BIOTIC STRESS MANAGEMENT IN TOMATO

Biotechnological Approaches

Innovations in Horticultural Science

BIOTIC STRESS MANAGEMENT IN TOMATO

Biotechnological Approaches

Edited by

Shashank Shekhar Solankey, PhD

Md. Shamim, PhD

AAP | APPLE ACADEMIC PRESS

First edition published 2022

Apple Academic Press Inc.
1265 Goldenrod Circle, NE,
Palm Bay, FL 32905 USA

4164 Lakeshore Road, Burlington,
ON, L7L 1A4 Canada

CRC Press
6000 Broken Sound Parkway NW,
Suite 300, Boca Raton, FL 33487-2742 USA

2 Park Square, Milton Park,
Abingdon, Oxon, OX14 4RN UK

Library and Archives Canada Cataloguing in Publication

Title: Biotic stress management in tomato : biotechnological approaches / edited by Shashank Shekhar Solankey, PhD, Md. Shamim, PhD.

Names: Solankey, Shashank Shekhar, editor. | Shamim, Md., 1985- editor.

Series: Innovations in horticultural science.

Description: First edition. | Series statement: Innovations in horticultural science | Includes bibliographical references and index.

Identifiers: Canadiana (print) 2021020513X | Canadiana (ebook) 20210205237 | ISBN 9781774630402 (hardcover) | ISBN 9781774639566 (softcover) | ISBN 9781003186960 (ebook)

Subjects: LCSH: Tomatoes—Diseases and pests. | LCSH: Tomatoes—Diseases and pests—Control. | LCSH: Tomatoes—Molecular aspects. | LCSH: Tomatoes—Biotechnology.

Classification: LCC SB608.T75 B56 2022 | DDC 635/.64293—dc23

Library of Congress Cataloging-in-Publication Data

CIP data on file with US Library of Congress

ISBN: 978-1-77463-040-2 (hbk)
ISBN: 978-1-77463-956-6 (pbk)
ISBN: 978-1-00318-696-0 (ebk)

INNOVATIONS IN HORTICULTURAL SCIENCE

Editor-in-Chief:

Dr. Mohammed Wasim Siddiqui, Assistant Professor-cum- Scientist
Bihar Agricultural University | www.bausabour.ac.in
Department of Food Science and Post-Harvest Technology
Sabour | Bhagalpur | Bihar | P. O. Box 813210 | INDIA
Contacts: (91) 9835502897
Email: wasim_serene@yahoo.com | wasim@appleacademicpress.com

The horticulture sector is considered as the most dynamic and sustainable segment of agriculture all over the world. It covers pre- and postharvest management of a wide spectrum of crops, including fruits and nuts, vegetables (including potatoes), flowering and aromatic plants, tuber crops, mushrooms, spices, plantation crops, edible bamboos etc. Shifting food pattern in wake of increasing income and health awareness of the populace has transformed horticulture into a vibrant commercial venture for the farming community all over the world.

It is a well-established fact that horticulture is one of the best options for improving the productivity of land, ensuring nutritional security for mankind and for sustaining the livelihood of the farming community worldwide. The world's populace is projected to be 9 billion by the year 2030, and the largest increase will be confined to the developing countries, where chronic food shortages and malnutrition already persist. This projected increase of population will certainly reduce the per capita availability of natural resources and may hinder the equilibrium and sustainability of agricultural systems due to overexploitation of natural resources, which will ultimately lead to more poverty, starvation, malnutrition, and higher food prices. The judicious utilization of natural resources is thus needed and must be addressed immediately.

Climate change is emerging as a major threat to the agriculture throughout the world as well. Surface temperatures of the earth have risen significantly over the past century, and the impact is most significant on agriculture. The rise in temperature enhances the rate of respiration, reduces cropping periods, advances ripening, and hastens crop maturity, which adversely affects crop productivity. Several climatic extremes such as droughts, floods, tropical cyclones, heavy precipitation events, hot extremes, and heat waves cause a negative impact on agriculture and are mainly caused and triggered by climate change.

In order to optimize the use of resources, hi-tech interventions like precision farming, which comprises temporal and spatial management of resources in horticulture, is essentially required. Infusion of technology for an efficient utilization of resources is intended for deriving higher crop productivity per unit of inputs. This would be possible only through deployment of modern hi-tech applications and precision farming methods. For improvement in crop production and returns to farmers, these technologies have to be widely spread and adopted. Considering the above-mentioned challenges of horticulturist and their expected role in ensuring food and nutritional security to mankind, a compilation of hi-tech cultivation techniques and postharvest management of horticultural crops is needed.

This book series, Innovations in Horticultural Science, is designed to address the need for advance knowledge for horticulture researchers and students. Moreover, the major advancements and developments in this subject area to be covered in this series would be beneficial to mankind.

Topics of interest include:

1. Importance of horticultural crops for livelihood
2. Dynamics in sustainable horticulture production
3. Precision horticulture for sustainability
4. Protected horticulture for sustainability
5. Classification of fruit, vegetables, flowers, and other horticultural crops
6. Nursery and orchard management
7. Propagation of horticultural crops
8. Rootstocks in fruit and vegetable production
9. Growth and development of horticultural crops
10. Horticultural plant physiology
11. Role of plant growth regulator in horticultural production
12. Nutrient and irrigation management
13. Fertigation in fruit and vegetables crops
14. High-density planting of fruit crops
15. Training and pruning of plants
16. Pollination management in horticultural crops
17. Organic crop production
18. Pest management dynamics for sustainable horticulture
19. Physiological disorders and their management
20. Biotic and abiotic stress management of fruit crops
21. Postharvest management of horticultural crops
22. Marketing strategies for horticultural crops
23. Climate change and sustainable horticulture
24. Molecular markers in horticultural science
25. Conventional and modern breeding approaches for quality improvement
26. Mushroom, bamboo, spices, medicinal, and plantation crop production

BOOKS IN THE SERIES

- **Spices: Agrotechniques for Quality Produce**
  Amit Baran Sharangi, PhD, S. Datta, PhD, and Prahlad Deb, PhD
- **Sustainable Horticulture, Volume 1: Diversity, Production, and Crop Improvement**
  Editors: Debashis Mandal, PhD, Amritesh C. Shukla, PhD, and Mohammed Wasim Siddiqui, PhD
- **Sustainable Horticulture, Volume 2: Food, Health, and Nutrition**
  Editors: Debashis Mandal, PhD, Amritesh C. Shukla, PhD, and Mohammed Wasim Siddiqui, PhD
- **Underexploited Spice Crops: Present Status, Agrotechnology, and Future Research Directions**
  Amit Baran Sharangi, PhD, Pemba H. Bhutia, Akkabathula Chandini Raj, and Majjiga Sreenivas
- **The Vegetable Pathosystem: Ecology, Disease Mechanism, and Management**
  Editors: Mohammad Ansar, PhD, and Abhijeet Ghatak, PhD
- **Advances in Pest Management in Commercial Flowers**
  Editors: Suprakash Pal, PhD, and Akshay Kumar Chakravarthy, PhD
- **Diseases of Fruits and Vegetable Crops: Recent Management Approaches**
  Editors: Gireesh Chand, PhD, Md. Nadeem Akhtar, and Santosh Kumar
- **Management of Insect Pests in Vegetable Crops: Concepts and Approaches**
  Editors: Ramanuj Vishwakarma, PhD, and Ranjeet Kumar, PhD
- **Temperate Fruits: Production, Processing, and Marketing**
  Editors: Debashis Mandal, PhD, Ursula Wermund, PhD, Lop Phavaphutanon, PhD, and Regina Cronje
- **Diseases of Horticultural Crops: Diagnosis and Management, Volume 1: Fruit Crops**
  Editors: J. N. Srivastava, PhD, and A. K. Singh, PhD

- **Diseases of Horticultural Crops: Diagnosis and Management, Volume 2: Vegetable Crops**
  Editors: J. N. Srivastava, PhD, and A. K. Singh, PhD
- **Diseases of of Horticultural Crops: Diagnosis and Management, Volume 3: Ornamental Plants and Spice Crops**
  Editors: J. N. Srivastava, PhD, and A. K. Singh, PhD
- **Diseases of Horticultural Crops: Diagnosis and Management, Volume 4: Important Plantation Crops, Medicinal Crops, and Mushrooms**
  Editors: J. N. Srivastava, PhD, and A. K. Singh, PhD
- **Biotic Stress Management in Tomato**
  Editors: Shashank Shekhar Solankey, PhD, and Md. Shamim, PhD
- **Medicinal Plants: Bioprospecting and Pharmacognosy**
  Editors: Amit Baran Sharangi, PhD, and K. V. Peter, PhD

ABOUT THE EDITORS

Shashank Shekhar Solankey, PhD
Assistant Professor-cum-Jr. Scientist (Horticulture: Vegetable Science) Dr. Kalam Agricultural College, Kishanganj Under Bihar Agricultural University, Sabour, India

Shashank Shekhar Solankey, PhD, is an Assistant Professor–cum–Jr. Scientist (Horticulture: Vegetable Science) at Dr. Kalam Agricultural College, Kishanganj under Bihar Agricultural University (BAU), Sabour (Bhagalpur), India. He has more than seven years of experience in teaching and research. He has previously worked as a Research Associate at ICAR-IIVR, Varanasi, India. His prime targeted area of research is biotic and abiotic stress resistance as well as quality improvement in vegetables, particularly solanaceous crops and okra. Dr. Solankey has supervised several MSc students and has handled several research projects. Among his many national and international awards are a Best Teacher Award (2016) and Best Researcher Award (2016) by BAU. He has published over 54 research papers as well as review papers, a souvenir paper, 9 books, 2 abstract books, over 45 book chapters, 23 popular articles, etc. He is a life-time member of several societies and is an editorial board member and reviewer for many reputed journals.

Dr. Solankey acquired a master's degree in Vegetable Science from Acharya Narendra Deva University of Agriculture and Technology, Kumarganj, Ayodhya, India, and a doctorate in Horticulture from Banaras Hindu University, Varanasi, India.

Md. Shamim, PhD

Assistant Professor cum Scientist, Department of Molecular Biology and Genetic Engineering of Dr. Kalam Agricultural College, Kishanganj (Bihar Agricultural University), India

Md. Shamim, PhD, is an Assistant Professor cum Scientist in the Department of Molecular Biology and Genetic Engineering of Dr. Kalam Agricultural College, Kishanganj (Bihar Agricultural University), India. He is the author or coauthor of over 30 peer-reviewed journal articles, over fifteen book chapters, and two conference papers. He has several authored and edited books to his credit along with several practical books. He is an editorial board member of several national and international journals. Recently, Dr. Shamim received the Young Faculty Award 2016 from the Venus International Foundation, Chennai, India. Before joining Bihar Agricultural University, Sabour. Dr. Shamim worked at the Indian Agricultural Research Institute, New Delhi, where he was engaged in heat-responsive gene regulation in wheat. Dr. Shamim also has working experience at the Indian Institute of Pulses Research, Kanpur, India, on molecular and phylogeny analysis of several *Fusarium* fungus and has also done research at the Biochemistry Department of Dr. Ram Manohar Lohia Institute on plant protease inhibitor isolation and their characterization. He is a member of the soil microbiology core research group at Bihar Agricultural University (BAU), where he helps with providing appropriate direction and assisting with prioritizing the research work on PGPRs. He has proved himself as an active scientist in the area of biotic stress management in rice and other crops, especially in yellow stem borer management by isolating protease inhibitor from jackfruit seeds and sheath blight resistance mechanism in wild rice, cultivated rice, and other hosts.

Dr. Shamim acquired his master's (Biotechnology) and PhD (Agricultural Biotechnology) degrees from Narendra Deva University of Agriculture and Technology, Kumarganj, Faizabad, India, with specialization in biotic stress management in rice through molecular and proteomics tools.

CONTENTS

CONTRIBUTORS

Anupam Adarsh
Krishi Vigyan Kendra, Saraiya, Dr. Rajendra Prasad Central Agricultural University, Pusa (Samstipur) – 848125, Bihar, India

Md. Feza Ahmad
Department of Horticulture (Fruit and Fruit Technology), Bihar Agricultural University, Sabour, Bhagalpur – 813210, Bihar, India

Shirin Akhtar
Department of Horticulture (Vegetable and Floriculture), Bihar Agricultural University, Sabour, Bhagalpur, Bihar – 813210, India, E-mail: shirin.0410@gmail.com

Bahadur Singh Bamaniya
Department of Horticulture, School of Life Sciences, Sikkim University, Tadong, Gangtok, East Sikkim, Sikkim, India

Sushmita Chhetri
Department of Horticulture, School of Life Sciences, Sikkim University, Tadong, Gangtok, East Sikkim, Sikkim, India

Kunvar Gyanendra
Faculty of Agriculture Sciences &Technology, Madhyanchal Professional University Ratibad, Bhopal-462044, Madhya Pradesh, India

Raja Husain
Department of Agriculture, Himalayan University, Jullang, Itanagar, Arunachal Pradesh – 791111, India, E-mail: rajahusain02@gmail.com

Dan Singh Jakhar
Department of Genetics and Plant Breeding, Institute of Agricultural Sciences, Banaras Hindu University, Varanasi – 221005, Uttar Pradesh, India, E-mail: dan.jakhar04@bhu.ac.in

V. B. Jha
Associate Dean cum Principal, Department of Plant Breeding and Genetics, Dr. Kalam Agricultural College, Kishanganj, Bihar Agricultural University, Sabour, Bhagalpur, Bihar – 855107, India

Ravi Keshri
Department of Molecular Biology and Genetic Engineering, Bihar Agricultural University, Sabour, Bhagalpur – 813210, Bihar, India

N. A. Khan
Department of Plant Molecular Biology and Genetic Engineering, Acharya Narendra Deva University of Agriculture and Technology, Kumarganj, Ayodhya – 224229, Uttar Pradesh, India

Anjani Kumar
Director, ICAR-ATARI, Zone-IV, Patna, Bihar, India

Arun Kumar
Department of Agronomy, Bihar Agricultural University, Sabour – 813210, Bihar, India

Deepak Kumar
Department of Manufacturing and Development, Nextnode Bioscience Pvt. Ltd.,
Kadi – 382725, Gujarat, India

Mahesh Kumar
Department of Molecular Biology and Genetic Engineering, Dr. Kalam Agricultural College (Bihar Agricultural University, Sabour), Kishanganj, Bihar – 855107, India,
E-mail: maheshkumara2z@gmail.com

Manoj Kumar
Division of Vegetable Science, Indian Institute of Horticultural Research, Hessarghatta,
Bangalore – 560089, Karnataka, India

Pankaj Kumar
Department of Agricultural Biotechnology and Molecular Biology,
Dr. Rajendra Prasad Central Agricultural University, Pusa (Samastipur) – 848125, Bihar, India,
E-mail: pankajcocbiotech@gmail.com

Rajesh Kumar
Department of Horticulture, School of Life Sciences, Sikkim University, Tadong, Gangtok,
East Sikkim, Sikkim, India, E-mail: rkumar@cus.ac.in

Ravi Ranjan Kumar
Department of Molecular Biology and Genetic Engineering, Bihar Agricultural University,
Sabour, Bhagalpur – 813210, Bihar, India

Santosh Kumar
Department of Plant Pathology, Bihar Agricultural University, Sabour, Bhagalpur – 813210,
Bihar, India, E-mail: santosh35433@gmail.com

Tribhuwan Kumar
Department of Molecular Biology and Genetic Engineering, Bihar Agricultural University,
Sabour – 813 210, Bihar, India

Vinod Kumar
Department of Molecular Biology and Genetic Engineering, Bihar Agricultural University,
Sabour, Bhagalpur – 813210, Bihar, India

Anupma Kumari
Krishi Vigyan Kendra, Saraiya, Dr. Rajendra Prasad Central Agricultural University,
Pusa (Samstipur) – 848125, Bihar, India

Meenakshi Kumari
Department of Vegetable Science, Chandra Shekhar Azad University of Agriculture and Technology,
Kanpur – 208002, Uttar Pradesh, India, E-mail: meenakshisinghupcs@gmail.com

Rima Kumari
Department of Agricultural Biotechnology and Molecular Biology,
Dr. Rajendra Prasad Central Agricultural University, Pusa (Samastipur) – 848125, Bihar, India,
E-mail: rimakumari1989@gmail.com

Jitendra Kumar Kushwah
Department of Agriculture, Government of Uttar Pradesh, India

Deepa Lal
Department of Horticulture, Babasaheb Bhimrao Ambedkar University, Lucknow – 226025, Uttar Pradesh, India

Mehi Lal
Plant Protection Section, ICAR-Central Potato Research Institute, Regional Station, Modipuram, Meerut – 250 110, Uttar Pradesh, India

Abhishek Naik
Product Development Manager, Nath Bio-Genes (I)Pvt. Ltd., Kalyani, Nadia - 741235, West Bengal, India

Md. Abu Nayyer
Integral Institute of Agricultural Science and Technology, Integral University, Dasauli, Lucknow – 226021, Uttar Pradesh, India, E-mail: nayyer123@gmail.com

M. S. Nimmy
NRC on Plant Biotechnology, IARI, Pusa Campus, New Delhi, India

Hari Om
Department of Agronomy, Dr. Kalam Agricultural College (Bihar Agricultural University, Sabour), Kishanganj, Bihar – 855107, India

Nagendra Rai
ICAR-Indian Institute of Vegetable Research (IIVR), Jakhini (Shahanshahpur), Varanasi – 221305, Uttar Pradesh, India, E-mail: nrai1964@gmail.com

Tushar Ranjan
Department of Molecular Biology and Genetic Engineering, Bihar Agricultural University, Sabour, Bhagalpur – 813210, Bihar, India

P. K. Ray
Krishi Vigyan Kendra, Saharsa (Bihar Agricultural University, Sabour), Bihar, India, E-mail: pankajveg@gmail.com

B. N. Saha
Department of Soil Science and Agricultural Chemistry, Dr. Kalam Agricultural College (Bihar Agricultural University, Sabour), Kishanganj, Bihar – 855107, India

Md. Shamim
Department of Molecular Biology and Genetic Engineering, Dr. Kalam Agricultural College (Bihar Agricultural University, Sabour), Kishanganj, Bihar – 855107, India, E-mail: shamimnduat@gmail.com

Md. Wasim Siddiqui
Department of Food Science and Post-Harvest Technology, Bihar Agricultural University, Sabour, Bhagalpur – 813210, Bihar, India

Ashutosh Kumar Singh
Department of Biotechnology and Crop Improvement, College of Horticulture and Forestry, Rani Lakshmi Bai Central Agricultural University, Jhansi, Uttar Pradesh, India

H. K. Singh
Krishi Vigyan Kendra, Kishanganj (Bihar Agricultural University, Sabour), Bihar, India

Major Singh
ICAR-Indian Institute of Vegetable Research (IIVR), Jakhini (Shahanshahpur), Varanasi – 221305, Uttar Pradesh, India

R. N. Singh
Associate Director Extension Education, Bihar Agricultural University, Sabour, Bhagalpur, Bihar, India

Rajesh Singh
Department of Genetics and Plant Breeding, Institute of Agricultural Sciences, Banaras Hindu University, Varanasi – 221005, Uttar Pradesh, India

Ramesh Kumar Singh
ICAR-Indian Institute of Vegetable Research (IIVR), Jakhini (Shahanshahpur), Varanasi – 221305, Uttar Pradesh, India

Umesh Singh
Associate Dean cum Principal, Mandan Bharti Agricultural College, Saharsa (Bihar Agricultural University, Sabour), Bihar, India

Vineeta Singh
Department of Plant Molecular Biology and Genetic Engineering, Acharya Narendra Deva University of Agriculture and Technology, Kumarganj, Ayodhya – 224229, Uttar Pradesh, India

S. S. Solankey
Department of Horticulture (Vegetable and Floriculture), Dr. Kalam Agricultural College (Bihar Agricultural University, Sabour), Kishanganj, Bihar – 855107, India, E-mail: shashank.hort@gmail.com

Deepti Srivastava
Integral Institute of Agricultural Science and Technology, Integral University, Dasauli, Lucknow – 226021, Uttar Pradesh, India

V. K. Tripathi
Department of Horticulture, Chandra Shekhar Azad University of Agriculture and Technology, Kanpur, Uttar Pradesh – 208002, India

B. Vanlalnehi
Division of Vegetable Science, Indian Institute of Horticultural Research, Hessarghatta, Bangalore – 560089, Karnataka, India

D. K. Verma
Department of Soil Science Agricultural Chemistry, Dr. Kalam Agricultural College (Bihar Agricultural University, Sabour), Kishanganj, Bihar – 855107, India

Rahul Kumar Verma
Krishi Vigyan Kendra, Madhepura, Bihar Agricultural University, Sabour, Bhagalpur, Bihar, India

Nitin Vikram
Department of Biochemistry, Zila Parishad Agriculture College, Banda, Uttar Pradesh, India

ABBREVIATIONS

°C	degree Celsius
°F	degree Fahrenheit
AChE	acetylcholinesterase
AFLP	amplified fragment length polymorphism
AMPs	antimicrobial peptides
APN	aminopeptidase N
atv	atroviolacea
BLAST	basic local alignment search tool
BOLD	barcode of life data systems
BSA	bulked segregant analysis
Bt	*Bacillus thuringiensis*
BTH	benzothiadiazole
bv	biovars
CAPS	cleaved amplified polymorphic sequence
CarE	carboxylesterase
Cas	CRISPR-associated
cfu/ml	colony-forming units per milliliter
CMV	cucumber mosaic virus
Cnr	colorless non-ripening
CP	capsid protein
CP	coat protein
CPB	Colorado potato beetle
CpTI	cowpea trypsin inhibitor
CRISPR	clustered regularly interspaced short palindromic repeats
Cry	crystal
DAS-ELISA	double-protein enzyme enzyme-linked immunosorbent research
DNA	deoxyribonucleic acid
DSB	double-strand break
dsRNA	double-stranded RNA
EB	early blight
ELISA	enzyme-linked immune sorbent assay
etc.	etcetera

FAO	Food and Agriculture Organization
FYM	farmyard manure
g	gram
GABA	gamma aminobutyric acid
GES	geraniol synthase
Gr	green-ripe
HaNPV	*Helicoverpa armigera* nucleopolyhedrovirus
HCl	hydrochloric acid
HPR	host plant resistance
HR	hypersensitive response
HRM	high-resolution melting
I.U.	international unit
ICP	insecticidal crystal protein
IDM	integrated disease management
ILs	introgression lines
INA	isonicotinic acid
ISR	induced systemic resistance
ISSR	inter-simple sequence repeat
ITmLCV	Indian tomato leaf curl virus
ITS	internal transcribed spacer
JA	jasmonic acid
kg	kilogram
LRR	leucine-rich repeat
lt	liter
MACE	modification of AChE
MAPK3	mitogen-activated protein kinase 3
MAS	marker-assisted selection
mg	milligram
Mi	*Meloidogyne incognita*
miRNAs	microRNAs
mm	millimeter
MP	movement protein
mRNA	messenger RNA
mtDNA	mitochondrial DNA
NBS	nucleotide-binding site
NGS	next-generation sequencing
NILs	near isogenic lines
NLSs	nuclear location sequences

NPV	nucleopolyhedrovirus
Nr	never-ripe
OAS	origin of assembly sequence
OP	organophosphate
OPDA	oxophytodianoic acid
ORF	open reading frame
PABP	poly(A)-bound protein
PAL	phenylalanine ammonia-lyase
PB	probenazole
PCR	polymerase chain reaction
PED	Pusa early dwarf
PGPR	plant growth-promoting rhizobacteria
pH	potential of hydrogen
PIs	protease inhibitors
PITGS	plant-induced cistron silencing
POX	peroxidase
PPO	polyphenol oxidase
PR	pathogenesis-related
PRP	proline rich protein
PVX	potato virus X
QTL	quantitative trait loci
R	races
RAPD	random amplified polymorphic DNA
rDNA	recombinant DNA technology
RdRp	RNA dependent RNA polymerase
RFLP	restriction fragment length polymorphism
RILs	recombinant inbred lines
rin	ripening-inhibitor
RKN	root-knot nematode
RNA	ribonucleic acid
RNAi	RNA interference
ROS	reactive oxygen species
Rt PCR	real-time PCR
SA	salicylic acid
SCAR	sequence characterized amplified region
SGA	steroidal glycol-alkaloid
SLFSR	fruit shelf-lifestyles regulator
SLMs	small lipophilic molecules

SLS	septoria leaf spot
SNP	single nucleotide polymorphisms
sq.m	square meter
SSCP	single-strand conformational polymorphism
SSR	simple sequence repeat
TBB	tomato big bud
TBTV	tomato bunchy top virus
TIR	toll interleukin receptor-like
TLCV	tomato leaf curl virus
TMV	tobacco mosaic virus
ToLCD	tomato leaf curl disease
ToMoV	tomato mottle virus
ToMV	tomato mosaic virus
TSP	total soluble macromolecule
TSWV	tomato spotted wilt virus
TSWV-BL	tomato spotted wilt tospovirus
TYLCSV	tomato yellow leaf curl Sardinia virus
TYLCV	tomato yellow leaf curl virus
UTR	untranslated region
VIP	vegetative insecticidal proteins
WP	wettable powder

PREFACE

Tomato is the world's most important vegetable crop consumed by us in our daily diet. Tomato shares a huge presence in the world as a fresh vegetable and is used as a variety of processed products, such as soups, juices, sauces, pastes, canned fruits, and dehydrated tomatoes. However, this important vegetable is vulnerable to more than 200 diseases that encounter many biotic agents. These diseases include viruses, fungi, viruses, phytoplasma, insects, and nematodes.

With the onset of climate change, global agriculture faces many challenges; biotic stress caused by pests and diseases continues to cause a serious problem in tomato production despite years of investment in research and development aimed at understanding plant interactions and finding effective ways to control them. Tomato plants are exposed to attacks from a broad spectrum of biotic stresses. The general introduction of major diseases of tomato and their probable management along with major biotic stress management through biotechnological approaches are systematically discussed in this volume. These stresses cause bacterial wilt, bacterial canker, damping-off seedlings, late blight, early blight (EB), fusarium wilt, septoria leaf spot (SLS), Cercospora leaf spot, verticillium wilt, tomato leaf curl virus (TLCV), tobacco mosaic virus (TMV), tomato spotted wilt virus (TSWV), root-knot nematode, fruit borer, sucking pests, gene stacking/pyramiding, and post-harvest management.

In tomato, there are very limited numbers of horizontal resistant varieties so far, and it is very difficult to manage many diseases during the same cropping season. However, pesticides are helpful in managing major diseases to some extent, but their adverse effects are harmful for us. Hence, it would be better to incorporate integrated approaches like grafting on resistant root stalks, crop rotation, summer plowing, removal of crop debris, cleaning of fields, use of bio-agents, etc., for effective and eco-friendly disease management. Moreover, in order to substantiate a sustainable production system along with high-quality products in terms of protective food, the use of resistant/tolerant cultivars becomes a principal tool to decrease the damages caused by these biotic factors. At present, advanced breeding tools, especially marker-assisted selection (MAS) and marker-assisted back cross-breeding, open new doors for efficient and significant disease management in tomatoes.

Other biotechnological tools, such as plant tissue culture, *in-vitro* mutagenesis, and genetic engineering, can help to restrict these biotic constraints.

In the present book, we explain the present status of biotechnological tools and their utilization for biotic stress management in tomatoes. In recent years, significant progress has been made towards understanding the interplay between tomatoes and their hosts, particularly the role of resistant varieties in regulating, attenuating, or neutralizing invading biotic stresses. The novel accomplishments and perspective on biotechnological approaches to combat biotic stresses, carried out gene expression, and analyses of genes targeted functional expressed against various biotic stresses have been described critically.

There is a great need to integrate conventional and biotechnology approaches for efficient disease management in tomatoes. Recent advancements in tomato breeding by using genomics tools such as structural and functional genomics and molecular breeding tools provide an efficient platform for tomato improvement. Genetic diversity in the tomato gene pool has indicated that various genes, expressed as QTL, are helpful in obtaining the resistance phenotype. Thus the biotechnological tools will help for the improvements of the tomato research program and the development of new tomato varieties/hybrids.

This book is the most up-to-date and comprehensive review and will be a greatly useful reference workbook. It contains basic facts and the new and recent discoveries on the biotechnological aspects of tomato for biotic stresses management. This book should find a remarkable place on the shelves of new vegetable scientists, plant breeders, biotechnologists, plant pathologists, virologists, and entomologists working in academic and commercial tomato research programs, and in the libraries of all research establishments and companies where this new exciting subject is researched, studied, or taught.

With very sincerely we wish to acknowledge the authors of the chapters who contributed in this endeavor from different states and who have assisted in the formulation and compilation of *Biotic Stress Management in Tomato: Biotechnological Approaches* by contributing the novel information and comprehensive scripts without which it would have been extremely complex to finish this herculean task. I crave to recognize all the theme specialists who were involved in the book to create this book a success.

We feel immense pleasure to express our heartfelt gratitude to Dr. Kirti Singh (President, International Society for Noni Science, Former Chairman, Agricultural Scientists Recruitment Board, New Delhi, Former Vice-Chancellor, NDUAT, Faizabad (UP), HPKV – Palampur (HP) & IGKV, Raipur, Former Secretary, National Academy of Agricultural Sciences, New Delhi) for his inspiring guidance, encouragement, and blessings. Recognitions are due to our academic team members, especially to Dr. Md. Wasim Siddiqui, Series Editor, Apple Academic Press, who generously supported the compilation and completion of this assignment. We are also thankful to our family members, whose sustained support and encouragement led us to complete this book.

We are also thankful to our teachers, seniors, colleagues, and friends for their direct and indirect support and valuable suggestions during the compilation of this book. We also invite fruitful suggestions and useful criticism from the book readers so we can improve the information in future on this subject.

The Editors sincerely acknowledge the help and support taken from the various resource persons, books, and journals, authors, and publishers whose publications have been used while preparing this manuscript. We thank Apple Academic Press for bringing out this publication in a very systematic manner.

Finally, we acknowledge 'Almighty God' who provided all the strength, inspirations, positive thoughts, insights, and possible way to complete this book in due time.

—***Shashank Shekhar Solankey, PhD***
Md. Shamim, PhD

CHAPTER 1

TOMATO DISEASES, THEIR IMPACT, AND MANAGEMENT

S. S. SOLANKEY,[1] P. K. RAY,[2] MEENAKSHI KUMARI,[3] H. K. SINGH,[4] MD. SHAMIM,[5] D. K. VERMA,[6] and V. B. JHA[7]

[1]*Department of Horticulture (Vegetable and Floriculture), Dr. Kalam Agricultural College (Bihar Agricultural University, Sabour), Kishanganj, Bihar – 855107, India, E-mail: shashank.hort@gmail.com (S. S. Solankey)*

[2]*Krishi Vigyan Kendra, Saharsa (Bihar Agricultural University, Sabour), Bihar, India*

[3]*Department of Vegetable Science, Chandra Shekhar Azad University of Agriculture and Technology, Kanpur – 208002, Uttar Pradesh, India*

[4]*Krishi Vigyan Kendra, Kishanganj (Bihar Agricultural University, Sabour), Bihar, India*

[5]*Department of Molecular Biology and Genetic Engineering, Dr. Kalam Agricultural College (Bihar Agricultural University, Sabour), Kishanganj, Bihar – 855107, India,*

[6]*Department of Soil Science and Agricultural Chemistry, Dr. Kalam Agricultural College (Bihar Agricultural University, Sabour), Kishanganj, Bihar – 855107, India*

[7]*Department of Plant Breeding and Genetics, Dr. Kalam Agricultural College (Bihar Agricultural University, Sabour), Kishanganj, Bihar – 855107, India*

ABSTRACT

Tomato is a well-known prime vegetable crop. It comes under a warm-season crop and is grown all over the world in warm and cool climatic conditions. The plant is extremely affected by adverse climatic conditions, i.e., responsible for promoting the disease infection. The warm and cool climatic conditions offer an optimal condition for the development of various foliar, stem, and soil-borne diseases in tomatoes. Diseases including fungal, viral, phytoplasma, and bacterial are the major limiting factors for its economic cultivation. The major diseases and its probable management in tomato are discussed here such as damping off, septoria leaf spot, early blight (EB), late blight, fusarium wilt, verticillium wilt, Cercospora leaf spot, anthracnose, powdery mildew, sclerotium collar rot, buckeye rot, bacterial wilt, bacterial canker, bacterial spot, tomato leaf curl virus (TLCV), tobacco mosaic virus (TMV), tomato spotted wilt virus (TSWV), tomato bunchy top virus (TBTV), and tomato big bud (TBB). For tomato growers, it would be better to incorporate integrated approaches for their management along with the use of resistant/tolerant high-yielding varieties; therefore, it could not economically affect the crop, as well as the recommended practices are feasible and easily understandable by the tomato growers.

1.1 INTRODUCTION

Tomato (*Solanum lycopersicum* L.) is one of the prime vegetable crops across the globe, shares a desirable position in India as a fresh vegetable, and also being used as a variety of processed products such as soup, juice, ketchup, sauce, puree, paste, canned fruits, and dehydrated tomatoes. It is susceptible to more than 200 diseases caused by pathogens, bacteria, viruses, and phytoplasma during its cultivation and postharvest management in existing varieties and hybrids (Abdel-Sayed, 2006; Singh et al., 2017). Outdoor production of tomato is seriously impaired due to increasing infections with evolving early blight (EB), late blight, and tomato leaf curl virus (TLCV) diseases (Scholthof et al., 2011), particularly in the Gangetic plains of India. Out of these bacterial wilts, bacterial canker, damping-off seedlings, fusarium wilt, septoria leaf spot, Cercospora leaf spot, verticillium wilt, tobacco mosaic virus (TMV) and tomato spotted wilt virus (TSWV) also causes the significant yield losses in tomato. Among all, TLCV is the most devastating disease in India. TLCV has become a major limiting factor in tomato cultivation, particularly during summer crop (February to May) in southern Indian states (Sadashiva et al., 2002) and autumn crop (August to December) in northern plains (Banerjee and Kalloo, 1987) and both early-autumn and autumn-winter (September to February) in Eastern India (Mandal et al., 2017). Disease management of tomatoes is essential for getting the optimal fruit yield. Not only the field-grown, even the crop grown under protected structures (poly houses, net houses, and polytunnels) and in kitchen gardens are also at high risk. For efficient management and to ensure timely control of diseases and its spreading, it is necessary to follow precautious measures along with early identification. The major tomato diseases, its diagnosis, impact, and its feasible control measures are as follows:

1.2 FUNGAL DISEASES OF TOMATO

1.2.1 DAMPING OFF (PYTHIUM, RHIZOCTONIA, PHYTOPHTHORA, ETC.)

It is a serious soil-borne pathogenic disease that chiefly affects seeds (pre-emergence) and new seedlings (post-emergence). Damping-off usually causes rotting of stems at the soil surface and root tissues below the soil surface. Sometimes, infected plants will germinate and look fine; however, within 2–4 days, they become water-soaked and squidgy, eventually fall over at the base and die. In the pre-emergence stage the seedlings are fall out just before they reach the soil surface or at seed emergence stage, while in the post-emergence stage the young, juvenile tissues of the collar at the ground level become softer and finally seedlings become fall off at collar region.

Hot water treatment of seeds (at 52°C for 30 minutes) is recommended as well as treats the seeds with fungicides like, Thiram/Captan/Metalaxyl-Mancozeb @ 2.5–3.0 g/kg of seeds. Drenching of nursery beds one week prior to sowing with Thiram/Captan or any copper-based fungicide @ 3 g/liter of water. Soil solarization of nursery area by cover the beds with transparent poly-sheets before sowing and left to open in direct sunlight for at least 10 days. Spraying the young seedlings with 0.2% Blitox or 0.25% Metalaxyl+Mancozeb, moreover 0.25% Fosetyl-Aluminum is found most effective against this pathogen. Provision of proper drainage to the beds, avoid dense planting, removal of infected seedlings from beds, and avoid flooding the beds as soon as the symptom arises to check spread of the disease.

1.2.2 SEPTORIA LEAF SPOT (SLS) (SEPTORIA LYCOPERSICI)

The characteristic symptom is small, water-soaked circular spots of 1.6 to 3.2 mm in diameter first appeared on the lower surface of older leaves. The circular spots have a dark brown margin with gray or tan heart. When spots reached to maturity, they enlarge to a diameter of about 6–7 mm and coalesce.

The center of the spots will show various dark brown, pimples-like structures that are known as pycnidia fruiting bodies of the pathogen. The structures can be seen with the necked eyes or by hand lens because they are large enough. At lateral stage, these spots may also appear on stems, calyxes, and blossoms, but rarely on fruits. Severely infected leaves will turn yellow, then dry up and finally drop off. This drop off of leaves will result in exposure of fruits in direct sunlight and cause physiological disorder, sun-scald.

Use disease-free seed or if the seed is suspected, then treat with hot water at 50°C for 25 minutes. Apply crop rotation with non-related crops for at least 3 years. Uprooting infected plants at the end of the growing season to prevent the spores from overwintering in the field. Plucked off and destroy infected leaves at first occurrence of the symptom and disinfect pruning tools before moving from one plant to another. Apply copper-based fungicides, e.g., copper oxychloride or bio-agent *Bacillus subtilis* for effective management of this disease, especially when used as a prophylactic measure for its management.

1.2.3 EARLY BLIGHT (EB) (ALTERNARIA SOLANI)

This is also a common disease of tomato, initially occurring on the leaves or foliage at any stage of the plant growth. The characteristic symptom is the appearance of small, black lesions predominantly on the older leaves. Day-by-day spots enlarge, and with time the diseased area expanded up to 0.25 inch in diameter or larger, concentric rings in a bull's eye pattern can be seen at the center. At the lower surface of the leaves just behind the bulls-eye-shaped brown spots, often the tissue around the spots will turn yellow. The fungus primarily attacks on the leaves/foliage, causing characteristic leaf spots and blight. Further, infected leaves will drop from the plant. In many cases, the tomato fruits will continue to ripe, as well as the disease symptoms will also appeared on the skin of the fruit. Under favorable conditions (hot and humid), much of the foliage will die. Lesions on the shoots are similar to those on leaves, occasionally girdling the plant, if they occur near the soil surface. In fruits, the infection transmitted by calyx or stem attachment. Lesions attain substantial size, usually on the entire fruit; concentric pitted spots are also seen on the surface of the fruit. Apart from the foliage symptoms that are known as EB, this disease generally causes less economically significant symptoms on tomato crop, including collar rot, stem lesions on the fully developed plant and fruit rot (Walker, 1952). Due to EB disease the yield losses up to 79% have been reported from India, Canada, United States, and Nigeria (Datar and Mayee, 1981).

The pathogen of EB lives in the soil and once a field has shown symptom of the EB pathogen, it will remain persist in the soil, because the fungus easily overwinters in the soil, even in cold months. Luckily, most tomatoes will continue to produce fruits even with moderately severe infestation of EB. Growing of resistant varieties like Arka Rakshak, Arka Samrat, Arka Vikas and Swarn Sampada are the best practice to prevent the tomato crop from this fungal disease, moreover mulching with a layer of paddy straw, leaf mold, or well-decomposed compost immediately after they are transplanted. Actually, this mulch creates a protective barrier that prevents the soil-borne spores from splashing up out of the soil and onto the plant. When the pathogen strikes, organic fungicides like *Bacillus subtilis* or copper-based fungicides can help to prevent or control the spread of this pathogen. Bicarbonate fungicides are also found effective (including Bi-carb, Green cure, etc.), against this disease. As prophylactic measure, spray the crop with 0.3% Diathane M-45 or 0.1% Carbendazim or 0.2% Chlorothalonil at 10–15 days intervals starting from 45 days after transplanting. Two sprays of 0.2% Chlorothalonil or Difenoconazole (0.5 g/lt) is also effective to control this disease.

1.2.4 LATE BLIGHT (PHYTOPHTHORA INFESTANS)

In the past one decade, the late blight is observed as the most destructive tomato disease. This plant pathogen is one of the most distressing organisms in human history, being responsible for the shocking Irish potato famine in the 1840s, and it is arguably the most important pathogen of tomatoes, causing yield loss of 91.80% (Byrne et al., 1997). The fungal spores create irregular-shaped, slimy, and water-soaked blotches on foliage. Often the blotches occur initially on the top leaves and stems. Later on, infection occurs on entire stems and it will "rot," turning black in color and slimy. Sometimes white spores' patches may also be appeared on the leaf undersides. In the northern plains of India, the pathogen overwinters in buried potato tubers after harvesting. However, in the southern parts of India, it easily survives during winter. Late blight disease commonly occurs in humid regions with cool temperatures ranging from 15–21°C and cloudy weather; however, prolonged hot dry days can check its spread. Under ideal condition, the disease development is rapid, causing serious economic losses. Lesions on the leaves initially produce irregular large shaped, greenish-black, and water-soaked spots. These spots enlarge rapidly, becoming brown, and under humid conditions, develop a white moldy growth close to the margins of the diseased area on the lower surface of the leaves and on stems also. The disease spread speedily under humid conditions and destroying quickly large areas of leaf as well as shoot tissues. Fruit lesions come with large, green to dark brown, principally on the upper half of the fruit, but they may also find on other parts. Under humid conditions, white moldy growth may also appear on fruits surface and thereafter converted into black color. The disease damaged the fruits as well as the foliage of the plant. Characteristic symptoms on the fruits generally begin on the shoulders of the fruit because spores land on fruits from the upper parts of the plant. However, it is a worldwide problem, but most severe epidemics occur in frequent cool and moist areas. In Bihar, the highest disease infestation and severity occurred during the month of January-February when the maximum temperature ranged from 10.4–10.8°C and maximum relative humidity ranged from 90–95% (Solankey et al., 2017).

The spores of this disease are easily transmitted by winds for miles that promote its fast infection. Late blight is an uncommon fungus, but once it occurs, there is a very limited option to prevent its infection because the spores spread so hastily. If it struck on few plants then uproot and destroy them immediately to check spores spreading. Few bio-agents like, *Bacillus subtilis* is effective to some extent in preventing this disease when it's first knocked in your area. Under prevailing weather condition, prophylactic spray with copper-based fungicides like, Copper oxychloride (0.3%)/Diathane M-45 (0.25%)/Chlorothalonil (0.2%) at 10 days interval. However, when it found in patches, without delay spray fungicides *viz*., Metalaxyl + Diathane M-45 (0.25%)/Cymoxanil+ Diathane M-45

(0.25%)/Dimethomorph (1 g) + Diathane M-45 (2 g)/liter of water, and alternate spray schedule with systemic fungicides at 7–10 days intervals depending upon weather condition. Careful irrigation or avoid sprinkler irrigation under late blight prevailing conditions. However, the varieties, Arka Rakshak, Arka Alok, Kashi Anupam, Arka Ananya, Azad T-5, Kashi Vishesh, Arka Saurabh, Pusa Rohini Azad T-6 have high yield potential along with lower incidence of late blight (Solankey et al., 2017; Ray et al., 2018).

1.2.5 FUSARIUM WILT (FUSARIUM OXYSPORUM F. SP. LYCOPERSICI)

Fusarium wilt habitually causes yellowing on one side of the leaves and plant. In general, yellowing starts from older leaves followed by wilting, browning, and finally defoliation. Plant growth is generally stunted, and little or no fruits will form. When the infected stem is cut at its base, then brown, vascular tissue can be found due to fusaric acid formation by fusarium. Infected plants often die before reaching maturity. It is basically a soil-borne disease that is found all over the humid and sub-humid warm regions of India and the world. Disease development is favored by warm and humid soils, and symptoms are most common when temperatures range from 26–32°C. Fusarium is specific for tomato and other solanaceous vegetables, and it is surprising that fusarium fungi survive in the soil or associated with plant debris for up to 10 years. In high nitrogen and low potassium soils, the disease develops more rapidly. However, plants grown in sandy soils lean-to contract this disease more easily.

As, the spores of pathogen live in the soil and can survive for many years and can frequently spread by garden tools, equipment, water, manures, plant debris, and sporadically by human and animals. The best method of avoidance is growing of resistant varieties like Pant Bahar. Moreover, disinfection of tomato debris with 10% bleaching powder solution at the time of crop ending. Once this disease occurs, there's little remedy to control it, and it is better to focus on preventive measures for future years by soil solarization to kill fungal spores in the topsoil as well as crop rotation with non-related crops. Drenching with biological fungicides like, *Streptomyces griseoviridis* and *Trichoderma virens* on topsoils are helpful to check the disease infection from colonizing the future crops roots.

1.2.6 VERTICILLIUM WILT (VERTICILLIUM SPP.)

Verticillium wilt is a true wilt occasionally occurs in tomato crop. It developed under good moisture and nutrition conditions in the field, initially older leaves show the yellow botches as a primary symptom of this disease subsequently, veins become brown, and thereafter chocolate brown dead spots on leaves. The spot sometimes seems like the EB, but they are not definite nor expressed concentric bull's eye rings symptom. The leaves will finally wilt, die, and drop off. The symptoms of disease progress up the stem and the plant goes stunted. In this disease, the top leaves may stay green, fruits remain small with yellow shoulders, and due to loss of leaves, the fruits may also sunburned.

Infection happens directly when the fungal threads penetrate the root hairs. Further, the rootlets become broken and nematodes may feed on the roots. The pathogen grows quickly up the xylem, or sap-conducting channels. Due to damage of xylem, the normal upward movement of water and nutrients is to be disturbed. This fungus also produces a toxin substance that leads to the spotting and wilting of the leaves. At advance stage, discoloration can be traced upwards and downwards into the roots. Differ to fusarium wilt, verticillium wilt discoloration seldom goes up to 10–12 inches above the soil, beside these its toxin may also increase.

Verticillium pathogen can survive in the soil and plants for many years (10 years or more) in the form of minute, black, seed-like structures known as microsclerotia. They thrive in moderate temperatures, i.e., 21–27°C. Once this disease occurs, there's little chance to control it in the existing year. Hence, focus must be given on preventing this disease in upcoming years. Soil Solarization is helpful in killing the microsclerotia in the topsoil. Destroying of crop debris and follow the crop rotations with non-host crop at least for 3–4 years after the infection.

1.2.7 CERCOSPORA LEAF SPOT (CERCOSPORA SPP.)

Cercospora leaf spot disease is characterized by small circular or angular brownish or purplish spots enclosed by yellow or red margins across the surface of infected leaves. Typical foliar symptoms are round spots of about 3.2 mm in diameter with ash gray centers and dark brown or reddish-purple margins. When the infection increases, many spots will appear until the leaf ceases to function as the site of the food production process by disturbing photosynthesis which needs sunlight to complete. Green leaves perform photosynthesis under sunlight and when they die, that usually signals the carbohydrate and energy deficiency in the plant and ultimately the death of the entire plant from starvation will occur. The disease often destroys entire leaves, leaving nothing behind except the shrunken stems and leaf veins before they fall off. Day temperatures of 26–32°C and night temperatures above 15°C favors disease expansion. Day temperature above 34°C is uncongenial for disease development. Under optimum conditions, the fungus may have 4–5 disease cycles during the season, and moreover, with each cycle, there is a substantial increase in the amount of inoculums. The best way of management is early control of disease when first symptoms occur. As the fungus damages the leaves and it adversely affect carbohydrate accumulation of plants that ultimately decreases plant yield.

Collection and burning of infected plant debris and follow crop rotation with non-solanaceous crops. Proper drainage of the field is also necessary because disease infection and severity rise under flooded conditions. Always use disease-free, healthy seeds that must be obtained from healthy disease-free plants. The chemical control measures for its management is, spray the crop with carbendazim (0.1%) or thiophanate methyl (0.1%) or Diathane M-45 (0.25%) or Redomil (0.25%) or copper oxychloride (0.3%) and repeat at 10–14 days interval.

1.2.8 ANTHRACNOSE (COLLETOTRICHUM PHOMOIDES)

It is a widespread disease of ripe or overripe tomato fruits and occasionally found on green fruits. The infected ripe fruits become small, concentric with sunken spots that may increase in size up to 0.5 inch in diameter. The core of older spots becomes blackish at later stage. Numerous spots may be visible under severe infection and resulting completely rotting of infected fruits. The anthracnose forms small, dark survival structures called sclerotia in the core of fruit spots. These sclerotia can survive up to 3 years in soil and cause infections in both ways, i.e., directly or by producing secondary spores. Green fruits are infected by this disease but show symptoms at the time of ripening. The disease is transmitted from infected to healthy fruit by spores that are splashed by rain or overhead irrigation, or by pickers working wet plants. Warm rainy weather favors this disease, moreover, overhead irrigation and heavy defoliation facilitates rapid disease transmission.

Infected plant debris should be destroyed or burn at the time of first appearance of the disease. Seed should be extracted from healthy fruits and treatment with Captan (0.3%) before sowing. Fungicides like, carbendazim (0.1%) or thiophanate methyl (0.1%) or combination of Diathane M-45 (0.25%) and carbendazim (0.1%) or copper oxychloride (0.3%) should be spread at the time of first appearance of disease symptom and further repeat it at 10–14 days intervals.

1.2.9 POWDERY MILDEW (LEVEILLULA TAURICA)

Powdery mildew is a dry season disease that makes a white powdery fungal coating on the leaf surface. Infected leaves become dwarf, narrow, and stiff. The pathogen gradually attacks new tender leaves, dispersing over leaf stems, twigs, and ultimately the fruit. Apical growth of the infected shoots becomes stunted. The patches turn dull, dirty white with age and contain small black embedded specks. Low light intensity along with dry weather and a dense cropping favors this disease infection and spreading. The affected fruits will not take optimum size and fruit yield will also decrease. Avoid plants overcrowding by using adequate spacing and thinning. Growing of tolerant variety, e.g., Arka Ashish. The effective management of this disease can be done by a single spray of propiconazole (Tilt 25EC@ 0.1%) or wettable sulfur (4 g/l of water) at the time of disease appearance.

1.2.10 SCLEROTIUM COLLAR ROT (SCLEROTIUM ROLFSII)

It is an important fungal disease that causes considerable yield losses up to 30% in tomato (Thiribhuvanamala et al., 1999) as well as can harm seedling losses of 20–40% in the field (Sherf and Macnab, 1986). It is reported that *Bacillus subtilis* controls *Sclerotium rolfsii* by 92% under greenhouse condition in peanut (Maienfisch et al., 2001). Bio-agents secreted an array of chemically diverse antimicrobial secondary metabolites and hydrolytic enzymes such as proteases, cellulases, chitinases, lipases, etc., which may have a probable function in enhancing the host development and vigor, increasing antagonistic microbial activity and enabling them to tolerate this pathogen (Mahato et al., 2017).

For its proper management, it is necessary to follow the adequate crop rotation practice with non-related crops such as maize, sorghum, small grains, or cotton because several tomato pathogens live in the soil for up to 3 years. Farmyard manure (FYM) of green manure should be fully decomposed before transplanting the tomato seedlings. Immediate clipping of leaves when any signs of disease appeared and then dispose them to check the disease spread. Sowing of resistant varieties, especially horizontal resistant varieties that have moderate and durable tolerance against multiple diseases including collar rot are to be preferred. Remove plant debris at the end of the growing season and destroy it carefully to check the future spread. Do not put infected plant debris in the compost pit. Adequate aeration should be provided around each plant by applying proper plant spacing followed by training and pruning. Two or three inches of compost can be applied for crop mulching or mulch by leaf mold, hay, or paddy straw serves to maintain soil cover and restrict disease transmission. Try to maintain the foliage dry whenever possible. Drip irrigation allows to target the water in the root zone only as well as fertigation makes more effective fertilizer management along with disease management and ultimately increases fruit yield. Application of fluxapyroxad + pyraclostrobin reduce incidence of Sclerotium collar rot in tomato. Soil application of *Trichoderma viride* and seedling dip with tebuconazole @ 0.05% followed by, soil drenching with tebuconazole @ 0.05% is also found effective to control collar rot of tomato (Banyal et al., 2008). By adoption of these techniques we can manage tomato diseases as well as enhance fruit yield and assure high economic return.

1.2.11 BUCK EYE ROT (PHYTOPHTHORA PARASITICA)

Buckeye rot or fruit rot is a severe disease in entire tomato growing areas. Initially, this disease affects the fruits near the ground level by rotting them. The disease is differed from the late blight in terms of the pathogen does not affect the foliage. Where the fruit touches the soil, grayish-green or brown water-soaked spots appears, i.e., the characteristic symptoms of this disease. With time, these spots enlarges, the surface of the lesion assumes a pattern of dark brown narrow concentric rings and the infected young green fruits become mummified.

For minimizing the disease infection, good drainage conditions should be provided in the field and staking plants, furthermore removing foliage along with fruits up to a height 15–30 cm from ground level. Apply fungicide Difolatan (0.3%) for 4 times at an interval of 10 days can control this disease more effectively.

1.3 BACTERIAL DISEASES OF TOMATO

1.3.1 BACTERIAL WILT (PSEUDOMONAS SOLANACEARUM)

This is one of the most dreaded soil borne disease of tomato, causing severe yield loss of 25–75% (Dutta et al., 2012). High soil moisture along with high temperatures (over 29°C) promotes disease development. The pathogen enters into plant roots through microscopic wounds (often injured by insects, intercultural operations, or at the time of transplanting) under favorable conditions. High soil moisture along with high soil temperature promotes disease development. The disease can infect a wide range of hosts like solanaceous crops (tomato, brinjal, chilies, potato, etc.), and a wide variety of ornamental plants (zinnia, marigold, hollyhock, dahlia, geranium, sunflower, nasturtium, etc.). At the initial stage, it shows wilted appearance on the youngest leaves, and with progress of disease, the foot of the plant may show root rot, brown cankers, and a cross-section of an infected stem sometimes represents a brown discoloration of the vascular tissue. The plant finally wilted permanently and then died. A freshly cut stem at the base of the plant dipped in clear water can also demonstrate a stream of a white slimy bacterial ooze that is a key indicator of the bacterium present in the vascular tissue.

Treat the seeds through hot water at 52°C for 20 minutes or with bactericides like streptomycin solution (0.01%) for 30 minutes. There are various cultural practices that reduce its incidence like, soil solarization, soil fumigation, removal of crop debris, drenching of soil with fungicides, etc., but these are not long-term control measure. The use of bacterial wilt resistant rootstocks can also been used in disease prone areas. Growing of commercially available resistant/tolerant varieties like Arka Rakshak, Shakthi, Sonali, Arka Samrat, Arka Alok, Arka Abha, Arka Ananya, Arka Shreshta, and Arka Abhijit are the best option for tomato growers in bacterial wilt prevalent regions. Adequate crop rotation with non-solanaceous crops for not less than 2 to 3 years reduces the chance of infestation. Application of bleaching powder @ 15–20 kg/ha and 4.5 quintal limes to the field and mix it properly to the soil at least 3–4 weeks prior to transplanting. Proper drainage should be maintained. A combination of Sun hemp as a green manure crop and chemical soil amendment with lime along with bleaching powder and application of bio-agents (*Pseudomonas fluorescens* and *Bacillus subtilis*) in the soil is the most efficient management method of wilt of solanaceous vegetables.

1.3.2 BACTERIAL CANKER (CLAVIBACTER MICHIGANENSIS PV. MICHIGANENSIS)

Tomato seedlings generally do not express any symptoms of bacterial wilt; however, the young plant show poor growth and temporary wilting of branches. At a later stage, lower leaves become yellow and shrivel, but symptoms perhaps not visible until flowering. When flowering occurs, the plants show two types of symptoms; first one is from systemic infections, i.e., the bacteria penetrate the vasculature and invade much of the plant, and then secondary infections occur, i.e., the bacteria cause local infections of leaves, stem, and fruits.

During primary infections on flowering plants, leaflets of the basal leaves curl, yellow, wilt, and finally turn brown and expressed burning symptoms. Sometimes, one side of the leaves is affected, and plants' growth is stunted, and after few days, these plants become wilted. Stem's pith becomes yellow and further turned in to reddish-brown color, particularly at the nodes shows a mealy appearance, and later the pith may become hollow. At an advanced stage of infection, cankers sometimes form at the nodes. Initially, light followed by dark streaks may form on stems, branches easily break off, and finally plant will die.

In secondary infections, leaves margins infection is common. Lesions of dark brown to black in color are formed, and later on, round to irregular spots occurred on leaves as well as on fruits near the calyx. Yellow to brown, slightly raised spots, surrounded by a persistent white halo (bird's eye spot) are appear on fruits. Spots are generally 3 mm in diameter; moreover, vascular tissue under the calyx scar, leading to seeds that possibly brown.

For its efficient management, use disease-free seeds or seedlings and soaking of seeds in hot water (56°C) for 30 minutes or in HCl (hydrochloric acid) (5%) for 5 hours to ensure disinfection. Soil, potting mixture and pots or beds should be sterilized before sowing. The pruning tools will also be disinfect before each use and clean them well after use. Destroy crop debris and follow crop rotation with non-host crop minimum for 3 years. Applications of Metalaxyl + Mancozeb @ 0.2% may reduce secondary spread of disease.

1.3.3 BACTERIAL SPOT (XANTHOMONAS CAMPESTRIS PV. VESICATORIA)

This disease is transmitted by seed, insects, irrigation water, infected plant debris in addition to *Solanaceous* crops and weeds. Humid weather and splattering rains promote disease progress. Most of the times, the outbreaks of this disease can be observed in the areas where heavy rainstorms occur. The bacteria enter the plant by transmitting agents through stomata and wounds. The bacterium affects the whole aerial part of the plant such as leaves, stems, and fruits. Initially, small circular brown spots appear on the leaves and fruits of infected plants, followed by yellowing of leaves and dropping off and formation of elliptical lesions on stems and petioles.

Selection of disease-free seeds and before sowing gives them hot water treatment by soaking in hot water (50°C) for 25 minutes. Follow crop rotation and remove weeds and other crops of *Solanaceae* family as well as remove crop debris and apply copper-based bactericides for its suitable management.

1.4 VIRAL DISEASES OF TOMATO

1.4.1 TOMATO LEAF CURL VIRUS (TLCV)

Tomato leaf curl virus (TLCV) is a viral disease and transmitted by whitefly (*Bemisia tabaci*). It is known as one of the most common and devastating diseases of tomato, covering the highest tomato disease area in India. TLCV disease is characterized by severe plant stunting along with downward rolling and crinkling of the leaves. The newly emerging leaves show discoloration and express pale yellow coloration along with curling symptoms (Mandal et al., 2017); however, older leaves turn into leathery and brittle. Nodes and internodes are radically reduced in size that gives infected plants a dwarf bushy appearance along with many lateral branches. Many studies verified that TLCV causing up to 100% yield loss under favorable condition (Green and Kalloo, 1994).

The prime strategy for its management is destroying of infected plant at the time of first appearance of disease symptom and destruction of all weeds around the field which serve as alternate host of the virus. Cover the nursery beds with 40 mesh fine nylon net and spray the seedlings with Imidacloprid (3.5 ml/10lt). Pre-sowing treatment of nursery beds with Furadon 3G @15 g/100 sq.m. Apply Thimet 10G @ 10–15 kg/ha 10 days after transplanting, subsequently 2–3 foliar sprays with 0.05% dimethoate (Rogor 1.5 ml/l) or Imidacloprid or Acetameprid or Thiomethoxam (3.5 ml/10lt) at 10 days interval. Use of yellow sticky traps for whiteflies trapping. Apply 2% power oil to the plants to prevent acquisition and inoculation of the virus by whiteflies. Sow 5–6 rows of border crops of maize, pearl millet and sorghum all-round the tomato plot at least 50–60 days prior to transplanting of tomato seedlings. Moreover, grow resistant varieties/hybrids *viz*., Kashi Abhiman, Kashi Aman, Kashi Amrit, Kashi Vishesh, Hisar Anmol, Arka Rakshak, Arka Samrat, Arka Ananya, Kalyanpur Angoorlata, BCTH-4 and TLBRH-6 (Mandal et al., 2017; Nagendran et al., 2017; Anonymous, 2019).

1.4.2 TOBACCO MOSAIC VIRUS (TMV)

The characteristic symptom of this disease is by light and dark green mottling on the leaves along with wilting of young leaves observed during sunny days at initial infection on plants. The leaflets of infected leaves are distorted, creased, and smaller in size than normal leaves. At an advanced stage, the leaflets become sunken and expressing the symptom of "fern leaf." The infected plant growth will be checked and become pale green color. The virus is transmitted by contact with hands when we touch infected plants with healthy ones, plant debris, implements, and by clothes also.

For its management, soaking of seeds in 10% trisodium phosphate (Na_3PO_4) solution for 15 minutes or heat dry the seeds at 70°C for 2–4 hrs. Always use disease-free seeds and destroy infected plants and plant debris. Avoid contact with infected plants as well as never smoke and allow anyone for smoking tobacco near the field because cigarette, tobacco, and its ash can also transmit infection. Sanitize your hand with sanitizers or wash gently with soap and water before visiting the tomato field. Do not grow other solanaceous crops near the field and growing of resistant varieties/hybrids is the best practice to get economical yield.

1.4.3 TOMATO SPOTTED WILT VIRUS (TSWV)

The TSWV is spreads through thrips (*Thrips tabaci, Frankliniella schultzi*, and *F. occidentalis*). There are numerous symptoms of TSWV out of which two most common and authentic symptoms are firstly bronzing of new leaves followed by development of numerous small, dark spots and second symptom is the dropping of infected leaves that creates a wilt-like appearance for which this disease was named. Moreover, sometimes plant show stunting, mottling, terminal stems dark streaking and dieback of the growing tips. One-sided growth habit is found in infected plants, and sometimes their growth may be completely checked. Early affected plants often do not able to produce any fruit, while those plants infected after fruit set produce few deformed fruits with striking symptoms with chlorotic concentric spots, raised bumps, and uneven ripening.

For its management, it is necessary to eliminate thrips and host plants populations from the field to prevent the disease. Grow tomato crops as far away as possible from flower crops field. Apply optimum nitrogen fertilizers, because excess nitrogen may lead to high thrips populations and TSWV levels. Talc based powder formulations of bio-agent *Pseudomonas fluorescens* strains to seed, soil, and foliage or as a seedling dip followed by three foliar sprays of 3% Neem oil reduced the thrips populations including nymphs as well as adults and also reduced the incidence of this disease (Vasanthi et al., 2017). Spray the seedlings with Imidacloprid (3.5 ml/10 lt). Treatment of nursery beds with Furadon 3G @15 g/100 sq.m. Apply Thimet 10G @ 10–15 kg/ha 10 days after transplanting, subsequently 2–3 foliar sprays with 0.05% dimethoate (Rogor 1.5 ml/l) or Imidachlorpid or Acetameprid or Thiomethoxam (3.5 ml/10 lt) at 10 days interval. Moreover, growing resistant varieties/hybrids is the best practice to harvest economical yield.

1.4.4 TOMATO BUNCHY TOP VIRUS (TBTV)

The bunchy top virus causes extensive abnormal plant growth with apical proliferation. The new leaves of axillary buds give dense bunchy appearance of leaves. The margins of leaflet curl towards the tips and the surface show puckered features. It causes necrosis of leaves followed by stems that is clearly seen in the infected plants. The diseased plants produce very few flowers as well as fruits of smaller in size and numbers.

When the first symptom arises on the plans in the field, then immediately removed and destroyed them to check the disease spreading. Alternate or collateral hosts are also removed from the field at the time of first weeding to minimize disease spreading.

1.5 PHYTOPLASMA DISEASE OF TOMATO

1.5.1 TOMATO BIG BUD (TBB)

This phytoplasma disease is transmitted by leafhopper (*Orosius argenatatus*) and infects all parts of the plant. The primary symptom of disease is the large (big), swollen green buds that fail to develop normally and do not set fruit. Apical stems become thick and show upright bushy growth because of shortened internodes and small leaves. The newest fruit truss, instead of becoming recurred like usual plants, assumes a vertical direction. The buds on the truss also direct in an upright position, the calyx segments stay united nearly tips, as well as the whole calyx blows up to look like a bladder with a toothed opening at the top. Especially in indeterminate varieties where training and pruning is important practice, the growing points fail to develop normally after pruning. In a short time, the axillary buds grow out, forming shoots affected in the same way as the main shoot. Concurrently, there is a continuing thickening of the stems of the infected parts because of the creation of an abnormal tissue. If the terminal buds' growth is completely ceases, the stems will be thicker and clearly marked in the plant. Young shots initially infected by this disease subsequently affected leaves become yellow-green and puckered. The leaves size reduces and fruits will remain green or colors extremely slowly or not at all as the disease advances. No resistant variety have been yet reported, so far, the best management practice is removal and destruction of the affected plant parts at the time of first symptom appearance over and above use of insecticides like, Pyrethrin or Azadirachtin can be apply for eco-friendly management of leafhoppers in disease-prone areas. Moreover, companion planting of marigolds or geraniums is also found helpful in repelling the leafhoppers.

1.6 INTEGRATED DISEASE MANAGEMENT (IDM) OF TOMATO

The major tomato diseases can be managed by using fungicides or bactericides or insecticides (for viruses/phytoplasma vectors) but it is expensive and not easily available in the local markets. However, these diseases can be restricted by several eco-friendly means like by grafting on resistant root stalks in case of soil-borne pathogens, use of resistant cultivars, crop rotation, summer plowing, removal of crop debris, cleaning of fields, etc. In view of the above facts, there is an utmost need to develop varieties/ hybrids of tomato that can resist/tolerate the ravage of major disease. Lack of horizontally resistant varieties of tomato is also a limiting factor for tomato growers. Therefore, the efficient management of these diseases can be done by employing IDM practices (Table 1.1). The IDM practices reduce the disease management cost along with efficiently secure the crop from these diseases along with reduce the chances of attack by the other pathogens (fungi, viruses, and bacteria); over and above, it saves the environment and good for our health.

TABLE 1.1 Integrated Management of Tomato Against Major Diseases

Diseases	Cultural Practices and Bio-Control	Use of Chemical Pesticides
Tomato leaf curl virus	Cover the nursery beds with 40 mesh fine nylon net and destroying of an infected plant at the time of the first appearance of the disease. Use of yellow sticky trap for whiteflies. Sow 5–6 rows of border crops of maize, pearl millet, and sorghum all-round the tomato plot at least 50–60 days prior to transplanting of seedlings.	Spray the seedlings with Imidacloprid (3.5 ml/10lt). Pre-sowing treatment of nursery beds with Furadon 3G @15 g/100 sq.m. Apply Thimet 10G @ 10–15 kg/ha 10 days after transplanting, subsequently 2–3 foliar sprays with 0.05% dimethoate (Rogor 1.5 ml/l) or Imidachlorpid or Acetameprid or Thiomethoxam (3.5 ml/10lt) at 10 days interval.

TABLE 1.1 *(Continued)*

Diseases	Cultural Practices and Bio-Control	Use of Chemical Pesticides
Tomato mosaic virus	Heat dries the seeds at 70°C for 2–4 hrs. Avoid contact with infected plants as well as never smoke and allow anyone for smoking tobacco near the field. Sanitize your hand with sanitizers or wash gently with soap and water before visiting the tomato field. Do not grow other solanaceous crops near the field and grow resistant varieties.	Soaking of seeds in 10% trisodium phosphate (Na_3PO_4) solution for 15 minutes.
Tomato spotted wilt virus	Grow tomato crops as far away as possible from flower crops field. Apply optimum nitrogen fertilizers, because excess nitrogen may lead to high thrips populations and TSWV levels.	Bio-agent *Pseudomonas fluorescens* and 3% Neem oil reduces the thrips populations that is vector of this disease. Spray the seedlings with Imidacloprid (3.5 ml/10lt).
Bacterial wilt	Seed treatment with hot water at 52°C for 20 minutes. Proper drainage should be maintained. Application of bio-agents (*Pseudomonas fluorescens, Streptomyces virginiae* (Y30 and E36) and *Bacillus subtilis*) in the soil.	Apply bleaching powder @ 15–20 kg/ha and 4.5 quintal lime to the field and mix it properly to the soil at least 3–4 weeks prior to transplanting. Soil drench with Plantomycin @ 100 ppm and Copper oxychloride @ 0.3% w/v.
Bacterial Canker	Use disease-free seeds or seedlings and soaking of seeds in hot water (56°C) for 30 minutes. Destroy crop debris and follow crop rotation with non-host crop minimum for 3 years. Use of bio-agent *Pseudomonas putida* in the soil.	Applications of Metalaxyl + Mancozeb @ 0.2% may reduce secondary spread of disease.
Damping-off seedlings	Seed treatment with bioagent, *Trichoderma viride* @ 4 g/kg of seeds or Thiram @ 3 g/kg of seeds is the most effective and safe preventive measure to check the disease at the initial stage. Use of bio-agents viz., *Pseudomonas corrugata, P. fluorescens, P. marginalis, P. putida, P. syringae, P. viridiflava* in the soil.	Drench of tomato seedlings with Copper oxychloride 0.2% or Bordeaux mixture 1%. Spraying of 0.2% Metalaxyl when cloudy weather seen.

TABLE 1.1 *(Continued)*

Diseases	Cultural Practices and Bio-Control	Use of Chemical Pesticides
Late blight	Overhead irrigation should be avoided.	Spraying with mancozeb 0.2% or captafol @ 0.2% or Metalaxyl 0.2% or copper oxychloride @ 0.2% or Tridemorph 80%EC 1.5 ml/L
Early blight	Use pathogen-free seeds, or collection of seeds from disease-free plants. Crop rotation with non-host crops minimum for 2 years. Use of bio-agents viz., *Pseudomonas, Bacillus, Azotobacter, Seeatia.*	Spraying of diathane M-45 @ 0.2% for effective control of this disease.
Fusarium wilt	Soil solarization during May-June. Optimum dose of nitrogenous fertilizer and use of well decomposed FYM/compost. Use of antagonistic organisms like, *Alcaligenes faecalis* (S18) and *Bacillus cereus* (S42).	Crop drenching with carbendazim 0.1% and follow crop rotation with a non-relative crop.
Septoria leaf spot	Collection and disposal of all foliage from infected plants. Improve air circulation around the plants. Do not use overhead watering. Nightshade and horse nettle host plants should be eradicated. Crop rotation with non-host crops, i.e., less than 2 years.	Treatment of tomato seeds with thiram or mancozeb @ 2 g/kg seeds is effective in managing seed-borne pathogens. On standing crop initial spraying with diathane M-45 @ 0.2% for timely controls of this disease.
Cercospora leaf spot	Collection and burning of infected plant debris and follow proper drainage and adopt crop rotation with non-solanaceous crops.	Spray the crop with carbendazim (0.1%) or thiophanate methyl (0.1%) or diathane M-45 (0.25%) or redomil (0.25%) or copper oxychloride (0.3%).
Verticillium wilt	Crop rotation to non-host crops, such as Poaceae family crops.	Spot drench with carbendazim (0.1%) or benomyl 0.05%.
Powdery mildew	The best option for its management is to protect plants and developing fruit.	Dinocap 48% EC 0.1% or propiconazol 25% EC @ 0.15% or tridemorph 80% EC @ 0.15% or tradimefon 25% WP 0.15% for effective disease control

TABLE 1.1 *(Continued)*

Diseases	Cultural Practices and Bio-Control	Use of Chemical Pesticides
Sclerotium collar rot	Rotate crop with non-related crops such as maize, sorghum, small grains, or cotton. FYM of green manure should be fully decomposed. Immediate clipping of leaves when any signs of disease appeared. Mulching by compost, leaf mold, hay, or paddy straw serves to maintain soil cover and restrict disease transmission. Use of antagonistic organisms like, *Burkholderia cepacia* (T1A-2B) and *Pseudomonas* sp. (T4B-2A).	Soil application of *Trichoderma viride* and seedling dip with tebuconazole @ 0.05% followed by, soil drenching with tebuconazole @ 0.05%. Fluxapyroxad + pyraclostrobin also reduce the incidence of *Sclerotium* collar rot in tomato.
Anthracnose	Enrich soil with organic mulches. Improve drainage protect against rain splash	Chlorothalanil 75% WP 0.2% or benomyl 50%WP 0.1% or thiopenate methyl 70% WP 0.2%
Buckeye fruit rot	Planting on raised beds and staking of the crop. Ensure proper soil drainage and aeration. Crop rotation with non-related crops for at least 2 years.	Spraying of fungicides like, redomil or metalaxyl + mancozeb or cymoxinal + mancozeb @ 0.25% can effectively manage this disease.

Source: Modified from Singh et al. (2017); Kumar et al. (2018).

1.7 CONCLUSION

Tomato is a versatile vegetable and affected by more than 200 diseases caused by fungi, bacteria, viruses, phytoplasma, etc., during its germination, growth, flowering, fruiting, harvesting, and postharvest handling. These diseases directly affected the quality and quantity of tomatoes and ultimately cause huge economical losses to tomato growers. Among all diseases, EB, late blight, and TLCV are responsible for huge losses in most of the tomato-growing regions of the world. Due to the lack of horizontal resistant varieties, it is difficult to manage many diseases that occur at different plant growth stages. By using pesticides, we can manage major diseases to some extent, but their residual effect may cause several severe diseases in our body as well as make environmental pollutions. Hence, for tomato growers and consumers, it would be better to incorporate integrated approaches *viz.*, use of resistant cultivars, grafting on resistant root stalks in case of soil-borne pathogens, crop rotation, summer plowing, removal of crop debris, cleaning of fields, use of bio-agents, etc., for its effective and eco-friendly disease management.

KEYWORDS

- **integrated disease management**
- **resistant/tolerant varieties**
- **symptoms**
- **tobacco mosaic virus**
- **tomato diseases**
- **tomato leaf curl virus**
- **tomato spotted wilt virus**

REFERENCES

Abdel-Sayed, M. H. F., (2006). *Pathological, Physiological and Molecular Variations Among Isolates of Alternaria Solani the Causal of Tomato Early Blight Disease* (p. 181). Faculty of Agriculture, Cairo University, Egypt.

Anonymous, (2019). *Annual Report 2018–19* (p. 188). ICAR-Indian Institute of Vegetable Research, Varanasi.

Banerjee, M. K., & Kalloo, G., (1987). Sources and inheritance of resistance to leaf curl virus in *Lycopersicon*. *Theor. and Appl. Genet., 73*(5), 707–710.

Banyal, D. K., Mankotia, V., & Sugha, S. K., (2008). Integrated management of tomato collar rot caused by *Sclerotium rolfsii*. *J. Myco. and Plant Patho., 38*(2), 164–167.

Byrne, J. M., Hausbeck, M. K., & Latin, R. X., (1997). Efficacy and economics of management strategies to anthracnose fruit rot in processing tomatoes in the Midwest. *Plant Dis., 81*, 1167–1172.

Datar, V. V., & Mayee, C. D., (1981). Assessment of loss in tomato yield due to early blight. *Indian Phytopatho., 34*, 191–195.

Dutta, P., (2012). Management of bacterial wilt of tomato through an innovative approach. *J. Biolog. Contr., 26*(3), 288–290.

Green, S. K., & Kalloo, G., (1994). Leaf curl and yellowing viruses of pepper and tomato: An overview. *Technical Bulletin, 21*. AVRDC, Taiwan.

Kumar, S. P., Srinivasulu, A., & Babu, K. R., (2018). Symptomology of major fungal diseases on tomato and its management. *J. Pharmacol. and Phytochem., 7*(6), 1817–1821.

Mahato, A., Biswas, M. K., & Patra, S., (2017). Eco-friendly management of collar rot disease of tomato caused by *Sclerotium rolfsii* (Sacc.). *Intel. J. Pure and Appl. Biosci., 5*(1), 513–520.

Maienfisch, P., Huerlimann, H., Rindlisbacher, A., Gsell, L., Dettwiler, H., Haettenschwiler, J., Sieger, E., & Walti, M., (2001). The discovery of thiamethoxam: A second generation neonictinoid. *Pesticide Sci., 57*, 165–176.

Mandal, A. K., Maurya, P. K., Dutta, S., & Chattopadhyay, A., (2017). Effective management of major tomato diseases in the Gangetic plains of eastern India through integrated approach. *Agri. Res. and Tech: Open Access J., 10*(5), 555796. doi: 10.19080/ARTOAJ.2017.10.555796.

Nagendran, K., Pandey, K. K., Rai, A. B., & Singh, B., (2017). Viruses of vegetable crops: Symptomatology, diagnostics and management. *IIVR Technical Bulletin No. 75* (p. 48). IIVR, Varanasi.

Ray, P. K., Verma, R. B., Solankey, S. S., & Chaudhary, A., (2018). Assessment of tomato advanced lines to resistance of late blight. *Intel. J. Current Microbio. and Appl. Sci., 7*(1), 2622–2629.

Sadashiva, A. T., Reddy, M., Reddy, K., Krishna, M., & Singh, T. H., (2002). Breeding tomato (*Lycopersicon esculentum* Mill.) for combined resistance to bacterial wilt and tomato leaf curl virus. *Proceed. of the Intl. Conference on Vegetables India* (pp. 125–133).

Scholthof, K. B. G., Adkins, S., Czosnek, H., Palukaitis, P., Jacquot, E., Hohn, T., Hohn, B., et al., (2011). Top 10 plant viruses in molecular plant pathology. *Mol. Plant Pathol., 12*(9), 938–954.

Sherf, A. F., & Macnab, A. A., (1986). *Vegetable Diseases and Their Control* (pp. 634–640). John Wiley and Sons, New York, USA.

Singh, V. K., Singh, A. K., & Kumar, A., (2017). Disease management of tomato through PGPB: Current trends and future perspective. *3 Biotech., 7*(4), 255. doi: 10.1007/s13205-017-0896-1.

Solankey, S. S., Akhtar, S., Neha, P., Ray, P. K., & Singh, R. G., (2017). Reaction of tomato genotypes for resistance to late blight (*Phytophthora infestans* Mont. de Bary) disease. *Indian J. Agril. Sci., 87*(10), 1358–1364.

Thiribhuvanamala, G., Rajeswar, E., & Doraiswamy, S., (1999). Inoculum levels of *Sclerotium rolfsii* on the incidence of stem rot in tomato. *Madras Agril. J., 86*, 334.

Vasanthi, V. J., Samiyappan, R., & Vetrivel, T., (2017). Management of tomato spotted wilt virus (TSWV) and its thrips vector in tomato using a new commercial formulation of *Pseudomonas fluorescens* strain and neem oil. *J. Entomol. and Zoology Stud., 5*(6), 1441–1445.

Walker, J. C., (1952). *Diseases of Vegetable Crops* (pp. 306–308). Mc Graw Hill Company. Inc. New York, USA.

CHAPTER 2

MOLECULAR TOOLS FOR BACTERIAL WILT RESISTANCE IN TOMATO

RAJESH KUMAR,[1] SUSHMITA CHHETRI,[1] BAHADUR SINGH BAMANIYA,[1] JITENDRA KUMAR KUSHWAH,[2] and RAHUL KUMAR VERMA[3]

[1]Department of Horticulture, School of Life Sciences, Sikkim University, Tadong, Gangtok, East Sikkim, Sikkim, India, E-mail: rkumar@cus.ac.in (Rajesh Kumar)

[2]Department of Agriculture, Government of Uttar Pradesh, India

[3]Krishi Vigyan Kendra, Madhepura, Bihar Agricultural University, Sabour, Bhagalpur, Bihar, India

ABSTRACT

Bacterial wilt is one of the known devastating diseases caused by *Ralstonia solanacearum* in tomato and other crops within the family. It is not only affecting yield but also the economy by limiting the crop production area. The bacterium causes yield reduction across the globe because of the availability of its wide range of host plants. Bacterial wilt pathogen is further classified as race, biovar, and sequevar based upon the carbohydrate and sugar oxidation. Identification and diagnosis of the pathogen in early stages based upon morphological, biochemical, and DNA markers, helps efficiently in control of further spread of pathogen. There are several partial resistant genotypes (or) varieties and efficient chemical control for the disease but efficient and sustainable control is not reported so far. Scientists from the plant science are still struggling to find sustainable management of the pathogen and to extract knowledge on the nature of attacking of pathogen. Marker (molecular) assisted selection in correlation with the traditional breeding may help in controlling the disease and development of the new cultivars with durable resistance. However, this chapter provides the various aspects of the bacterial wilt of tomato and achievements made so far to control this widespread pathogen.

2.1 INTRODUCTION

Solanaceous vegetables are also termed as "nightshade family," consists of about 75 genera and 2000 species (Rubatzky and Yamaguchi, 1997). Solanaceous vegetables include several edible crops such as tomato (*Solanum lycopersicum* L.), brinjal (*Solanum melongena* L.), potato (*Solanum tuberosum* L.), hot pepper (*Capsicum annuum* L.), and sweet pepper (*Capsicum annuum var. grossum* L.). Among these, Tomato (*Lycopersicon esculentum* Miller) is utmost important crop cultivated globally. The crop was renamed as *Solanum lycopersicum* L. (Peralta et al., 2008). The leading tomato growing countries are the USA, Japan, China, and several European countries (Ram, 2017) and worldwide production of tomatoes accounts for 170.8 million tons and China being the leading producer of tomato (31% share) (Sen Nag, 2017). In India, it ranks 2nd among all the vegetables in terms of area and production. However, Andhra Pradesh (14.42%), Madhya Pradesh (12.70%), Karnataka (9.48%), Gujarat (7.36%), Odisha (6.77%), West Bengal (6.53%), Maharashtra (5.88%), Chhattisgarh (5.34%), Bihar (5.26%), Telangana (4.60%), Uttar Pradesh (4.34%), Haryana (3.97%), and Tamil Nadu (3.76%) account for 91% of the total output production of the country (Monthly Report Tomato, 2018).

Tomato is extensively being grown globally as an annual plant. It is treated as "protective food" universally as it provides plenty of essential vitamins, minerals, and organic acids. The total sugar percent present in ripe fruit is 2.5%. It also consists of ascorbic acid which varies from 16–65 mg/100 g of total fruit weight. The total amino acid present in the fruit is 100–350 mg/100 g of total fruit weight. However, the amount of vitamin A present in tomato accounts for 320 I.U/100 g (Thamburaj and Singh, 2014). Tomatoes are usually consumed as raw vegetables in salads, etc. However, processed products like puree, juice, paste, syrup, ketchup, whole peeled tomato, etc., are prepared from tomato extensively. Consuming tomato helps in curing constipation and it is considered as a good appetizer as well. However, the seeds of tomato consist of 24% of oil which is used in the manufacture of margarine (Muthukumar and Selvakumar, 2017).

Tomato is said to have originated in the Peru-Ecuador area of the Andes (South America). Existence of tomato was first recorded in South America in the year 1554. Later on, it was moved to Mexico and Europe for its domestication and cultivation in the same year. However, the ancestor of tomato is *Lycopersicon esculentum* var. *cerasiformae.* Tomato is an herbaceous, annual to perennial, prostrate, and sexually propagated and self-pollinated crop. It has an identical genome formula as 2n = 2x = 24. The tomato plant is considered as indeterminate and determinate based upon growth habit. In the determinate types, the growth of plants is restricted by the initiation of terminal flowers and therefore, plants are dwarf. Whereas, in indeterminate types, the growth of the plant is continued and there is less appearance of flowers and fruits on the stem (Thamburaj and Singh, 2014).

However, every year the yield of tomato as affected by an incidence of a wide range of fungal, bacterial, and viral diseases. Plant disease may be defined as a deviation from the normal growth and functioning of the plant due to biotic and abiotic factors (Manoharachary and Kunwar, 2014). As a result, there is a reduction in economic value, yield loss and degradation in quality. The most common diseases of solanaceous vegetables are damping-off (*Pythium aphanidermatum*), early blight (EB) (*Alternaria solani*), bacterial leaf spot (*Xanthomonas campestris PV. vesicatoria*), late blight (*Phytophthora infestans*), bacterial wilt (*Ralstonia solnacearum*), etc. Among all these diseases the bacterial disease *viz.* bacterial wilt (*R. solanacearum)* is the most destructive ones across the world (Jan et al., 2010). The characteristic symptoms of bacterial wilt include initial wilting of upper leaves followed by wilting completely in the later stage. The vascular tissue with brown discoloration of the infected stem and the cross-section cut gives white and yellowish bacterial ooze (Tans-Kersten et al., 2001). Furthermore, it is reported that the management of this disease is difficult due to the high pathogenic diversity.

According to The United Nation's Food and Agriculture Organization (FAO), the harm caused by pests and diseases towards the yield every year is 20–40% (Terra et al., 2015). As a result, the increasing population and decreasing production of the crops will give rise to food shortage and will challenge food security in the days to come.

In today's scenario, the basic method adopted for the management of the disease is by applying fungicides, antibiotics, and chemical fertilizers. But these fungicides and chemical fertilizers have proved to cause several health hazards and environmental pollution (Nicolopoulou-Stamati et al., 2016). Moreover, the application of these harmful chemicals has led to degradation of soil, pesticide residue in the crops and resistivity by the causal organisms. It was reported that bacterial diseases are resistant to conventional antibiotics (Lee Ventola, 2015). Even the resistant cultivars have been evolved from the different resistant sources to bacterial wilt in tomato (Scott et al., 2005). Hence bacterial wilt is location-specific, therefore it is very difficult breeding for durable as well as stable resistance. According to carbohydrate utilization, *Ralstonia solanacearum* is categorized into five and six based upon their pathogenicity (Hayward, 1964; Hayward et al., 1991; Xue et al., 2011). Race 1 affects Solanaceous and other vegetables.

Therefore, as a part of the solution, biotechnological aspects can be employed in combination with conventional breeding methods. For instance, the application of molecular markers can be implemented for mapping and tagging of markers for desirable traits, which further can be utilized to derive resistance genes for the management of various diseases in the crops (Bahadur et al., 2015).

2.2 PHYLOGENY, CLASSIFICATION, AND DIVERSITY OF RALSTONIA SOLANACEARUM

Ralstonia solanacearum, formerly called *Pseudomonas solanacearum*, the causal organism of bacterial wilt, is a most destructive pathogen cause harm to more than 200 plant species, including a wide range of vegetables, ornamental crops, and weeds. The genus *Ralstonia* falls within the β-subdivision of Proteobacteria and includes other species such as *R. insidiosa, R. picketii, R. syzygii, R. mannitolilytica* and *R. solanacearum* (Yabuuchi et al., 1995). The strains of pathogens are classified as biovars and races. Based upon the host range *R. solanacearum* has five races and six biovars as well based on their differential capability to produce acid from the panel of carbohydrates (Denny, 2006). However, all the five races of *R. solanacearum* comprise of wide host ranges and vast degree of geographic distributions. Race 1 is prevalent in the southern United States as well as Asia, Africa, and South America. While race 2 is mainly found in Southeast Asia and Central America and its main host is banana. Race 3 is mainly linked with potato and has a wide range of hosts. Meanwhile, race 4 is associated with ginger and abundantly found in Asia and Hawaii.

Furthermore, in China race 5 affects the mulberry (Kelman, 1997; Denny, 2006).

2.2.1 CAUSAL ORGANISM OF THE DISEASE

Ralstonia solanacearum is a gram-negative plant pathogenic bacterium which doesn't form spore. The cells of *Ralstonia solanacearum* is non-encapsulated and its size is 0.5–0.7 × 1.5–2.0 μm (Olson, 2005). The optimum temperature for the growth of bacteria is 27 to 32°C (Patrice et al., 2008). It is the second most important pathogens known to affect over 200 species (Yuliar et al., 2015). It is also known by the name Southern bacterial blight. The pathogens are an aerobic obligate organism and thrives well in the temperature ranging from 10 to 41°C (Tahat and Sijam, 2010). The bacterium is mainly cultured in liquid and solid (agar) media. When cultured in solid agar medium, two colonies differing in their morphology can be observed after 36 to 48 hours of growth at a temperature of 32.4°C. However, the colonies which are cream-colored or white, irregularly shaped or round, opaque, and fluidal are virulent types while, colonies that are round uniformly, small, and butyrous are the non-virulent types. The occurrence of virulent and non-virulent types is due to oxygen stress in liquid media during storage conditions. It was reported that virulent types appeared as white with pink centers while non-virulent types appeared as dark red when the strains were subjected to (Patrice et al., 2009).

2.2.2 CLASSIFICATION OF R. SOLANACEARUM

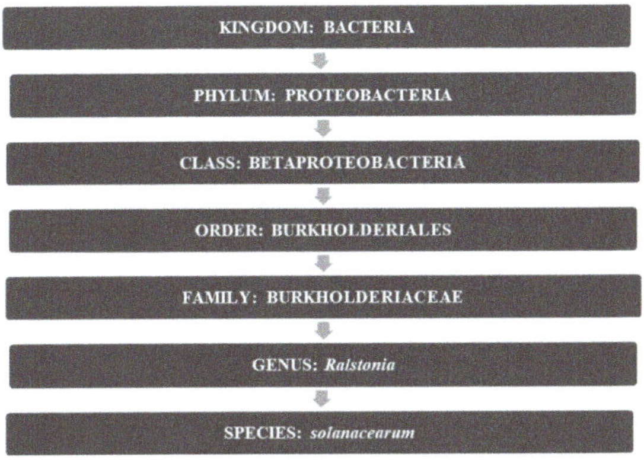

2.2.3 OTHER NAMES OF R. SOLANACEARUM

i. *Bacillus solanacearum*;
ii. *Burkholderia solanacearum*;
iii. *Pseudomonas solanacearum*;
iv. *Ralstonia solanacearum*.

2.2.4 DISEASE SYMPTOMS

- During the warmest period of the day, the initial sign of disease shows wilting of younger leaves at the terminal end of the branches. However, the wilted plant may look towards recover when the temperature lowered down.
- As the favorable condition prevails, the disease develops and causes the whole plant to wilt quickly. Further, the yellowing of plant occurs, and the plant eventually dies.
- Another common symptom that appears is stunting of plants which may occur at different stage of plants.
- The vascular bundles of young stem become long, narrow with dark brown streaks. However, in highly susceptible varieties, the young and succulent stem also may collapse down.
- The cross-cut section of badly infected plants may show brown discoloration of infected tissues.
- The appearance of sticky and milky white exudates from freshly cut infected stem revealed the presence of masses of bacterial cells densely in the vascular tissues, particularly with xylem.
- Another common diagnosis of the disease can be revealed by placing the infected cut stem in water which shows the streaming of viscous white spontaneous slime. The streaming occurs due to the presence of bacterial ooze colonized in the vascular bundles. This "stem-streaming" is used to detect the diseased plant in the field.

2.2.5 EPIDEMIOLOGY AND DISEASE CYCLE

Ralstonia solanacearum is soil and water-borne pathogen. The soil and water form a reservoir for inoculum; therefore, the bacterium survives and disperse for periods of time in infested water or soil. Tomato plants are infected by the bacterium through the plant roots. The bacterium penetrates through plant roots favored by another organism, such as root-knot nematode. However, the infection may also occur through injuries caused either during cultural practices or through insect damage. In some cases, the infection may also occur through irrigation practices, where bacteria can move from the roots of infected plants to healthy ones. Moreover, the factors that favor the survival of bacteria are soil type and structure, soil moisture content, high temperature, organic matter in soil, water salt content and pH, and the existence of other antagonist microbes (Coutinho, 2005). The bacterium can reside outside of the plants termed as 'exterior phase.' However, the bacteria do not survive for longer period epiphytically when exposed to high temperature or relative humidity below 95%.

However, infected plant material in the soil, surface water irrigation and diseased weeds provide a medium for survival of *R. solanacearum* for days to years. From these sources, bacteria may disperse from infected to healthy fields or by transfer of soil on farm machinery and surface runoff water after rainfall or irrigation water (Elphinstone, 2005). The bacterium can also be multiplied in infected water bodies and spread to other fields through water flows. At low temperatures like 4°C, bacterial population can decrease rapidly, but bacteria may survive often in the latent state too. In natural habitat, bacterium *R. solanacerum* race 3 biovar 2 might survive during winter in the semi-aquatic weeds, in crop debris or in the rhizosphere of non-host plants. Whenever the temperatures start to increase the bacterium from semi-aquatic weeds was increasingly released after winter (Fegan, 2005).

2.2.6 DIAGNOSIS AND IDENTIFICATION

The early diagnosis of bacterial wilt of tomato as a first step is symptom identification. However, several microbiological and molecular methods are adopted for accurate identification of *R. solanacerum* from water or soil samples or symptomatic and asymptomatic plants. The early detection as well as identification of bacteria from contaminated soil or infected plants and water samples is facilitated by screening test. However, the screening test include stem stream test, modified (plating on semi-selective medium), *R. solanacerum* specific antibodies test as immunodiagnostic assays, DNA identification using specific primers, and pathogenicity assessment through susceptible plant hosts such as seedlings of tomato. But they cannot identify the race or biovar detection of the organism. In addition, for rapid and field detection of *R. solanacearum* immunostrips (Agdia) rapid screening tests are available commercially (Patrice et al., 2009).

The biovar determination of *R. solanacearum* is facilitated by a biochemical growth test. This test depends upon the differential ability of the strains of pathogen to produce acid from many carbohydrate sources, including sugar alcohols and disaccharides. At the sub-species level, several nucleic-acid-based methods with specific primers can be used to identify the strains of *R. solanacearum* (Pradhanang, 2005).

2.3 INTEGRATION OF BIOTECHNOLOGY IN BREEDING PROCEDURES

Several molecular markers such as RAPD, AFLP, RFLP, CAPs, and SSR are mapped on the tomato genetic map. Molecular markers basically consist of specific molecules, which show easily detectable differences among different species. These markers may be based on proteins, e.g., isozymes, or DNA (Singh, 2018). According to the published classical data on linkage map, tomato comprising of 285 phenological and other physiological traits, disease resistance traits, and isozymes markers for which corresponding genes were mapped on 12 chromosomes. However, with the DNA marker technology, the development of high-density tomato maps has been possible today. The first molecular linkage map of tomato comprised 18 isozymes and 94 DNA markers. High-density molecular linkage maps exploited the identification of the DNA markers which are tightly linked to gene of interest (Ram, 2017).

2.3.1 MOLECULAR MARKERS LINKED TO BACTERIAL WILT RESISTANCE OF TOMATO

2.3.1.1 SINGLE NUCLEOTIDE POLYMORPHISMS (SNP)

It arises when different nucleotide occurs at the same genomic site of different strains of a species. The analysis of genome sequence information data of few to several lines/individuals leads to the identification of SNPs initially. However, after the identification of SNP loci, the genome sequence is utilized to design a suitable assay based on, usually, PCR amplification of the concerned genomic segment. Among different SNP genotyping platforms, the Illumina Golden Gate SNP genotyping platform is the most sophisticated and highly sensitive one. The next generation sequencing (NGS) methods have led to the development of SNP genotyping methods based on whole-genome sequencing. Further, due to this approach genotyping can be done in a single step and can detect all the alleles present at a given locus.

2.3.1.2 AMPLIFIED FRAGMENT LENGTH POLYMORPHISM (AFLP)

It utilized restriction fragments towards PCR amplification and generates highly polymorphic DNA marker system by combining the restriction digestion of DNA step of RFLP system with the PCR technique. In this method, 100–500 ng DNA is digested properly with two restriction enzymes, suitable adapters are then ligated at the ends of the restriction fragments resulting, and a quite smaller set of these fragments is amplified selectively by the PCR. An AFLP marker is detected as the presence of an amplified DNA fragment. However, AFLP's can detect a large number of markers and are codominant.

2.3.1.3 SIMPLE SEQUENCE REPEATS (SSR)

SSR's are microsatellites-based markers that are analyzed by PCR amplification of small genomic sites contain the repeated sequence. A pair of specific primers amplifies SSR markers which is located in a unique position in the genome. However, variation produced by differences in the number of repeats present in the different alleles can be easily detected by SSR markers. These are simple, PCR-based, codominant, extremely polymorphic, and highly informative and has the ability to distinguish between closely related individuals.

2.3.1.4 SEQUENCE CHARACTERIZED AMPLIFIED REGIONS (SCAR)

They are developed from desirable selected RAPD markers. However, SCAR are dominant and polymorphism markers and can be developed into array (plus/minus) to eliminate the necessity for electrophoresis. Thus, SCAR is similar to sequence-tagged sites in the construction and application. However, SCAR is dominant marker but theoretically can be converted to codominant markers.

2.3.1.5 IDENTIFICATION OF MOLECULAR MARKERS LINKED TO BACTERIAL WILT RESISTANCE OF TOMATO

According to the experiment, "identification of a molecular marker tightly linked to bacterial wilt resistance in tomato by genome-wide SNP analysis" conducted by Boyoung et al., (2018), the functional SNP within a putative leucine-rich repeat (LRR) receptor-like gene that developed an effective SNP marker in *Solyc12g009690.1* gene on the chromosome 12 that might efficiently differentiate the resistant genotypes of tomato by high-resolution melting (HRM) analysis was identified by using whole-genome sequencing again of tomato varieties. According to the obtained results, the *Solyc12g009690.1* gene is linked tightly to the *Bwr-12*, which has been found in the variety 'Hawaii 7996.' This gene was present in other tomato cultivars resistant to the disease bacterial wilt such as '10-BA-3-33,' 'BWRs,' and '10-BA-4-24' and contain the same SNP (G/A) at 699 bp position. For the development of the SNP markers, seven resistant germplasm 'Hawaii7998,' 'Hawaii7996,' '10-BA-3-33,' 'BWR-22,' 'BWR-1,' 'BWR-23' and '10-BA-4-24' with two susceptible genotypes 'BWS-3' and 'Heinz1706' were taken.

In an experiment, "an identification of two AFLP markers are linked to bacterial wilt resistance in tomato and their conversion to SCAR markers" conducted by Miao et al. (2009). Molecular markers which are linked to the resistance of bacterial wilt in tomato was identified using bulked segregant analysis (BSA) in combination with AFLP methods. The two markers identified were E-AAT/MCGA and E-AAG/M-CAT with a size of 200 bp and 300 bp respectively from tomato cultivar T51A categorized as resistant parent, while T9230 as a susceptible. However, the markers were linked to the gene TRSR-1. Since, the AFLP markers are dominant and for the breeding purpose it needs to be converted; therefore, these markers were converted into codominant SCAR markers as TSCARAAT/CGA and TSCARAAG/CAT.

Miao et al. (2009) reported in an experiment, "identification of AFLP markers linked to bacterial wilt resistance gene in tomato." AFLP marker RRS-342 was found closely linked to one of the bacterial wilt resistant gene, with a genetic distance of 6.7 cm with the crosses made between two tomato cultivars like resistant tomato variety T51A and susceptible variety T9230. Later it was reported that the AFLP marker was converted to a SCAR marker.

In a study, "genetic studies of yield and its component traits and identification of Molecular markers linked to bacterial wilt resistance in tomato (*Solanum lycopersicum* L.)" conducted by Ramesh (2013) reported that the F_2 individual derived from crosses made between Pusa Ruby and two other genotypes named CLN2768A and CLN2777H when subjected to genotyping with the SSR 20 marker indicated to be associated with bacterial wilt resistance. However, the derived marker is SLM6-15 using BSA.

A study, "screening of tomato genotypes against bacterial wilt (*Ralstonia solanacearum*) and validation of resistance linked DNA markers" was conducted by (Kumar et al., 2018) using 57 different tomato germplasm to screen against bacterial wilt under greenhouse conditions using inoculation technique. Morphological and molecular screening of the bacterial strains isolated from affected plants was also carried out. Therefore, the experiment revealed the identification of the SCAR marker, SCU176-534 using the technique of phyllo type-specific multiplex PCR found to be related with bacterial wilt resistance that may be used in marker-assisted breeding programs (Table 2.1).

TABLE 2.1 Identification of Molecular Markers Linked to Bacterial Wilt Resistance of Tomato

SL. No.	Sources	Technique	Molecular Markers	Genes
1.	Hawaii7996 (Boyoung et al., 2018)	HRM	SNP	Solyc12g009690.1
2.	T51A (Miao et al., 2007)	Bulked segregant analysis	AFLP markers (EAAT/MCGA and E-AAG/M-CAT)	TRSR-1

TABLE 2.1 *(Continued)*

SL. No.	Sources	Technique	Molecular Markers	Genes
3.	T51A AND T9230 (Shao et al., 2006)	–	AFLP marker RRS-342	–
4.	Pusa Ruby × CLN2768A and Pusa Ruby × CLN2777H (Ramesh, 2013)	Bulked segregant analysis	SSR 20 marker (SLM6-15)	–
5.	EC802398 (Kumar et al., 2018)	Phylotype-specific multiplex PCR	SCAR marker (SCU176-534)	–

2.4 METHOD OF SCREENING OF GENOTYPES FOR BACTERIAL WILT RESISTANCE OF TOMATO

The different screening methods were employed, among which the initial ones include clipping of leaves using scissors and dipping it into a bacterial suspension. This method was developed in 1974 and was used up to the year 1982. However, another method commonly known as stem puncture method was also tested for selecting resistance but the method being too rigorous was rejected later on. In the year 1983 to 1986, an accurate method using infectivity titrations was developed. In this method, the number of bacteria injected into the stem of seedlings is controlled using micropipette 0.1 ml (100 ul) of inoculum with one of six levels of bacterial concentrations (10^3–10^8 cfu/ml). In recent years other methods are adopted like the root dip and soil drenching methods, injection methods and pinprick methods (Tables 2.2 and 2.3).

TABLE 2.2 Varieties/F_1 hybrids resistance to bacterial wilt in tomato

SL. N.	Varieties/F_1 Hybrids	Breeding Method	Source
1.	BWR 5 (Arka Alok)	Pureline selection	IIHR-719-1-6 (CL114-5-1-0) from AVRDC, Taiwan
2.	BWR 1(Arka Abha)	Pureline selection	IIHR 663-12-3-SB-SB (VC-8-1-2-1) from AVRDC, Taiwan

TABLE 2.2 *(Continued)*

SL. N.	Varieties/F_1 Hybrids	Breeding Method	Source
3.	Arka Abhijit (BHR 2)	Hybridization	15 SBSB × IIHR 1334
4.	BT-1	–	–
5.	Arka Ananya	Hybridization	TLBR-6 × IIHR-2202
6.	BT-10	–	–
7.	Arka Shreshta	Hybridization	15 SBSB × IIHR 1614
8.	BT-12	–	–
9.	Arka Rakshak	Hybridization	IIHR-2834 × IIHR-2833
10.	Arka Samrat	Hybridization	IIHR-2835 × IIHR-2832
11.	Swarna Sampada	Hybridization	EC-339074 × EC-386021
12.	Swarna Lalima	Selection	EC 339074 as exotic collection
13.	Swarna Naveen	Selection	EC-369060 as exotic collection
14.	Sakthi	–	–
15.	Mukthi	–	–
16.	Anagha	–	–
17.	Sonali	Hybridization	VC 48-1 × Tamu chico III

Source: Modified by Swarup (2014); and Kumar et al. (2018b).

TABLE 2.3 Genetic Linkage Maps of Tomato (*Lycopersicon* spp.) Developed based Upon Intra- and Interspecific Crosses

Linkage Map	Population Type[a])	Population Size	Number of Markers	Type of Markers[b]
L. esculentum × *L. esculentum* var. *cerasiformae.*				
1. Cervil × Levovil	F_7-RIL	153	377	AFLP, RFLP, RAPD
L. esculentum × *L. pimpinellifolium*				RAPD, RFLP
1. M82-1-7 × LA1589	BC_1	257	120	
				Morphological
2. NC84173 × LA722	BC_1	119	151	RFLP
3. Giant Heirloom × LA1589	F_2	200	90	CAPS, RFLP

TABLE 2.3 *(Continued)*

Linkage Map	Population Type[a]	Population Size	Number of Markers	Type of Markers[b]
4. E6203 × LA1589	BC_2F_6-BIL	196	127	RFLP
5. NC84173 × LA722	F_{10}RIL	119	191	RGA, RFLP
6. NCEBR1 × PSLP125	F2	172	256	EST, RGA, RFLP
7. NCEBR1 × PSLP125	F_8-RIL	172	255	RFLP, EST
L. esculentum × *L. cheesmanii*				
1. UC204B × LA483	F_2	350	71	RFLP
2. UC204B × LA483	F_7-RIL	97	132	RFLP
L. esculentum × *L. parviflorum*				Morphological, RFLP, SCAR
1. E6203 × LA2133	BC_2	170	133	
L. esculentum × *L. chmielewskii*				
1. UC82B × LA1028	BC_1	237	70	RFLP, Isozyme
L. esculentum × *L. hirsutum*				
1. E6203 × LA1777	BC_1	149	135	RFLP
2. E6203 × LA1777	NIL, BIL	111	95	RFLP
3. NC84173 × PI126445	BC_1	145	171	RFLP, RGA
L. esculentum × *L. pennellii*				
1. VF36 *Tm2*[a] × LA716 (high-density map of tomato)	F_2	67	1050	Isozyme, RFLP, morphological
2. Vendor *Tm2* (a) × LA716	F_2	432	98	
3. M82 × LA716	IL	50	375	RFLP
4. VF36 *Tm2* (a) × LA716	F_2	67	1242	AFLP, RFLP
5. E6203 (LA925) × LA716	F_2	83	1500	COS
6. E6203 × LA1657	BC_2	175	110	RFLP
7. E6203 × LA716	F2	83	152	SSRs, CAPs
L. esculentum × *L. peruvianum*				
1. E6203 × LA1706	BC_3	241	177	RFLP, SCAR

TABLE 2.3 *(Continued)*

Linkage Map	Population Type[a])	Population Size	Number of Markers	Type of Markers[b]
L. esculentum var. *cerasiformae.* × *L. pimpinellifolium*				
1.E9 × L5	F_6-RIL	142	132	SSR, SCAR
L. esculentum var. *cerasiformae* × *L. cheesmanii*				
1.E9 × L3	F_6-RIL	115	114	SSR, SCAR
L. peruvianum × *L. peruvianum*				
1. LA2157 × LA2172	BC_1	152	73	RFLP

Note: (a) RIL: recombinant inbred line; NIL: near isogenic line; BIL: backcross inbred line.

(b) RFLP: restriction fragment length polymorphism; RAPD: randomly amplified polymorphic DNA; AFLP: amplified fragment length polymorphism; CAPS: cleaved amplified polymorphic sequence; RGA: resistance gene analog; EST: expressed sequence tag; SCAR: sequence characterized amplified region; SSR: simple sequence repeat.

Source: Adapted from Foolad (2007).

2.5 CONCLUSION

Though the molecular approaches of disease resistance can effectively replace the time-consuming breeding methods. Yet, there are certain limitations, such as the high cost incurred for large-size population marker-based genotyping in many breeding programs. The meager availability of PCR-based molecular markers reliability for many simple as well as complex traits. Genetic polymorphism is hardly sufficient for the markers used presently in many inter varietal populations. The scarce knowledge of the particular locations of QTLs because most of the identified QTL intervals are relatively large and also the limited knowledge of the reliable identified QTLs across the genetic backgrounds. The unfamiliarity of many breeders with molecular marker technology or laboratories brings the situation of lack of inherent interest in the identification of genes or QTLs for agricultural traits.

Despite of many disadvantages and limitations yet there lies some future perspective of marker-assisted selection (MAS). In the near future, it will emerge as a promising method for improving many simple horticultural traits controlled by one or oligo-genes like disease resistance. Further, with the advancement of in the development of high-resolution and high-throughput markers such SSRs, SNPs, MAS will become available to many tomato improvement programs. For quantitative traits, rather than the whole traits, QTLs will be available, which will be more reliable for marker-assisted breeding. Furthermore, problems associated with linkage drag will be avoided with the identification of QTLs. Therefore, more focus and efforts should be made on the research based on marker-assisted breeding.

KEYWORDS

- **amplified fragment length polymorphism**
- **DNA markers**
- **high-resolution melting**
- ***Ralstonia solanacearum***
- **resistant genotypes**
- **sequence characterized amplified regions**
- **sustainable management**

REFERENCES

Bahadur, B., Kadirvel, P., Senthilvel, S., Sujatha, M., & Varaprasad, K. S., (2015). *Plant Biology and Biotechnology: Volume II: Plant Genomics 65 and Biotechnology.* Springer, India.

Boyoung, K., In Sun, H., Hyung, J. L., Je Min, L., Eunyoung, S., Doil, C., & Chang-Sik, O., (2018). Identification of a molecular marker tightly linked to bacterial wilt resistance in tomato by genome-wide SNP analysis. *Theor. and Appl. Gene.*

Coutinho, T. A., (2005). *Introduction and Prospectus on the Survival of R. Solanacearum in Bacterial wilt Disease and the Ralstonia solanacearum Species Complex* (pp. 28–30). APS, St Paul.

Denny, T. P., & Gnanamanickam, S. S., (2006). Plant pathogenic *Ralstonia* species. In: *Plant Associated Bacteria*. Springer, Netherlands.

Elphinstone, J. G., (2005). The current bacterial wilt situation: A global overview. In: *Bacterial Wilt Disease and the Ralstonia solanacearum Species Complex* (pp. 9–28). APS Press, St-Paul.

Fegan, M., & Prior, P., (2005). How complex is the "*Ralstonia solanacearum*" complex? In: *Bacterial Wilt Disease and the Ralstonia solanacearum Species Complex* (pp. 449–461). APS Press, St. Paul.
Foolad, M. R., (2007). Genome mapping and molecular breeding of tomato. *Inter. J. Pl. Geno., 2007*, 1–52.
Hayward, A. C., (1964). Characteristics of *Pseudomonas solanacearum. J. Appl. Bacteriolo., 27*, 265–277.
Hayward, A. C., Sequeira, L., French, E. R., El-Nashaar, H., & Nydegger, U., (1991). Tropical variant of biovar 2 of *Pseudomonas solanacearum. Phytopath., 82*, 607.
Jan, P. S., Huang, H. Y., & Chen, H. M., (2010). Expression of a synthesized gene encoding cationic peptide cecropin B in transgenic tomato plants protects against bacterial diseases. *Appl. Environ. Microb., 76*(3), 769–775.
Kelman, A., (1997). One hundred and one year of research bacterial wilt. *Bacteri. Wil. Dis., 14*, 1–5.
Kumar, M., Srinivas, V., & Kumari, M., (2018b). Screening of tomato line/varieties for bacterial wilt (*Ralstonia solanacearum*) resistance in hill zone of Karnataka, India. *Int. J. Curr. Microbiol. App. Sci., 7*, 1451–1455.
Kumar, S., Gowda, P. H. R., Saikia, B., Debbarma, J., Velmurugan, N., & Chikkaputtaiah, C., (2018). Screening of tomato genotypes against bacterial wilt (*Ralstonia solanacearum*) and validation of resistance linked DNA markers. *Australas. Pl. Path., 47*, 365–374.
Lee, V. C. M. S., (2015). The antibiotic resistance crisis: Causes and threats. *Pharmc. and Therap., 40*(4), 277–283.
Mane, R. S., (2013). *Genetic Studies of Yield and Its Component Traits and Identification of Molecular Markers Linked to Bacterial Wilt Resistance in Tomato (Solanum lycopersicum L.).* Doctoral degree thesis, University of Agricultural Sciences, Dharwad, Bangalore.
Manoharachary, C., & Kunwar, K., (2014). *Host-Pathogen Interaction, Plant Diseases, Disease Management Strategies, and Future Challenges* (pp. 185–229). New York: Springer.
Miao, L., Shou, S., Cai, J., Jiang, F., Zhu, Z., & Li, H., (2009). Identification of two AFLP markers linked to bacterial wilt resistance in tomato and conversion to SCAR markers. *Mol. Biol. Rep., 36*(3), 479–486.
Monthly report Tomato, (2018). *Horticultural Statistics Division.* Department of Agriculture, Cooperation, Farmer's Welfare. Ministry of Agriculture and Farmer's Welfare. Government of India, New Delhi.
Muthukumar, P., & Selvakumar, R., (2017). *Glaustas Horticulture.* New Vishal publications, New Delhi.
Nicolopoulou-Stamati, P., Maipas, S., Kotampasi, C., Stamatis, P., & Hens, L., (2016). Chemical pesticides and human health: The urgent need for a new concept in agriculture. *Fron. Pub. Health., 4*, 148.
Olson, H. A., (2005). *Soil Borne Pathogens.* https://projects.ncsu.edu/cals/course/pp728/profile.html (accessed on 1 March 2021).
Patrice G. C., & Timur M., (2009). Bacterial Wilt of Tomato. USDA-NRI Project: *R. solanacearum* race 3 biovar 2: detection, exclusion and analysis of a Select Agent. Educational modules. pp. 1–11. https://plantpath.ifas.ufl.edu/rsol/Trainingmodules/RalstoniaR3b2_Sptms_Module.html (accessed on 1 March 2021).

Peralta, I. E., Spooner, D. M., & Knapp, S., (2008). Taxonomy of wild tomatoes and their relatives (Solanum sect. Lycopersicoides, sect. Juglandifolia, sect. Lycopersicon; Solanaceae). *Systematic Botany Monographs* (Vol. 84, p. 186). The American Society of Plant Taxonomists.

Pradhanang, P. M., Ji, P., Momol, M. T., Olson, S. M., Mayfield, J. L., & Jones, J. B., (2005). Application of acibenzolar-*S*-methyl enhances host resistance in tomato against *Ralstonia solanacearum*. *Plan. Dis., 89*, 989–993.

Ram, H. H., (2017). *Vegetable Breeding Principle and Practices*. Kalyani, New Delhi.

Rubatzky, V. E., & Yamaguchi, M., (1997). Tomatoes, peppers, eggplants, and other solanaceous vegetables. *World Vegetables*, 532–576. Springer, Boston.

Scott, J. W., Wang, J. F., & Hanson, P., (2005). Breeding tomatoes for resistance to bacterial wilt, a global view. *Acta. Hort., 695*, 161–168.

Sen, N. O., (2017). *The World's Leading Producers of Tomatoes*. World Atlas.

Singh, B. D., (2018). *Plant Breeding Principles and Methods*. Kalyani Publishers, New Delhi.

Swarup, V., (2014). *Vegetable Science and Technology in India* (p. 314). Kalyani Publishers, New Delhi.

Tahat, M. M., & Sijam, K., (2010). *Ralstonia solanacearum*: The bacterial wilt causal agent. *Asia. J. Plan. Scie., 9*(7), 385–393.

Tans-Kersten, J., Huang, H., & Allen, C., (2001). *Ralstonia solanacearum* needs motility for invasive virulence on tomato. *J. Bacteriology., 183*(12), 3597–3605.

Terra, I. A., Portugal, C. S., & Becker-Ritt, A. B., (2015). Plant antimicrobial peptides. In: Mendez-Vilas, A., (ed.), *The Battle Against Microbial Pathogens: Basic Science, Technological Advances and Educational Programs*. Formatex Research Center., 199–207.

Thamburaj, S., & Singh, N., (2014). *Textbook of Vegetables, Tuber Crops and Spices* (p. 489). Indian Council of Agricultural Research, New Delhi.

Xue, Q. Y., Yin, Y. N., Yang, W., Heuer, H., Prior, P., Guo, J. H., & Smalla, K., (2011). Genetic diversity of *Ralstonia solanacearum* strains from China assessed by PCR-based fingerprints to unravel host plant- and site-dependent distribution patterns. *FEMS Microbiol. Eco., 75*, 507–519.

Yabuuchi, E., Kosako, Y., Yano, I., Hota, H., & Nishiuchi, Y., (1995). Transfer of two *Burkholderia* and *Alcaligenes* species to *Ralstonia* gen. nov.: Proposal of *Ralstonia picketti* comb. nov., and *Ralstonia eutropha* comb. nov. *Microbiol. Immune., 39*, 897–904.

Yuliar, N. Y. A., & Toyota, K., (2015). Recent trends in control methods for bacterial wilt diseases caused by *Ralstonia solanacearum*. *Microb. Enviro., 30*(1), 1–11.

CHAPTER 3

MOLECULAR METHODS FOR THE CONTROLLING OF DAMPING OFF SEEDLINGS IN TOMATO

DAN SINGH JAKHAR,[1] RIMA KUMARI,[2] PANKAJ KUMAR,[2] and RAJESH SINGH[1]

[1]*Department of Genetics and Plant Breeding, Institute of Agricultural Sciences, Banaras Hindu University, Varanasi – 221005, Uttar Pradesh, India, E-mail: dan.jakhar04@bhu.ac.in (D. S. Jakhar)*

[2]*Department of Agricultural Biotechnology and Molecular Biology, Dr. Rajendra Prasad Central Agricultural University, Pusa – 848125, Bihar, India*

ABSTRACT

Damping-off is a significant tomato disease that decreases the germination of seeds and young plants, one of the key restrictions on farmers' returns. More than a dozen fungi and fungal organisms are reported to be induced by the soil, but just a handful of them. The disorder is also related to organisms of Pythium, Phytophthora, and Rhizoctonia. The seedlings either fail to come out of the soil or die when they appear. Conventional fungicides, among many biotic strains, are widely used to manage the disease with two main impacts. On the one hand, fungicide overuse threatens human health and causes environmental problems. This activity has also contributed to the presence in the atmosphere of fungicidal resistant microorganisms. Sustainable and robust maintenance damping methods, less reliant on traditional fungicides, are therefore progressively involved. To achieve this objective, new, and robust molecular approaches are required to control damping of tomato seedlings. Several molecular devices for damping seedlings can be defined and managed including PCR-RFLP, BLAST (basic local alignment search tool), PCR-ELISA, SSCP, and macro arrays.

3.1 INTRODUCTION

Tomato (*Solanum lycopersicum* L.) is a nightshade horticultural crop and an essential supply for all plants in the country. This developed in tropical America and was cultivated thousands of years before the colonization of Europe in Mexico and Peru. It is the only source of supply in late autumn and early winter and has a significant consumer position in India. In India, yields per hectare are low due to attacks by various viral and soilborne diseases which cause damage to crops each year. Seedlings are caused by a number of Pythium, Phytophthora, and Rhizoctonia species among soilborne fungal diseases (Kaprashvili, 1996; Lucas et al., 1997) and are widely dispersed all over the world.

Damping is a general word for the decay, before or after its growth, of seedlings under humid conditions. It is mainly an early-season challenge and causes the greatest losses in cold, wet soils. Damping fungi appear in all soils in which tomatoes are produced, and they invade tomatoes when the soil is warm. The infection is most widespread under cool conditions, while phytophthora and Rhizoctonia can also infect soils of seedlings. Tomato seedlings are not prone to Pythium or Rhizoctonia infections, but may be affected by Phytophthora at any point after 2- or 3-blade point. Pythium-related damping can increase if green dust is processed into the soil just before planting, such as voluntary grain. Damping is typically not moved from season to season in the same places, but only where and when circumstances promote infection. The Pythium genus is one of the most significant classes of soilborne pathogens in nearly any agricultural world, assaulting, and the crop yield and efficiency in the roots of thousands of hosts (Schroeder et al., 2013).

There are several diseases caused by *Pythium* spp. Wet cold environments are preferred, but in hot weather, some of the plants (*P. aphanidermatum* and *P. myriotylum*) are more popular. It can live under adverse conditions as thick-walled oospores. In a few species (*P. tracheiphilum* and *P. dimorphum*), the thick-walled chlamydospores can also be found in this region. The pathogen penetrates epidermal cells by releasing decaying enzymes from the cell-wall and destroys tissue by developing enzymes and toxins ahead of advancing hyphae (Schroeder et al., 2013).

3.2 TYPES OF DAMPING-OFF DISEASE

There are two types of damping-off disease (Figure 3.1):

1. **Pre-Emergent Damping-Off:** Seeds rot in the soil or seedlings decay before they push through the soil.
2. **Post-Emergent Damping-Off:** Seedling's sprout, but then pale, curl, wilt, or collapse at the soil line. The stem is water-soaked and turns gray, brown or black before disintegrating.

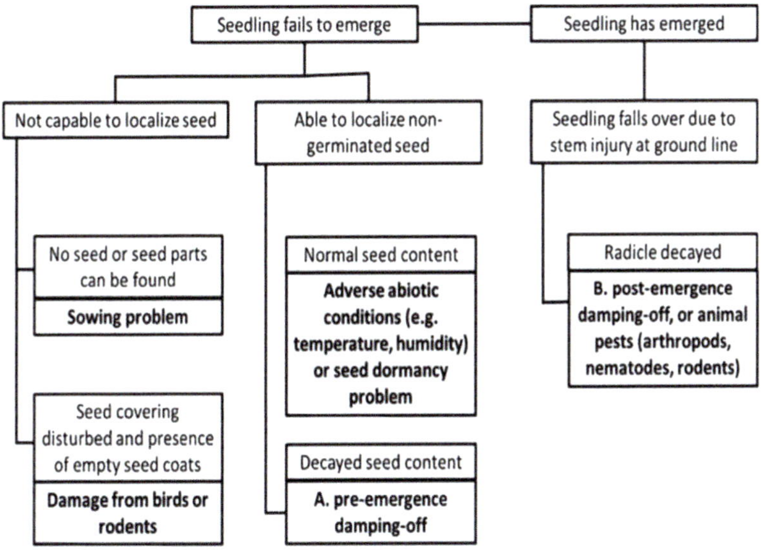

FIGURE 3.1 Damping-off is either a disease of germinating seeds (pre-emergence-A) or young seedlings (post-emergence-B).
Source: Adapted from Landis (2013).

3.3 SYMPTOMS

Damping-off disease of tomato occurs in two stages, i.e., the pre-emergence and the post-emergence phase. The seedlings are killed in pre-emergence phase just before they reach the soil surface. The young radical and the plumule are killed, and the seedlings are completely rotted. The post-emergence phase is characterized by infection of the collar's young, juvenile tissues at the ground level. The infected tissues are soft and soaked with water. The seedlings either collapse or overturn (Figure 3.2).

FIGURE 3.2 Damping-off disease symptoms in tomato.
Source: TNAU Agritech Portal.

3.4 FAVORABLE CONDITIONS FOR DISEASE DEVELOPMENT

The optimum for bacteria and illness growth is high humidity, high ground dust, thick vegetation, low pH, cloudlessness, and low temperatures for a few days. Dense planting, dehydration due to high runoff, slow drying and soil solutes interferes with the growth of the plant and enhances pathogenic damping.

3.5 TAXONOMY AND CLASSIFICATION

In the past of Pythium organisms, asexual, and sexual structures have been identified and categorized. The shape of sporangia and the ornamentation of oogonia are key characters. Four types of sporanges were described by Van der Plaats-Niterink (1981), either filamentous or non-inflationary and globose, either inter- or non-interesting (Figure 3.3). Moreover, some species do not sporangy; hyphal swelling often occurs.

The five sporangium forms for class D, contiguous sporangia consisting of (sub) globose elements connected in hyphal segments were observed by Lévesque and De Cock (2004). Oogonia is either smooth walled or lined of spines of various shapes in Pythium (Figure 3.4).

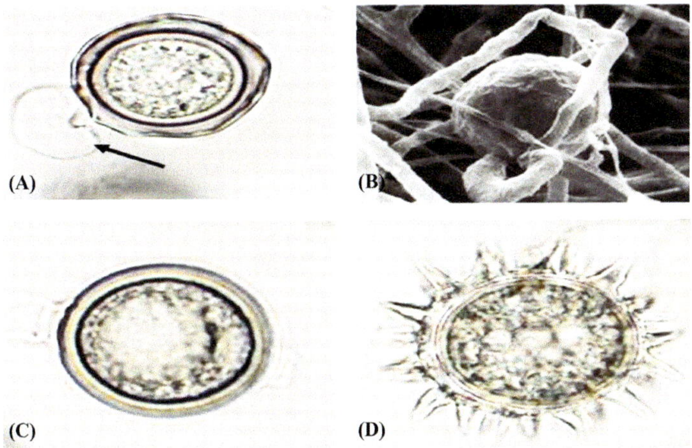

FIGURE 3.3 Sporangia of *Pythium*. (A) Filamentous sporangia of *Pythium dissotocum*; (B) Inflated filamentous (lobed) sporangia of *Pythium torulosum*; (C) Globose hyphal swellings of *Pythium ultimum*; (D) Internally proliferating sporangium of *Pythium middletonii*.
Source: Reprinted with permission from Schroeder et al. (2013).

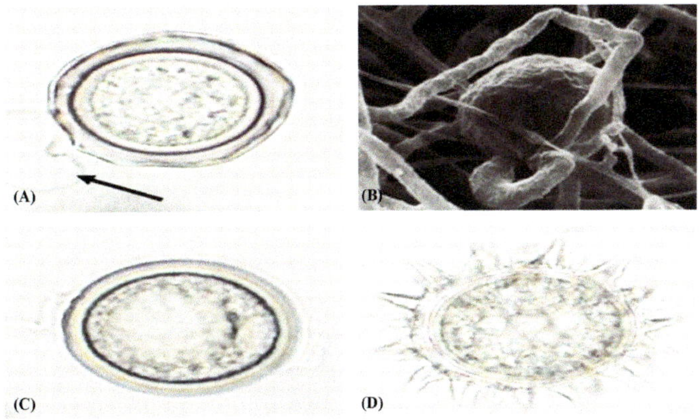

FIGURE 3.4 Oogonia and oospores of *Pythium*. (A) Smooth oogonium of *Pythium ultimum*; (B) Oogonium and antheridia of *P. ultimum*; (C) Smooth intercalary oogonium of *Pythium rostratifingens*; (D) Spiny oogonium of *Pythium oligandrum*.
Source: Reprinted with permission from Schroeder et al. (2013).

3.6 MOLECULAR PHYLOGENY

Pythium was based on a ribosomal DNA cistron restriction map for early research on Molecular Phylogeny (Klassen et al., 1987). CsCl ultra-centrifugation filtered CsCl rDNA, and data on RFLP were analyzed to generate Pythion species restriction maps. DNA amplification has been somewhat easier, and several Pythium species have provided many studies of oomycetes phylogeny (Dick et al., 1999; Riethmüller et al., 1999), but their samples in these experiments were not broad enough to include a clear explanation of the phylogenetic structures. The PCR amplify and the creation of ribosomal DNA primers that might work with oomycetes (White et al., 1999). The two Pythium species with filamentous sporangia varied from the five with globose spores, and the *P. vexans* varied from those of Pythium and Phytophthora, respectively (Briard et al., 1995) focused on only Pythium and Phytophthoras with the ribosomal small subunits (LSU). Cooke et al. (2000), along with three filamentous and three globose Pythium and *P. vexans*, reported identical findings in their first detailed Phytophthora research. Matsumoto et al. (1999) demonstrated that species of filamentous and globose sporangia have been phylogenetically separated using the internal transcribed spacer (ITS) of rDNA. Martin (2000) has shown that there were sixty isolates of 24 organisms developed into three phenotypes, utilizing the cytochromic oxidase 2 (cox 2) gene; one clade-shaped organism of filamentous sporangia, while two different clades developed species of globosis sporangia. Bedard et al. (2006) studied organizing the 5S gene family in a very broad range of Pythium species and found that the cladius A to K, including the very unique status of Clada K, was generally supported.

3.7 IDENTIFICATION THROUGH DNA SEQUENCING AND BARCODES

A sampling of different sites by utilizing a set of sources for BLAST (basic local alignment search tool) research by sequence alignments is the most accurate method to species level recognition. By utilizing a sequence-based method for isolation identification, certain concepts should be recognized. BLAST research involves an e-value score that indicates the e-value identity with a comparison list, depending on the corresponding percentage and list identity duration, which will also be analyzed in order to maintain the quality of the series in favor of making assumptions based on the maximum e-value score. The reference sequence database used for BLAST is also important for the use of curated isolate sequences which have their identification checked, and for the duration of the inquiry to not extend beyond the area sequenced in the reference datasets. For example, a BLAST search for an ITS series, which covers large parts of the flanking subunits which give back oomycete sequences with the same flanking regions as their highest e-value rather than those sequences which have an ITS match 100% without any flanking subunit regions.

3.8 MOLECULAR RECOGNITION TOOLS, DETECTIONS, AND QUANTIFICATIONS

Multiple molecular tools are available for monitoring, detecting, and calculating seedlings. The definitions below are as follows.

3.8.1 SEROLOGICAL METHODS AND ISOZYMES

Until the widespread use of DNA techniques dependent on antigens in the mycelia and fungal cell walls, serological approaches had been established for some Pythium spp. Polyclonal antibodies have little diagnostic species precision, however, they are beneficial as a method of Pythium quantifying in root and colonization studies in guided studies (Kyuchukova et al., 2006; Rafin et al., 1994). Infield experiments on early soybeans infections (Murillo-Williams, 2008) an enzyme-dependent immunosorbent assay (ELISA) unique to the Pythum form was developed and implemented. This tool is useful in certain cases that include essential Pythium identification in greenhouses or nurseries. It is sold as an ELISA package by Neogen Europe Ltd.-Adgen Phytodiagnostics. The monoclonal antibody may be more sensitive than polyclonal one. To usage of the ELISA dual-antibody conditional sandwich (Avila et al., 1995; Yuen et al., 1998) a mixture of the polyclone capture antibody and a monoclonal antibody unique to the *Pythium ultimum*. The product is extremely sensitive and is sold as a three-piece indirect ELISA kit by Agdia, Inc. (Elkhart, IN). Compared to DNA-based approaches, the benefits of ELISA kits include expense, simple farming usage and responsiveness to circumstances in which species recognition is no longer required.

3.8.2 RESTRICTION FRAGMENT LENGTH POLYMORPHISM (RFLP)

This may help to classify isolates at species level by matching digested DNA banding patterns to specific limiting enzymes, although the method is not as precise as the sequence review. Banding profiles useful for the identification of organisms were established for the digestion of purified mitochondrial DNA (mtDNA) with multiple restriction enzymes (Huang et al., 1992; Martin, 1990; Martin and Kistler, 1990). This method does not advocate the usage of restriction enzymes that accept 4 GC base pairs, large copies of the mtDNA, separated with the strength of unique lines, to digest complete DNA (Lévesques, 1993) from nuclear DNA. Nevertheless, depending on the restriction enzymes used, distinct bands can often be used for rDNA. RFLP research used as cloned sequences in Southern blots has also been used for population analyzes of *P. ultimum* (Francis and St-Clair, 1993; Francis et al., 1994) and *P. irregulare* (Harvey et al., 2000). When utilizing some PCR-RFLP analysis to make sure that the methodology will provide the extent of the species specificity that is required, it is crucial to test the species with phylogenetic relations.

3.8.3 SINGLE-STRAND CONFORMATIONAL POLYMORPHISM (SSCP)

Like RFLP, the SSCP study differentiates between species according to banding patterns, instead of producing them by digestion with restricting enzymes; it denatured and separated the amplified prototype with non-denatural gel that is distinguished by its secondary structure of the single-stranded DNA that defines the rate of migration. This technique was described by Kong et al. (2004) as a test for *Pythium* utilizing ITS and a direct amplification procedure of the settlement was recorded, removing the need for DNA extraction (Kong et al., 2005). It is used successfully to classify 7 000 isolates obtained in a broad field program in Ohio (Broders et al., 2007) and to model the existence of multiple organisms in a community survey (Vallance et al., 2009). This method is focused on a wide-scale field system. This approach also helps *Phytophthora spp.* to be identified.

3.8.4 SPECIES-SPECIFIC PROBES

Southern work utilizing species-specific samples was a tool used before the advent of PCR techniques to classify various organisms in a bigger sample. While DNA cultivation and extraction have always included, there have been few successful attempts to use *in situ* colony lysis on a nylon membrane. A species-specific sample of *P. ultimum* was developed by Lévesque et al. (1994) from the ITS region while *P. irregulare* was developed by Matthew et al. (1995) for a species-specific sample of repetitive genomic clone identification. A more comprehensive approach to species-specific studies was introduced by Klasse et al. (1996). The 5S rRNA gene is found in the IGS area of rDNA in species with inflated/filamentous sporangia and is used as a tandem series, isolated by spacer sequences (Bedard et al., 2006; Belkhiri et al., 1992), for species with spherical/globose sporangia. Klassen et al. (1996) found that the spacers are sufficiently different in order to design specimens for eight specific species (several cross-reactivities have been identified between *P. ultimum* var. ultimum and *P. ultimum* var. *sporangiiferum*). Even in the nature of PCR research for organisms with spherical/globo sporangia, the sequences of these spacer regions are likely to be included.

3.8.5 PCR-ELISA FOR ISOLATE IDENTIFICATION

The technique for identifying *Pythium* isolates at the species level, using both ELISA and PCR has been described by Bailey et al. (2002). The ITS region of the extracted DNA was enhanced by highly conserved primers in an amplification master blend containing digoxygenin-11-UTP. The amplicon was denatured, and to the well-containing streptavidin-coated microtiter was added a biotin-labeled collection sensor specific to a species. The digoxygenin-11-UTP is measured and quantified immunologically. This approach can also be utilized widely by the development of a wider variety of ITS sequence details and special oligonucleotide macroarray samples (Perneel et al., 2006).

3.8.6 ANALYSIS OF INTRASPECIFIC VARIATION AND PATHOGEN POPULATIONS

Intraspecific variance analyzes and demographic analyzes utilizing methods such as random amplified polymorphic DNAs (RAPDs) and amplified polymorphic fragment lengths (AFLP) were performed. These approaches have the benefit that genomic sequence details from the examined organisms are not required for the generation of the markers, but the markers are not co-dominant, which is essential when dealing with a diploid. Nonetheless, cloning, and design of a new key amplification pair to produce SCAR markers can mitigate this (Ahonsi et al., 2010). Intraspecific *P. ultimum* variability (Kageyama et al., 1998; Tojo et al., 1998), *P. irregulare* (Matsumoto et al., 2000), *P. porphyrae* (Park et al., 2003), *P. spinosum* (Gherbawy et al., 2002), and *P. insidiosum* (Pannanusorn et al., 2007) was assessed with RAPD. The *P. irregulare* Complex (Garzón et al., 2005) limit of species and the population systems of *P. aphanidermatum*, *P. irregulare*, and *P. ultimum* have been studied using AFLP analyzers (Lee et al., 2010).

The SSRs can be used in the genome to distinguish isolates without having knowledge of genomic sequences for priming. Inter-simple sequence repeat (ISSR) study utilizes a single prime generated from the SSR sequence pair, and a brand that is in near contact with an allele in the opposite orientation can be amplified. Vasseur et al. (2005) used this method in order to separate *Pythium* group F (filamentous sporangia) isolates from *P. oligandrum* (Vallance et al., 2009).

3.8.7 ARRAY-BASED DETECTION

Using arrays is a strategy which can be incredibly successful in recognizing all organisms that can be contained in a given sample (also known as reverted dot blot). A blueprint for the configuration of hybridization probes mounted to a nylon membrane from isolated DNA is supplemented. The DNA is labeled when processed and then used in southern testing with the sample as a study. The benefits of this approach include the potential to use Hybridization Sonder unique to a range of organisms in the collection, thus offering a wider study in the population. The usage of Hybridization Sonden built from a number of loci, thereby verifying the subsequent detection. Lévesque et al. (1994) found out that a macroarray utilizing ITS1, as a hybridization test, could classify *Pythium spp.* efficiently, but did not have the specificities required to assess the environmental sample.

3.8.8 PCR-BASED DETECTION AND QUANTIFICATION

Conventional diagnosis set limitations provided an opportunity to use PCR-based approaches for defining and detecting *Pythium*. Classic (endpoint) PCR, which has been accessible for industry since the mid-1980s, has been used to correctly classify at least 23 plant-pathogenic pythium organisms from different farms. In most cases, researchers have been trying to classify cultured isolates in soils or contaminated plant tissues or to detect known organisms. Although other insightful and unique DNA sequences were used (Bent et al., 2009), the field of ITS was commonly utilized for priming designs. Objective DNA may be quantified with a titration of target-DNA enhancement with established quantities of a rival with a near-sequence name, however distinguishable from that amplicon, for example, by hosting a special endonuclease cleavage site. The elimination by the endonuclease procedure from the competitor before the gel electrophoresis allows the target amplicon to be noticeable. Because the molar volume of the competitor is larger than that of the goal, the color strength of the target DNA strip approached the horizon.

Incorporated into the PCR platforms in the 1990s, fluorescence detection technology-enabled product amplification to be tracked in real-time following each PCR phase. The number of target DNA in a sample is inversely proportional to how much the amplicon is first detected in the fractional cycle (Ct) (its fluorescence goes beyond the baseline, or fluorescence, of the base). Scientists may remove gel electrophoresis as a drug testing tool by utilizing PCR technique in actual or quantitative time, instead of using melting temperature as an amplicon-specific feature. Over five to seven log units of target concentration and high sample performance (up to 384 per run) are other advantages of real-time technology. In eight real-time PCR (Rt PCR) studies, researchers used the SYBR Green in which this fluorescent dye binds to double-sided DNA during each of the stages of clogging and expansion. Fourteen Pythium plant pathologies were quantified using real-time PCR. The forward and reverse PCR primers and a fluorescent-labeled study (Okubara et al., 2005) was used for an alternate fluorescence chemistry, Taq-Man, in three papers. Taq-Man Chemistry gains from the ability to discern a single nucleotide polymorphism inside the sample binding region, which allows it a higher degree of precision and the potential to amplify multiplex.

3.8.9 EXAMPLES OF QUANTIFICATION USING REAL-TIME PCR-DETECTION SYSTEMS

A major challenge for *Pythium* spp. study is the large range of organisms typically contained in the soil or tissue (Spies et al., 2011) are locally present (Paulitz et al., 2003). Spp. Python. The assessment of the importance of rising causative agent may also confuse co-infectant of plant-pathogenic fungi (Atallah et al., 2006; Cullen et al., 2007). Further, *Pythium* is not uniformly distributed in the soil like many pathogens that are in the soil, such as *Rhizoctonia*, which makes it difficult to quantify. In the isolation of certain associations and the selection of isolates, conventional plating approaches along with microscopic species recognition have proved useful. Nevertheless, real-time PCRs have not only made possible the detection of specific species but also the quantification of the communities of such creatures. The use of real-time PCRs to quantify *Pythium* was used in a wide range of studies. It has been used to measure *Pythium* in the earth (Kertallah et al., 2008) and *P. oligandrum* biochemical regulation isolates (Le Floch et al., 2007) in mixed infections with certain forms of species (Atallah and Stevenson, 2006).

3.9 CONCLUSION

The ability to examine the community structure of *Pythium* species within an ecosystem would improve our understanding of the ecological niche that specific species occupy and the influence of the environment on their survival. Techniques such as PCR-RFLP, BLAST, PCR-ELISA, SSCP, and macroarrays have been successfully used for this purpose, but there are limitations with sample throughput and identification of unknown or new species when investigating large replicated trials. With the improvement of DNA sequencing technology, cost reduction, and improvement of computational support for the management of large data sets, an alternative approach is to use highly conserved primers to amplify a specific region with sufficient sequence divergence to identify species.

KEYWORDS

- **amplified polymorphic fragment lengths**
- **damping-off seedlings**
- **enzyme-dependent immunosorbent assay**
- **inter-simple sequence repeat**
- **molecular approaches**
- ***Pythium* spp.**
- **soilborne diseases**

REFERENCES

Ahonsi, M. O., Ling, Y., & Kageyama, K., (2010). Development of SCAR markers and PCR assays for single or simultaneous species-specific detection of *Phytophthora nicotianae* and *Pythium helicoides* in ebb-and- flow irrigated Kalanchoe. *J. Microbiol. Methods, 83*, 260–265.

Atallah, Z. K., & Stevenson, W. R., (2006). A methodology to detect and quantify five pathogens causing potato tuber decay using real-time quantitative polymerase chain reaction. *Phytopathology, 96*, 1037–1045.

Avila, F. J., Yuen, G. Y., & Klopfenstein, N. B., (1995). Characterization of a *Pythium ultimum*-specific antigen and factors that affect its detection using a monoclonal antibody. *Phytopathology, 85*, 1378–1387.

Bailey, A. M., Mitchell, D. J., Manjunath, K. L., Nolasco, G., & Niblett, C. L., (2002). Identification to the species level of the plant pathogens Phytophthora and *Pythium* by using unique sequences of the ITS1 region of ribosomal DNA as capture probes for PCR ELISA. *FEMS Microbiol. Lett., 207*, 153–158.

Bedard, J. E. J., Schurko, A. M., Cock, A. W. A. M., & Klassen, G. R., (2006). Diversity and evolution of 5S rRNA gene family organization in *Pythium*. *Mycol. Res., 110*, 86–95.

Belkhiri, A., Buchko, J., & Klassen, G. R., (1992). The 5S ribosomal RNA gene in *Pythium* species: Two different genomic locations. *Mol. Biol. Evol., 9*, 1089–1102.

Bent, E., Loffredo, A., Yang, J. I., McKenry, M. V., Becker, J. O., & Borneman, J., (2009). Investigations into peach seedling stunting caused by a replant soil. *FEMS Microbiol. Ecol., 68*, 192–200.

Briard, M., Dutertre, M., Rouxel, F., & Brygoo, Y., (1995). Ribosomal RNA sequence divergence within the *Pythiaceae*. *Mycol. Res., 99*, 1119–1127.

Broders, K. D., Lipps, P. E., Paul, P. A., & Dorrance, A. E., (2007). Characterization of *Pythium* spp. associated with corn and soybean seed and seedling disease in Ohio. *Plant Dis., 91*, 727–735.

Cooke, D. E. L., Drenth, A., Duncan, J. M., Wagels, G., & Brasier, C. M., (2000). A molecular phylogeny of *Phytophthora* and related oomycetes. *Fungal Genet. Biol., 30*, 17–32.

Cullen, D. W., Toth, I. K., Boonham, N., Walsh, K., Barker, I., & Lees, K., (2007). Development and validation of conventional and quantitative polymerase chain reaction assays for the detection of storage rot potato pathogens, *Phytophthora erythroseptica*, *Pythium ultimum*, and *Phoma foveata*. *J. Phytopathol., 155*, 309–315.

Dick, M. W., Vick, M. C., Gibbings, J. G., Hedderson, T. A., & Lopez- Lastra, C. C., (1999). 18S rDNA for species of *Leptolegnia* and other *Peronosporomycetes*: Justification for the subclass taxa *Saprolegnio-mycetidae*, and *Peronosporomycetidae* and division of the *Saprolegniaceae sensu lato* into the *Leptolegniaceae*, and *Saprolegniaceae*. *Mycol. Res., 103*, 1119–1125.

Francis, D. M., & St-Clair, D. A., (1993). Outcrossing in the homothallic oomycete, *Pythium ultimum*, detected with molecular markers. *Curr. Genet., 24*, 100–106.

Francis, D. M., Gehlen, M. F., & St. Clair, D. A., (1994). Genetic variation in homothallic and hyphal swelling isolates of *Pythium ultimum* var. *ultimum* and *P. ultimum* var. *sporangiferum*. *Mol. Plant-Microbe Interact., 7*, 766–775.

Garzón, C. D., Geiser, D. M., & Moorman, G. W., (2005). Amplified fragment length polymorphism analysis and internal transcribed spacer and *coxII* sequences reveal a species boundary within *Pythium irregulare*. *Phytopathology, 95*, 1489–1498.

Gherbawy, Y. A. M. H., & Abdelzaher, H. M. A., (2002). Using of RAPD-PCR for separation of *Pythium spinosum* Sawada into two varieties: Var. *spinosum* and var. *sporangiiferum*. *Cytologia, 67*, 83–94.

Harvey, P. R., Butterworth, P. J., Hawke, B. G., & Pankhurst, C. E., (2000). Genetic variation among populations of *Pythium irregulare* in southern Australia. *Plant Pathol., 49*, 619–627.

Huang, H. C., Morrison, R. J., Muendel, H. H., Barr, D. J. S., Klassen, G. R., & Buchko, J., (1992). *Pythium* sp. "group G", a form of *Pythium ultimum* causing damping-off of safflower. *Can. J. Plant Pathol., 14*, 229–232.

Kageyama, K., Uchino, H., & Hyakumachi, M., (1998). Characterization of the hyphal swelling group of *Pythium*: DNA polymorphisms and cultural and morphological characteristics. *Plant Dis., 82*, 218–222.

Kernaghan, G., Reeleder, R. D., & Hoke, S. M. T., (2008). Quantification of *Pythium* populations in ginseng soils by culture dependent and real-time PCR methods. *Appl. Soil Ecol., 40*, 447–455.

Klassen, G. R., McNabb, S. A., & Dick, M. W., (1987). Comparison of physical maps of ribosomal DNA repeating units in *Pythium*, *Phytophthora* and *Apodachlya*. *J. Gen. Microbiol., 133*, 2953–2959.

Klassen, G., Balcerzak, M., & De Cock, A. W. A. M., (1996). 5S ribosomal RNA gene spacers as species-specific probes for eight species of *Pythium*. *Phytopathology, 86*, 581–587.

Kong, P., Richardson, P. A., & Hong, C., (2005). Direct colony PCR-SSCP for detection of multiple pythiaceous oomycetes in environmental samples. *J. Microbiol. Methods, 61*, 25–32.

Kong, P., Richardson, P. A., Moorman, G. W., & Hong, C. X., (2004). A molecular fingerprint for identification of multiple *Pythium* species. (Abstr.) *Phytopathology, 94*, S55.

Kuprashvili, T. D., (1996). The use phytocides for seed treatment. *Plant Prot. Quar., 55*, 31pp.

Kyuchukova, M. A., Büttner, C., Gabler, J., Bar-Yosef, B., Grosch, R., & Kläring, H. P., (2006). Evaluation of a method for quantification of *Pythium aphanidermatum* in cucumber roots at different temperatures and inoculum densities. *J. Plant Dis. Prot., 113*, 113–119.

Landis, T. D., (2013). Forest nursery pests: Damping-off. *For Nurs. Notes, 2*, 25–32.

Le Floch, G., Tambong, J., Vallance, J., Tirilly, Y., Lévesque, A., & Rey, P., (2007). Rhizosphere persistence of three *Pythium oligandrum* strains in tomato soilless culture assessed by DNA macroarray and real-time PCR. *FEMS Microbiol. Ecol., 61*, 317–326.

Lee, S., Garzón, C. D., & Moorman, G. W., (2010). Genetic structure and distribution of *Pythium aphanidermatum* populations in Pennsylvania greenhouses based on analysis of AFLP and SSR markers. *Mycologia, 102*, 774–784.

Lévesque, C. A., & De Cock, A. W. A. M., (2004). Molecular phylogeny and taxonomy of the genus *Pythium*. *Mycol. Res., 108*, 1363–1383.

Lévesque, C. A., Beckenbach, K., Baillie, D. L., & Rahe, J. E., (1993). Pathogenicity and DNA restriction fragment length polymorphisms of isolates of *Pythium* spp. from glyphosate-treated seedlings. *Mycol. Res., 97*, 307–312.

Lévesque, C. A., Vrain, T. C., & De Boer, S. H., (1994). Development of a species-specific probe for *Pythium ultimum* using amplified ribosomal DNA. *Phytopathology, 84*, 474–478.

Lucas, G. B., Campbell, C. L., & Lucas, L. T., (1997). *Introduction to Plant Disease Identification and Management* (p. 364). CBS Pub. and Distributors, New Delhi.

Martin, F. N., & Kistler, H. C., (1990). Species-specific banding patterns of restriction endonuclease-digested mitochondrial DNA from the genus *Pythium*. *Exp. Mycol., 14*, 32–46.

Martin, F. N., (1990). Taxonomic classification of asexual isolates of *Pythium ultimum* based on cultural characteristics and mitochondrial DNA restriction patterns. *Exp. Mycol., 14*, 47–56.

Martin, F. N., (2000). Phylogenetic relationships among some *Pythium* species inferred from sequence analysis of the mitochondrially encoded cytochrome oxidase II gene. *Mycologia, 92*, 711–727.

Matsumoto, C., Kageyama, K., Suga, H., & Hyakumachi, M., (1999). Phylogenetic relationships of *Pythium* species based on ITS and 5.8S sequences of the ribosomal DNA. *Mycoscience, 40*, 321–331.

Matsumoto, C., Kageyama, K., Suga, H., & Hyakumachi, M., (2000). Intraspecific DNA polymorphisms of *Pythium irregulare*. *Mycol. Res., 104*, 1333–1341.

Matthew, J., Hawke, B. G., & Pankhurst, C. E., (1995). A DNA probe for identification of *Pythium irregulare* in soil. *Mycol. Res., 99*, 579–584.

Murillo-Williams, A., & Pedersen, P., (2008). Early incidence of soybean seedling pathogens in Iowa. *Agron. J., 100*, 1481–1487.

Okubara, P. A., Schroeder, K. L., & Paulitz, T. C., (2005). Real-time polymerase chain reaction: Applications to studies on soilborne pathogens. *Can. J. Plant Pathol., 27*, 300–313.

Pannanusorn, S., Chaiprasert, A., Prariyachatigul, C., Krajaejun, T., Vanittanakom, N., Chindamporn, A., Wanachiwanawin, W., & Satapatayavong, B., (2007). Random amplified polymorphic DNA typing and phylogeny of *Pythium insidiosum* clinical isolates in Thailand. *Southeast Asian J. Trop. Med. Publ. Health, 38*, 383–391.

Park, C., Kakinuma, M., Sakaguchi, K., & Amano, H., (2003). Genetic variation detected with random amplified polymorphic DNA markers among isolates of the red rot disease fungus *Pythium porphyrae* isolated from *Porphyra yezoensis* from Korea and Japan. *Fish. Sci., 69*, 361–368.

Paulitz, T. C., & Adams, K., (2003). Composition and distribution of *Pythium* communities in wheat fields in eastern Washington state. *Phytopathology, 93*, 867–873.

Perneel, M., Tambong, J. T., Adiobo, A., Floren, C., Saborio, F., Lévesque, A., & Hofte, M., (2006). Intraspecific variability of *Pythium myriotylum* isolated from cocoyam and other host crops. *Mycol. Res., 110*, 583–593.

Plaats-Niterink, A. J. V. D., (1981). Monograph of the genus *Pythium*. *Studies in Mycology No. 21*. Centraalbureau Voor Schimmel cultures, Baarn, The Netherlands.

Rafin, C., Nodet, P., & Tirilly, Y., (1994). Immuno-enzymatic staining procedure for *Pythium* species with filamentous non-inflated sporangia in soilless cultures. *Mycol. Res., 98*, 535–541.

Riethmüller, A., Weiß, M., & Oberwinkler, F., (1999). Phylogenetic studies of Saprolegniomycetidae and related groups based on nuclear large subunit ribosomal DNA sequences. *Can. J. Bot., 77*, 1790–1800.

Schroeder, K. L., Frank, N. M., Arthur, W. A. M. D. C., André, L. C., Christoffel, F. J. S., Patricia, A. O., & Timothy, C. P., (2013). Molecular detection and quantification of *Pythium* species: Evolving taxonomy, new tools, and challenges. *Plant Disease, 97*(1), 4–20.

Spies, C. F. J., Mazzola, M., & McLeod, A., (2011). Characterization and detection of *Pythium* and *Phytophthora* species associated with grapevines in South Africa. *Eur. J. Plant Pathol., 131*, 103–119.

Tojo, M., Nakazono, E., Tsushima, S., Morikawa, T., & Matsumoto, N., (1998). Characterization of two morphological groups of isolates of *Pythium ultimum* var. *ultimum* in a vegetable field. *Mycoscience, 39*, 135–144.

Vallance, J., Le Floch, G., Déniel, F., Barbier, G., Lévesque, C. A., & Rey, P., (2009). Influence of *Pythium oligandrum* biocontrol on fungal and oomycete population dynamics in the rhizosphere. *Appl. Environ. Micro-Biol., 75*, 4790–4800.

Vasseur, V., Rey, P., Bellanger, E., Brygoo, Y., & Tirilly, Y., (2005). Molecular characterization of *Pythium* group F isolates by ribosomal- and intermicrosatellite-DNA regions analysis. *Eur. J. Plant Pathol., 112*, 301–310.

White, T. J., Bruns, T., Lee, S., & Taylor, J., (1990). Amplification and direct sequencing of fungal ribosomal RNA genes for phylogenetics. In: Innis, M. A., Gelfand, D. H., Sninsky, J. J., & White, T. J., (eds.), *PCR Protocols, a Guide to Methods and Applications* (pp. 315–322). Academic Press, San Diego.

Yuen, G. Y., Xia, J. Q., & Sutula, C. L., (1998). A sensitive ELISA for *Pythium ultimum* using polyclonal and species-specific monoclonal antibodies. *Plant Dis., 82*, 1029–1032.

CHAPTER 4

MOLECULAR METHODS FOR THE CONTROLLING OF LATE BLIGHT IN TOMATO

P. K. RAY,[1] S. S. SOLANKEY,[2] R. N. SINGH,[3] ANJANI KUMAR,[4] and UMESH SINGH[5]

[1]*Krishi Vigyan Kendra, Saharsa (Bihar Agricultural University, Sabour), Bihar, India, E-mail: pankajveg@gmail.com (P. K. Ray)*

[2]*Department of Horticulture (Vegetable and Floriculture), Dr. Kalam Agricultural College, Kishanganj (Bihar Agricultural University, Sabour), Bihar, India*

[3]*Associate Director Extension Education, Bihar Agricultural University, Sabour, Bhagalpur, Bihar, India*

[4]*ICAR-ATARI, Zone-IV, Patna, Bihar, India*

[5]*Mandan Bharti Agricultural College, Saharsa (Bihar Agricultural University, Sabour), Bihar, India*

ABSTRACT

Molecular science is also a science that focuses on the involvement of possible genomic and ecological risk aspects, known at the molecular level, to the etiology, distribution, and bar of unwellness. Molecular provides the tools that have prophetical significance which epidemiologists will use to higher outline the etiology of specific diseases and work towards their management. The application of molecular methods has redoubled the thoughtful of the medicine of the foremost vital transmittable agents, fungus. Current evolution in *P. infestans* biology is provided that the knowledge used for such strategies and new biomolecular markers unit of measurement presently being technologically advanced that have incredible potential within the study of *P. infestans*. Molecular techniques facilitate to stratify and improve knowledge by provided that plenty of complex and exact measurements that help goings-on along with unwellness investigating, happening investigations, characteristic transmission forms, and risk features between seemingly different cases illustrating host agent relations and provided that higher understanding of unwellness organic process at the molecular level.

4.1 INTRODUCTION

Tomato (*Solanum lycopersicum* L.) is the leading common agricultural crops and grown all over the globe and local to the Andean vicinity of South America. It will be grownup an exceedingly big selection of tropical to temperate environments; when outdoor temperatures aren't favorable, then it can also be cultivated undercover conditions. Among vegetable crops, tomato is that the world's second most fed on vegetable after potato (Foolad, 2007). Fresh, mixed in other food items or processed and canned as sauce, ketchup, juice, salsa, paste, soup, and pickled are tomato products consumed in several ways. Tomato supplies an adequate quantity of the antioxidant lycopene that supports to protection the body against cancer and heart condition and richest source of fat-soluble vitamin and C (Bohm et al., 2012). There's a greater request used for fresh market and processed market for its wider adaptability, uses, and nutritional values. Due to increasing the demand of tomato, it is necessary to produce good quality tomato.

Though, several fungus, bacteria, viruses, and nematodes as well as blight hindered the commercial production of tomato. These species are more susceptible to disease epidemics due to narrow genetic diversity that lead to its strong choice and inbreeding throughout development and domestication. Cultivated tomato classes are extra susceptible to diseases against wild species. The foremost common process of governing late blight in tomato is through cultural methods and heavy practice of preventative fungicides because pathogen's quick lifetime and therefore the difficulty of detecting pathogen with low inoculum levels within the field (Foolad, 2008). Molecular techniques don't additional used for traditional approaches. They report epidemiologic difficulties that can't be advanced or would remain extra employment, luxurious in addition to time period towards handle through traditional methods. Today's molecular practice converted tomorrow's traditional investigative implement; otherwise may be given towards dustbin. Molecular techniques are accustomed study and solve the epidemiological problems that the normal epidemiological methods cannot.

4.2 SYMPTOMS AND SIGNS OF DISEASE

4.2.1 ON TOMATO LEAVES

Water-soaked patch that expands quickly in light green towards the brownish-black wound, lesions begin as indefinite and might cover huge parts of the foliage. Scratches arranged on the abaxial surface of the foliage could also remain enclosed through a grayish in the direction of whiteness moldy growing during wet weather. The leaf goes yellowish so brownish, curly, shrinks, and perishes and bottoms of more scratches, a hoop of moldy progress of fungus is usually noticeable in moist climate, because the fungus progress. The disease signs remain different and will not remain confused through signs of mildew disease, the fungus spores typically seem on the upper foliage of tomato.

4.2.2 ON TOMATO PETIOLES AND STEMS

Water-soaked patch that expand speedily in brownish to blackish scratches, lesions begin as indefinite and might cover huge parts of the petioles and branches. Scratches is also enclosed through grayish to whitish moldy growing of the fungus during wet weather conditions. Influences branches and petioles might ultimately fail through the purpose of infection, resulting in death of entirely distal portions of the plant.

4.2.3 ON TOMATO FRUITS

A skinny coating of whitish mycelium is also existing in rainy climate condition, shady, olivaceous grayish patch progress on green fruit.

4.3 THE INFECTION CYCLE

Around 120 recognized types of *Phytophthora* species, and everyone is a fungus of plants. The colonies diverse host materials, similar to root, bulbs, herbaceous branches, woody trunks, leaf, and berry. In their infection cycle *Phytophthora* species progress diverse cellular phases. The first dispersal stages are multinucleate sporangia and uninucleate motile zoospores represent. *P. infestans* could also be puffed or marked to novel hosts wherever they'll also release zoospores or propagate unswervingly to develop infection while sporangia arranged on foliage plant portions. Forming a cyst when synthesize a plasma membrane and zoospores discard flagella. These germinate within hours which dissemination of host epidermal cells arises and zoospores enter through natural openings of host tissue like stomata otherwise form an appressorium-like puffy germ duct. Hyphae develop towards branch over plant portion a spherical primary infection vesicle is created upon host cell penetration. In host cells, *P. infestans* hyphae raise intercellularly, prominent digit-like haustoria (Whisson et al., 2007; Avrova et al., 2008). Haustoria remain structures that arrangement a close collaboration through host cells. Haustoria eradicating the plant cell membrane then exit the skin integral to help molecular conversation among the fungus and alive plant cell. The way transition is regulated from one stage to the next and it's of attention to expose by what means these specialized cell categories vary as of each other then: sporangium → zoospore → cyst → germination → appressorium → host dispersion and contamination vesicle → intercellular hyphal growth → haustorium formation → initiation of sporulation. To manage contagion through targeted inhibition of controlling procedures inside the fungus information of the variances and likenesses among these progressive phases will help innovative means. As per these phases remain in nearby connection through plant cells, and this can be wherever the result of contagion is decided, contagion vesicles, haustoria, and intercellular hyphae remain of specific attention.

4.4 THE DISEASE CYCLE

Contagion potato tubers in scrap heaps; otherwise, those left adjacent or taking place on the earth surface remain the primary source of initial inoculum. When the tubers sprout by infecting the seedlings, the pathogens lie dormant in tubers in storing or within the ground and continues growing. The fungus produces sporangia is under favorable environmental conditions. Sporangia infect the prone host and induced additional pathogen sequence and the sporangia are blowout by air or splashed rain to adjacent prone host. On *Solanum tuberosum*, *Solanum lycopersicum*, *Solanum melongena*, *Petunia integrifolia* and various other crops within the family of Solanaceae, the pathogen remains accomplished active throughout minor wintertime in Southern conditions. The sporangia in northward is transported by prevailing winds. Inoculum source in North parts of India are *Lycopersicon esculentum* nursery grownup in these parts turn out to be infected and function like transportation of disease. During the winter on greenhouse tomatoes the pathogen also might persist vigorous in north parts of India. Sporangia are carried within the spring from the greenhouse to the sphere. Hot times 21°C to 29°C and a ratio nearly 100%, followed by calm nighttime 7°C to 15°C with heavy precipitation, mist or a light drizzle rain that continues over the before noon are ideal environmental conditions that favor epidemic development.

The pathogen hyphae raise quickly among the cells of *Solanum lycopersicum*. Haustoria enter the *Solanum lycopersicum* cells and suck nutrients and obtained food. Preliminary signs could seem in two or three days but frequently remain evident later five to seven days that fungus has begun to colonize in the plant. Function inoculum for repeating, secondary disease cycles, the sporangiophores develop over the stomates then develop sporangia after symptoms appear.

The extent of the epidemic was determined by situations that regulate spore creation, propagation, and contagion. The fungus forms sporangia at a ratio near one hundred pc for quite 10 to 15 hours and temperatures of 7°C to 30°C with an optimum of 18°C to 21°C. The speed of spore production is determined by temperature and quantity of periods of basically 100% relative humidity. Temperature may regulate the sporangia germination in one in all two ways. At one hundred pc ratio and with temperatures of 45°C to about 7°C to 21°C, each sporangium creates eight to twelve or additional motile zoospores in one to three hours. Around 12°C temperature is the optimum for zoospores development.

Each zoospore propagates to provide a germination tube in addition to an appressorium, after which a penetration pin enters the foliage swims for some minutes in an exceedingly moisture film, encysts. Around 12°C to 15°C temperature is the optimum for germination of zoospores and for tube development of germ is 21°C to 23°C. At about 21°C to 30°C temperature without producing zoospores sprouting of sporangia arises in a straight line via individual germination pipe. Around 25C temperature is the optimum temperature for direct penetration and 8 to 48 hours takes place. At overlapping temperatures both varieties germination occurs. Contagion and fungus growth remain utmost quick at 22–24°C temperature and penetration has occurred. The connecting pathogen occurrences *Solanum melongena, Petunia integrifolia, Physalis spp.* and supplementary wild plant in the *Solanum tuberosum* and *Lycopersicon esculentum* family in addition tomato and potato.

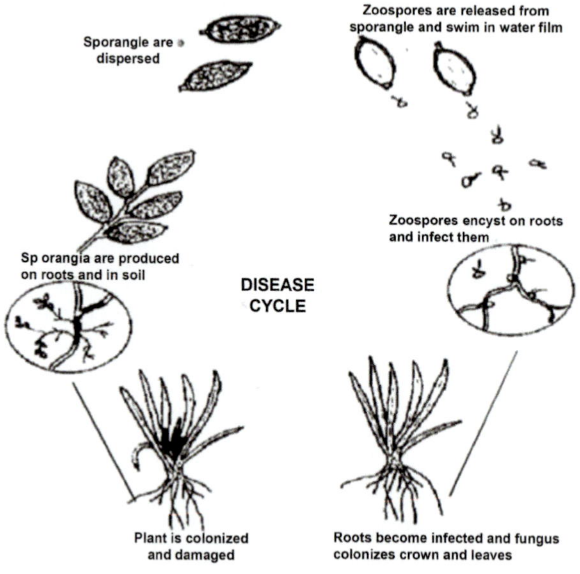

4.5 APPLICATIONS OF MOLECULAR MARKERS

4.5.1 INDIVIDUAL

1. Hybridization;
2. PCR-based:
   i. Classical PCR;
   ii. Nested PCR;
   iii. Real-time PCR.

4.5.2 POPULATION

1. PCR-based assay;
2. RFLP hybridization;
3. DNA sequence.

The efficacy and sturdiness of the latest managing methods evolution is a vital think about predicting. To attain this goal, a variety of phenotypic as well as genotypic assessments has remained practical but to each takes restrictions and innovative approaches are required. New high-through co-dominant biomolecular indicators are now being technologically advanced that have fabulous possible within the study of late blight population biology, epidemiology, ecosystem, inheritances, and progress as well as current improvement in genomics of late blight is provided that the statistics used for such procedures. Genetics, genomics, and evolutionary processes is additionally authoritative to better understanding of *P. infestans* infection biology. The generation and management of dissimilarity in inhabitants is vital. Transmutation, recombination, normal process, inheritable factor, accidental heritable drift and immigration contributions and rates (Burdon and Silk, 1997) so far, such reasons persist slight considered and poorly unstated. An absence of understanding of the impact of selection pressure and broadly implemented and unbiased means of assessing late blight inhabitant's variability also are hampering scientific progress. The practical investigation of genetic factor elaborates in progress, growth, and plant contagion (Torto et al., 2003) and current advances in *P. infestans* genomics (Bos et al., 2003), genetic mapping (van der Lee et al., 1997) and physical (Whisson et al., 2001), also are developing the scope of *Phytophthora* study. Simplifying reasonable, high-throughput investigation of various DNA-based markers to advances in laboratory automation and fingerprinting technology, valued DNA-based markers will be made. In key fields like epidemiology, genetics, population biology, and also the functional and mapping analysis of novel genes for biomolecular markers in evolving *P. infestans* study. Particular emphasis is going to be placed on the potential of the latest procedure for investigative *P. infestans* populations and also the efficacy of prevailing genotypic and phenotypic markers due to the pressures after varying fungus inhabitants in several countries of the global. It's proposed that innovative approaches and methods are required to inspire advances during this area.

4.5.3 MARKER TECHNOLOGY AND THEIR APPLICATIONS

It's clear that used for altogether characteristics of *Phytophthora infestans* examine there is not any single marker system (Milbourne et al., 1997) are acceptable. Primarily, it studies the investigative wants of every kind of study and primary applications of the latest marker technology. Some key thoughts in choosing a proper marker are resources available, run-in time and depth of taxonomic resolution, material required, running budgets and projected variation by other study groups.

4.5.4 POPULATION GENETICS AND POPULATION DIVERSITY

In the current pathogen population, the tomato breeding approaches, prediction tools, and management practices are suitable and possibly the leading common objective in the study of *Phytophthora infestans* populations. Thus, the danger of long-lived oospores serving as primary inoculum sources, the intensive care of A1 and A2 mating-type proportions is very imperative to assist estimates of the amount of sexual recombination. Furthermore, sexual recombination is possibly going to spread the speed of pathogen variation (Barton and Charlesworth, 1998) to its epidemiological effect.

4.6 AN IDEAL MARKER FOR GENETIC ANALYSIS OF *P. INFESTANS*

High throughput uses the foremost widespread and affordable technology available, capable of being multiplexed, robust, optimized protocols for running and objective scoring of the assays to encourage widespread adoption of a typical marker system, flexible, are often applied to both pure *P. infestans* DNA samples and infected leaf material or spore washings, may be modified to the resolution appropriate to the study, e.g., from the study of closely related species to intra-population diversity, suitable for rigorous genetic analysis. Markers unlinked, simply inherited and, ideally, mapped to every linked genes co-dominant. A mix of nuclear and mitochondrial targets, broadly applied, widely disseminated protocols leading to its universal adoption, safe, doesn't involve hazardous procedures or chemicals. Investigations are directed by gathering isolates that denote a snap of the inhabitants in period and space.

Variations of topographical and chronological in diversity of genotypic and phenotypic are then inspected and construed in relevance to the technical goals of the study. The advantage of the investigation has innovative from genotypic to phenotypic approaches and there are several samples of study within this kind, like isozymes analysis, mtDNA, and RG57 fragment length polymorphism (RFLP) patterns, amplified fragment length polymorphisms (AFLPs), and further newly, SSRs. In the general tendency of skyrocketing multiplicity in *Phytophthora infestans* has remained detected in various tomato-rising provinces of the biosphere through the exemption of the now various inhabitants at its center of derivation. An additional recent study highlights the expressions of the various novel genotypes via immigration and sexual recombination while primary studies defined inhabitants that remained clonal or liberated by some separate ancestries (Cooke et al., 2003).

4.6.1 AFLP

1. Genomic DNA is absorbed with both a restriction nuclease that cuts regularly and one that cuts fewer regularly.
2. The resultant fragments are ligated to finish exact connector fragments.
3. A pre-selective PCR amplification is completed using primers complementary to every of the two adaptor sequences, apart from the presence of one extra-base at the 3' end, which base is elected by the user. Amplification of the only $1/16^{th}$ of EcoRI-MseI fragments occurs.

AFLP fingerprinting, as a sample, differentiates isolates measured undistinguishable supported RG57 fingerprint and SSR markers (Knapova and Gisi, 2002). An endlessly increasing list of distinct genotypes, somewhere a high number of isolates in an inhabitant take sole genomic fingerprints. With subgroups for isolates supposed to possess arisen in a genotype, the presently assumed system of designating genotypes trusts arranged a rustic code tracked by a singular figure for every novel genotype. The quantity of genotypes that prerequisite towards remain defined in this method is possibly going to spread exponentially and within the long period, as an increasing traits of *Phytophthora infestans* inhabitants may be a distorting of the limits of inherently dissimilar subpopulations, this might not be a supportive method. Nowadays, there are various variants of the US1 lineage and a minimum of 19 US genotypes, nearly possibly produced as recombinants of prevailing lineages. To accommodate this growing diversity, a recognized identification arrangement is obviously required for leading subsections of the population.

4.6.2 RFLP

- Use the restriction endonucleases to acknowledge the precise DNA sequences;
- Hybridize to probe DNA or amplify by PCR;
- Analyze the variation of amplified bands.

4.6.3 REAL-TIME PCR (RT PCR)

1. Onward and inverse primers remain prolonged through Taq polymerase as in an exceedingly traditional PCR response. An exploration with two fluorescent dyes involved strengthens to the gene arrangement between the two primers.
2. Because the polymerase outspreads the primer, the probe is displaced.
3. An integral nuclease action inside the polymerase slashes the reporter dye from the probe.
4. Subsequently issue of the reporter dye from the quencher, a fluorescent signal is created.

4.6.4 SSR-PCR

Bands are generated by a primer of easy sequence repeats. The best grouping of essential features for inhabitants' study and their probable should be discovered more completely by SSRs. A change removed from only categorization of *Phytophthora infestans* dissimilarity and to researches with sample approaches planned to checked exact theories, the practice of such biomolecular markers takes excessive probable via such markers inside a hypothetical background of inhabitant's inheritances, is required. Helping to investigate the balance and evaluation gene flow among the forces of existence of the fittest and chance properties of genetic drift and migration likewise in the coming centuries, the distributions and tracing of allele regularities finished will development the thoughtful of the temporal and spatial dynamic forces of populations of *Phytophthora infestans*. From these statistics, the procedures driving inhabitant's variation and the way it should be succeeded to the good thing about upcoming control of disease will be measured. The strong point of the disciplines of epidemiology, molecular ecology, inhabitants' heredities and plant pathology are united for a synchronized method.

4.7 PHENOTYPIC MARKERS: POTENTIAL AND IMPORTANCE

There are comparatively rare consistent morphological traits through which to differentiate *Phytophthora infestans* isolates from the species of *Phytophthora*. The foremost considered of those phenotypic characters that persist utmost useful are mating type, virulence as well as fungicide resistance.

4.7.1 MATING TYPE

Late blight may be a disturbing potato blight produced by *P. infestans*. *P. infestans* produce sexually and asexually among the strains of dual mating types called A1 and A2. Mating type is a vital strain disturbing the pathogen's inhabitant's construction. Through the paired test results for twenty-six isolates together universal and used for a gaggle of one forty-six refine isolates confirm diverse PCR markers for *Phytophthora infestans* mating type. This study recognizes an encouraging type of the isolates of genotype US-1. Aimed at A1 mating type isolates, the merchandise exact for A2 isolates is improved with the marker W16. The US-1 isolates through current A2 mating type isolates indicate high similarity through the investigation of orders of W16 PCR products. Around 95% and 96% of isolates remain properly allocated by markers W16 and S1 separately, After US-1 isolates remain left out from investigation. Around 14% of discrepant results through the pairing test of Marker PHYB. Molecular markers may be suitable tools used for *Phytophthora infestans* mating-type purpose, then their application would become first by authentication to each local inhabitant later their effectiveness may differ reckoning on a fungus genotype.

4.7.2 VIRULENCE

The oomycete, *P. infestans*, is unique amongst the foremost significant plant pathogens wide-reaching. Considerable of the pathogenic achievement of *Phytophthora infestans* depend on its capacity to originate with huge quantities of sporangia on or after mycelia, which issue zoospores that encyst and form contagion structures. The avirulence fragments that are supposed by host barricades, little remained recognized about the molecular source of oomycete pathogenicity by. To know the molecular mechanisms relationship within the fungus and host relations, information of the genome assembly was utmost significant, which is obtainable today later genome sequencing. Potato and *P. infestans* mechanism of biotrophic interaction may be determined by thoughtful the effector ecology of the fungus, which is as yet lesser understood. Understanding the signal transduction pathways followed by apoplastic and cytoplasmic effectors for translocation into host cell by the availability of oomycete genome. Finally supported genomics, innovative approaches might be settled for actual managing of the crop sufferers because of the blight pathogen.

4.7.3 FUNGICIDE RESISTANCE

Phytophthora infestans is the entity that recently blights disease of *Solanum tubrosam* and *Lycopersicon esculentum*. This pathogen causes more financial losses every year then management is especially attained by the utilization of fungicides. Regrettably, inhabitants of *Phytophthora infestans* immune to fungicides are recognized. Additionally, studies have testified that delicate isolates to the phenylamide fungicide, mefenoxam, subsided delicate *in vitro* later one passageway over sublethal concentrations of fungicide-amended medium. The primary objective was to analyze if isolates of *Phytophthora infestans* are accomplished of obtaining resistance to 2 extra systemic fungicides, fluopicolide, and cymoxanil. In contrast to true through mefenoxam, contact of isolates to sublethal absorptions of fluopicolide and cymoxanil failed to encourage reduced sensitivity to those two fungicides. All species of *Phytophthora* measured (*P. infestans*, *P. betacei,* and *P. pseudocryptogea*) further as species of *Phytopythium* developed struggle to mefenoxam later earlier contact to source holding 1 μg ml per liter of mefenoxam. Remarkably, isolate sixty-six species of *Phytopythium* and also the isolate of *Phytophthora pseudocryptogea* tried don't look to be attaining resistance to mefenoxam later contact to medium comprising 5 μg ml per liter of the fungicide. The tried isolates of *Phytophthora palmivora* and *Phytophthora cinnamomi* remained very sensitive to mefenoxam and hence it absolutely remained uphill to achieve an additional transmission to access achievement of resistant to the current fungicide.

4.8 GENOTYPIC MARKERS: POTENTIAL AND IMPORTANCE

Though phenotypic behaviors remain significant for thoughtful the choice compressions on *Phytophthora infestans* inhabitants, in separation they are doing not fulfill several factors in various genotypic markers are accustomed study *Phytophthora infestans* and now the position and upcoming submissions of every are measured.

4.8.1 ISOZYMES

Isoenzyme difference of one hundred ninety-eight isolates of *Phytophthora infestans* collected from various places in 1987 to 1990 and of 4 pre-1987 isolates was inspected by means of starch gel electrophoresis. An earlier unreported allele at malic isoenzyme locus/ME was detected. A relation among mating types and isoenzyme genotypes at 3 isoenzyme loci, glucose phosphate isomerase, peptidase, and ME was observed. At PEP-1, the A2 isolates from Japan had an earlier unreported genotype. Typical segregation at the malic isoenzyme locus happened in an exceedingly crossing of parents of Mexican and Japanese. The outcomes offer a signal of a modification within the inhabitant's genetic structure of *Phytophthora infestans*.

4.8.2 RFLP

The moderately repetitive RFLP probe RG57 yields a biometric authentication of 25–29 bands and has evidenced a valued tool in nursing *Phytophthora infestans* genetic variability. Countless isolates wide-reaching are fingerprinted and a world catalog of the outcomes created. The dataset has remained significant in defining and nursing ancestries of *Phytophthora infestans* and tracing inoculum causes (Zwankhuizen et al., 2000). The strategy does have drawbacks, nevertheless; huge quantities of unadulterated DNA are mandatory, it's period intense, the banding designs may be tough to interpret, and therefore the resulting figures are important. Additionally, slight or no is understood approximately the distinct loci that frame the fingerprint, so evaluating the prospect of homoplasy is tough.

4.9 CONCLUSION

Since the recognition of *P. infestans* as a plant pathogen and also because the rationale for population displacement, the biology, genetics, and pathogenic strategies are studied. However, relatively little research has explored Phytophthora diseases at the molecular level. Establishing expression profiles to know mechanisms underlying different phenomena helps within the wealth of information sequences comprising involved in numerous developmental stages. There is a necessity, however, to develop systematic gene disruption techniques to help elucidate the functional role of these genes. A valuable resource will prove evidently sequencing the genome of *P. infestans*. Large UniGene set generated from the ESTs are facilitated by annotation of *Phytophthora infestans* genome also as available physical and genetic maps. For comparative analysis amongst *Phytophthora* species or with other organisms to decipher a whole-genome sequence could also be used. To understand mechanisms underlying pathogenicity and host responses and ultimately end in the event of improved strategies to manage *P. infestans*, there's much hope that within the near future, these resources will help us. Undoubtedly, the potentiality of the latest genetic markers is appreciated and as they're technologically advanced and subjugated by the global investigate community to nearly long-standing and imperative queries is essential and useful *Phytophthora infestans* study will develop. SSRs seem to produce the most effective possible across a decent series of submissions of the methods discussed and can be developed further. Thus, the equivalent tracing of impartial and purposeful markers will support to detect the forces driving fungus growth and genomics is additionally describing the role of the various innovative *Phytophthora infestans* genes. Marker potential is exploited by the quick community issue of protocols and applications, whichever marker systems are advanced, preferably collected into a catalog together with evidence on their map sites. A comparison of newer DNA-based markers and the resolution and suitability of existing is additionally being undertaken to recount the old and novel datasets on standard isolate collections. In accomplishing the critical mass of complete evidence essential to expose the driving forces and practical consequences of inhabitant's fluctuations on this scale cooperative approaches are visiting be important. Specialists within the areas of pathology, epidemiology, population genetics, molecular ecology, *Phytophthora infestans* biology and plant breeding are closer collaborations promoted to empower such improvement. To cut back agrochemical contributions, upcoming sustainable managing tactics should place additional importance on host resistance with increasing ecological and financial burden. Their success will, however, forecasting upcoming responses of *Phytophthora infestans* inhabitants to such resistant deployment and activate understanding current diversity. A population genetics approach that regulates the degree of gene flow among inhabitants and also the stability among the forces of action and unintended effects of genetic drift and migration is crucial to this thoughtful and exposes the genetic structure of inhabitants at both global and field scales. The method will interchange to the synchronized investigation of the various markers across subsets or perhaps the total genome which has the ability to discrete locus-specific properties from those affecting the complete genome and as extra markers are advanced and so the genome saturated a successive investigation of linkage disequilibrium (Luikart et al., 2003).

KEYWORDS

- **agricultural crops**
- **amplified fragment length polymorphisms**
- **etiology**
- **late blight**
- **molecular epidemiology**
- ***P. infestans***

REFERENCES

Avrova, A. O., Boevink, P. C., Young, V., Grenville-Briggs, L. J., Van, W. P., Birch, P. R., & Whisson, S. C., (2008). A novel *Phytophthora infestans* haustorium-specific membrane protein is required for infection of potato. *Cell Microbiol., 10*, 2271–2284.

Banke, S., Peschon, A., & McDonald, B. A., (2004). Phylogenetic analysis of globally distributed *Mycosphaerella graminicola* populations based on three DNA sequence loci. *Fungal Genet. and Biol., 41*, 226–238.

Barton, N. H., & Charlesworth, B., (1998). Why sex and recombination? *Science, 281*, 1986–1990.

Bohm, F., Edge, R., & Truscott, G., (2012). Interactions of dietary carotenoids with activated (singlet) oxygen and free radicals: Potential effects for human health. *Mol. Nutr. Food Res., 56*, 205–216.

Bos, J. I., Torto, T. A., Ochwo, M., Armstrong, M., Whisson, S. C., Birch, P. R. J., & Kamoun, S., (2003). Intraspecific comparative genomics to identify virulence genes from *Phytophthora. New Phytologist, 159*, 63–73.

Brumfield, R. T., Beerli, P., Nickerson, D. A., & Edwards, S. V., (2003). The utility of single nucleotide polymorphisms in inferences of population history. *Trends in Ecol. and Evol., 18*, 249–256.

Burdon, J. J., & Silk, J., (1997). Sources and patterns of diversity in plant pathogenic fungi. *Phytophthora., 87*(7), 664–669.

Chauvet, S., Van, D. V. M., Imbert, E., Guillemin, M. L., Mayol, M., Riba, M., Smulders, M. J. M., et al., (2004). Past and current gene flow in the selfing, wind-dispersed species *Mycelia muralis* in Western Europe. *Mol. Ecol., 13*, 1391–1407.

Cooke, D. E. L., Young, V., Birch, P. R. J., Toth, R., Gourlay, F., Day, J. P., Carnegie, S., & Duncan, J. M., (2003). Phenotypic and genotypic diversity of *Phytophthora infestans* populations in Scotland (1995–1997). *Plant Patho., 52*, 181–192.

Dowley, L. J., Griffin, D., & O'Sullivan, E., (2002). Two decades of monitoring Irish populations of *Phytophthora infestans* for metalaxyl resistance. *Potato Res., 45*, 79–84.

Flier, W. G., Grünwald, N. J., Kroon, L. P. N. M., Sturbaum, A. K., Van, D. B. T. B. M., Garay-Serrano, E., Lozoya-Saldana, H., et al., (2003). The population structure of *Phytophthora infestans* from the Toluca Valley of Central Mexico suggests genetic differentiation between populations from cultivated potato and wild *Solanum spp. Phytophthora., 93*, 382–390.

Foolad, M. R., Merk, H. L., & Ashrafi, H., (2008). Genetics, genomics and breeding of late blight and early blight resistance in tomato. *Critic. Rev. Plant Sci., 27*, 75–107.

Foolad, M., (2007). Genome mapping and molecular breeding of tomato. *Int. J. Plant Genom.*, 1–52.

Fry, W. E., Goodwin, S. B., Matuszak, J. M., Spielman, L. J., Milgroom, M. G., & Drenth, A., (1992). Population genetics and intercontinental migrations of *Phytophthora infestans*. *Annual Review of Phytopatho., 30*, 107–129.

Gavino, P. D., & Fry, W. E., (2002). Diversity in and evidence for selection on the mitochondrial genome of *Phytophthora infestans*. *Mycologia., 94*, 781–793.

Gisi, U., Sierotzki, H., Cook, A., & McCaffery, A., (2002). Mechanisms influencing the evolution of resistance to Qo inhibitor fungicides. *Pest Management Sci., 58*, 859–867.

Goodwin, S. B., (1997). The population genetics of *Phytophthora*. *Phytophthora., 87*, 462–473.

Hermansen, A., Hannukkala, A., Naerstad, R. H., & Brurberg, M. B., (2000). Variation in populations of *Phytophthora infestans* in Finland and Norway: Mating type, metalaxyl resistance and virulence phenotype. *Plant Patho., 49*, 11–22.

Hurst, L., (2002). The Ka/Ks ratio: Diagnosing the form of sequence evolution. *Trends in Genet., 18*, 486, 487.

Knapova, G., & Gisi, U., (2002). Phenotypic and genotypic structure of *Phytophthora infestans* populations on potato and tomato in France and Switzerland. *Plant Patho., 51*, 641–653.

Li, Y., Korol, A. B., Fahima, T., Beiles, A., & Nevo, E., (2002). Microsatellites: genomic distribution, putative functions and mutational mechanisms: A review. *Mol. Ecology, 11*, 2453–2465.

Luikart, G., England, P. R., Tallmon, D., Jordan, S., & Taberlet, P., (2003). The power and promise of population genomics: From genotyping to genome typing. *Nature Reviews Genet., 4*, 981–984.

Maggioni, R., Rogers, A. D., & Maclean, N., (2003). Population structure of *Litopenaeus Schmitt* (Decapoda: Penaeidae) from the Brazilian coast identified using six polymorphic microsatellite loci. *Mol. Ecol., 12*, 3213–3217.

Milbourne, D., Meyer, R., Bradshaw, J. E., Baird, E., Bonar, N., Provan, J., Powell, W., & Waugh, R., (1997). Comparison of PCR based marker systems for the analysis of genetic relationships in cultivated potato. *Mol. Breed., 3*, 127–136.

Mizubuti, E. S. G., & Fry, W. E., (1998). Temperature effects on developmental stages of isolates from three clonal lineages of *Phytophthora infestans*. *Phytophthora., 88*, 837–843.

Tang, S., Kishore, V. K., & Knapp, S. J., (2003). PCR-multiplexes for a genome-wide framework of simple sequence repeat marker loci in cultivated sunflower. *Theor. and Appl. Genet., 107*, 6–19.

Torto, T. A., Li, S. A., Styer, A., Huitema, E., Testa, A., Gow, N. A. R., Van, W. P., & Kamoun, S., (2003). EST mining and functional expression assays identify extracellular effector proteins from the plant pathogen *Phytophthora*. *Genome Res., 13*, 1675–1685.

Van, D. H. R. A. L., De Wit, P. J. G. M., & Joosten, M. H. A. J., (2002). Balancing selection favors guarding resistance proteins. *Trends in Pl. Sci., 7*, 67–71.

Van, D. L. T., De Witte, I., Drenth, A., Alfonso, C., & Govers, F., (1997). AFLP linkage map of the oomycete *Phytophthora infestans*. *Fungal Genet. and Bio., 21*(3), 278–291.

Wattier, R. A. M., Gathercole, L. L., Assinder, S. J., Gliddon, C. J., Deahl, K. L., Shaw, D. S., & Mills, D. I., (2003). Sequence variation of intergenic mitochondrial DNA spacers (mtDNA-IGS) of *Phytophthora infestans* (oomycetes) and related species. *Mol. Ecology Notes, 3*, 136–138.

Whisson, S. C., Boevink, P. C., Moleleki, L., Avrova, A. O., Morales, J. G., Gilroy, E. M., Armstrong, M. R., et al., (2007). A translocation signal for delivery of oomycete effector proteins into host plant cells. *Nature, 450*, 115–118.

Whisson, S. C., Van, D. L. T., Bryan, G. J., Waugh, R., Govers, F., & Birch, P. R. J., (2001). Physical mapping across an avirulence locus of *Phytophthora infestans* using a highly representative, large-insert bacterial artificial chromosome library. *Mol. Genet. and Genom., 266*, 289–295.

Zwankhuizen, M. J., Govers, F., & Zadoks, J. C., (2000). Inoculum sources and genotypic diversity of *Phytophthora infestans* in Southern Flevoland, the Netherlands. *European J. of Plant Patho., 106*, 667–680.

CHAPTER 5

MOLECULAR APPROACHES OF EARLY BLIGHT RESISTANCE BREEDING

MEENAKSHI KUMARI,[1] B. VANLALNEHI,[2] MANOJ KUMAR,[2] and S. S. SOLANKEY[3]

[1]*Department of Vegetable Science, Chandra Shekhar Azad University of Agriculture and Technology, Kanpur – 208002, Uttar Pradesh, India, E-mail: meenakshisinghupcs@gmail.com (Meenakshi Kumari)*

[2]*Division of Vegetable Science, Indian Institute of Horticultural Research, Hessarghatta, Bangalore – 560089, Karnataka, India*

[3]*Department of Horticulture (Vegetable and Floriculture), Dr. Kalam Agricultural College, Kishanganj – 855107 (Bihar Agricultural University, Sabour), Bihar, India*

ABSTRACT

Tomato (*Solanum lycopersicum* L.) is one of the chief vegetable crop grown throughout the world. Both fresh and processed fruits are used, which contain various vitamins, minerals, etc. The production of tomatoes is affected by several diseases, insects, and pests. Among them, early blight (EB) is one of the most serious diseases caused by *Alternaria solani.* EB resistance in tomatoes is difficult due to the polygenic nature of inheritance, which causes difficulties in understanding and screening of the disease. Several chemical pesticides were used for the management of this disease. But these chemical has shown not much effective measures leads to the use of crop-related species which is best alternatives. Two wild species namely, *S. habrochaites* and *S. pimpinellifolium* have inadequate level of resistance as an individual and the undesirable horticultural traits from the introgression of the wild species genes results in limited success for EB resistant breeding program.

5.1 INTRODUCTION

The achievement in tomato improvement for fungi resistance/tolerance is due to the major gene effect. Nevertheless, the early blight (EB) of tomato is one of the hurdles in resistance breeding due to its polygenic nature of inheritance (Foolad et al., 2000). The ineffective nature of biological and chemical management measures has led to the use of crop-related species which is the best alternatives. The resistance of EB identified in some breeding lines 71B2 and C1943 (resistant to collar rot and leaf blight) were limited to success as a result of the susceptibility to collar rot and low level of resistance to leaf blight, respectively. However, the development of many EB resistant tomato breeding lines *viz*., NCEBR-2, NCEBR-3, NCEBR-4, NCEBR-5, NCEBR-6 NC63EB, NC870 and several hybrids were still dependent on 71B2 and C1943 (Gardner, 1988, 2000; Gardner and Shoemaker, 1999). The wild species viz., *S. habrochaites* and *S. pimpinellifolium* has inadequate level of resistance as an individual and the undesirable horticultural traits (indeterminate, late maturing, and low yielding) from the introgression of the wild species genes results in limited success for EB resistant breeding program. Additionally, the high sugars as well as glycoalkaloids (e.g., solanine, chaconine, and solanidine) contribute toward EB resistant to younger leaves which in fact can lead to disease escape in late-maturing tomato varieties (Sinden et al., 1972; Johanson and Thurston, 1990; Rotem, 1994; Chaerani and Voorrips, 2006). The *Alternaria solani* triggered the metabolic changes and activates steroidal glycol-alkaloid (SGA), lignin with flavonoid biosynthetic pathways in *S. arcanum* accessions (Shinde et al., 2017). The intricate nature of EB resistance under the influence of multiple genes interacting in additive and non-additive manner coupled with high environmental effect is the other obstacle (Nash and Gardner, 1988).

5.2 RESISTANCE SOURCES AVAILABLE

EB is a crucial disease of tomato however; there is a lot of resistance sources available across the globe (Table 5.1) to help in the resistant breeding program.

TABLE 5.1 Worldwide Resistance Genetic Resources of Tomato Early Blight

Source of Resistance	Early Blight Resistant Accessions	Nature of Resistance	References
	Accessions: Solanum lycopersicum		
Unknown source	C1943	–	Barksdale (1971)
68B134	71B2	–	Barksdale (1969)
C1943	NC63EB	–	Gardner (1988)
C1943	NC870	–	Gardner (1988)
C1943	NC EBR-2	Quantitative	Gardner (1988)
71B2	NCEBR-5, NCEBR-6	Quantitative	Gardner (2000)
Unknown accessions	HRC90.145, HRC90.158, HRC90.159	–	Poysa and Tu (1996)
NCEBR-1	NC EBR-4, IHR1816	Quantitative	Gardner and Shoemaker (1999); Thirthamallappa and Lohithaswa (2000)
NCEBR-1, NCEBR-2	NCEBR-3	–	Gardner and Shoemaker (1999)
NCEBR-3, NCEBR-4	Mountain supreme	–	Gardner and Shoemaker (1999)
NCEBR-5, NCEBR-6	Plum Dandy	–	Gardner (2000)
NCEBR-6	Mountain magic	–	Gardner and Panthee (2012)
PI 406758	–	–	Martin and Hepperly (1987)
	Accessions: Solanum pimpinellifolium		
PI 365912, PI 390519	–	–	Martin and Hepperly (1987)
A 1921	P-1	–	Kalloo and Banarjee (1993)
L4394 (IHR1939)	–	–	Thirthamallappa and Lohithaswa (2000)
Accessions: *Solanum habrochaites*			
PI 127827	–	–	Locke (1949)
PI 390514, PI 390662	–	–	Martin and Hepperly (1987)
PI126445	NC EBR-1	–	Gardner (1988)

TABLE 5.1 *(Continued)*

Source of Resistance	Early Blight Resistant Accessions	Nature of Resistance	References
PI 1390662	87B187	–	Maiero, Ng, and Barksdale (1990)
B 6013	H-7, H-22, H-25	–	Kalloo and Banarjee (1993)
Unknown accessions	HRC90.303, HRC91.279, HRC91.341	–	Poysa and Tu (1996)
LA 2100, LA 2124, LA 2204, PE 36	–	–	Poysa and Tu (1996)
PI 126445	NC39E	–	Foolad et al. (2002)
PI 390513, PI 390516, PI 390658 and PI 390660	–	–	Martin and Hepperly (1987)
LA 2650	–	–	Poysa and Tu (1996)
Accessions: Solanum peruvianum			
PE 33, LA 1292, LA 1365, LA 1910, LA 1983 and PI 270435	–	–	Poysa and Tu (1996)

Source: Modified from Adhikari et al. (2017).

S. habrochaites have the most important QTLs on chromosomes 5, 8, 9, 10, 11, and 12 (Foolad et al., 2002), whereas *S. pimpinellifolium* have the most promising QTLs on chromosomes 1, 3, 4, 5, and 6. Hence, it is suggesting that accessions of *S. habrochaites* and *S. pimpinellifolium* probably have useful genes for resistance against EB. So far, there is an opportunity of pyramiding the QTLs of above both sources of resistance followed by introgression into elite tomato varieties (Foolad et al., 2005).

5.3 MARKER ASSISTED SELECTION (MAS)

DNA markers are increasingly being recognized as useful tools for disease resistant breeding since these are least influenced by the environment, unlike the traditional selection. A genetic marker is perhaps associated with a specific phenotype that can be viewed as a tag attached to a specific segment of a chromosome. To pace the breeding program, several molecular markers (RFLP, AFLP, RAPD, CAPs, SSRs, RGAs, ESTs, and COS) are employed and densely mapped on all the chromosomes of tomato that serves in mapping QTLs for agricultural relevant traits (Foolad, 2007). The EST database of tomato are screened against Arabidopsis genomic sequence and a set of 1025 genes/COS markers with single copy in both genomes are identified (Fulton et al., 2002b). The molecular markers linked to major genes are ease in transferring as compared to molecular markers link to QTLs. The next-generation sequencing (NGS) is a throughput genotyping platforms enable to identify single nucleotide polymorphisms (SNPs) (Yang et al., 2004; Yamamoto et al., 2005) that differentiate a minute changes between species and among populations (Foolad and Panthee, 2012). The SNPs marker has become a desirable marker in molecular breeding for identifying individual genes by dissecting a complex quantitative EB resistance, seedling stage selection of EB resistant genotype, and the genetic basis of negative correlation between EB resistance and economically important traits can be comprehended (Ashrafi and Foolad, 2015). In case of EB, the lack of incomplete resistance from individual source can be overcome by pyramiding of targeted genes from diverse genetic backgrounds using molecular markers. Moreover, the undesirable linkage drag and quick recurrent parent genome rediscovery in case of wild donors can be achieved through MAS. Despite the linkage drag from donor *S. habrochaites* (Ashrafi and Foolad, 2015) developed the *S. pimpinellifolium* (LA2093) and *S. lycopersicum* (NCEBR-1) derived recombinant inbred lines (RILs) which possessed EB resistance along with desirable characteristics such as early maturity, high yield, and dwarf plant type. Henceforth, the mapping of quantitative trait loci (QTL) is carried out to identify linked markers to EB resistance QTL. The markers linked to QTLs are less used due to the unavailability of tight linkage that can eliminate the horticultural value-related problems. Extensive research and further fine mapping is obligatory to identify exact locations of QTLs in order to maximum utilization of MAS.

5.4 EARLY BLIGHT (EB) RESISTANCE LINKED QTLS

The QTLs which are linked to EB are described in Table 5.2.

TABLE 5.2 The QTLs Conferring Tomato Early Blight Resistance and Linked Molecular Marker

Mapping Population	Molecular Markers	QTL Location	LOD Scores	R^2 Value (%)	References
145 BC1 *S. lycopersicum* NC84173 *S. habrochaites* PI126445	TG559-TG208A	chr1	7	–	Foolad et al. (2002)
–	TG337-CT59	chr2	2.9	15.3	–
–	XLRR.370-CT202	chr5	2.6	8.4	–
–	CD40-TG16	chr8	2.3	7.4	–
–	RLRR.130-CLRR.950	chr9	4.2	14.9	–
–	TG424-TG429	chr9	5.1	16.2	–
–	TG241-TG403	chr10	6.8	20.2	–
–	CT168-TG508	chr11	3.8	13.2	–
–	TG147-A41.3	chr11	2.3	7.4	–
–	CT100-TG68	chr12	3.1	10.3	–
–	AN23.390-TG180	chr12	4.1	12.9	–
145 BC_1S_1 *S. lycopersicum* NC84173 *S. habrochaites* PI126445	TG559-TG208A	chr1	3.6	11.9	–
–	TG337-CT59	chr2	2.8	15.9	–
–	TG411-TG214	chr3	2.9	9.1	–
–	TG441-CT242	chr5	2.6	7.9	–
–	XLRR.370-CT202	chr5	3.7	11.3	–
–	CD40-TG176	chr8	3	10.3	–
–	TG330-TG294	chr8	5.4	21.0	–
–	RLRR.130-CLRR.950	chr9	8.2	25.9	–

TABLE 5.2 *(Continued)*

Mapping Population	Molecular Markers	QTL Location	LOD Scores	R^2 Value (%)	References
–	TG424-TG429	chr9	3.7	16.2	–
–	TG241-TG403	chr10	5.6	16.3	–
–	TG508-TG651	chr11	3.8	11.5	–
–	CT55-CD17	chr11	3	9.9	–
–	CT100-TG68	chr12	2.5	8.3	–
Selective genotyping of 820 BC1 *S. lycopersicum* NC84173 *S. habrochaites* PI126445	TG66, TG621	chr3	–	≤–0.22	Zhang et al. (2003)
–	TG652, PK34-340	chr4	–	≤ 0.28	–
–	XLRR-370, RLRR-220, S23-300, PK12-340, CT202, TG318, CT172, CT118A, TG351, CT80A, TG185	chr5	–	≤ 0.32	–
–	AN23-410, CT825	chr6	–	≤ 0.23	–
–	TG46, TG330, TG36C, CT265, CT68, TG294	chr8	–	≤ 0.22	–
–	CT57, CEL79	chr10	–	0.21	–
–	TG497	chr11	–	0.19	–

TABLE 5.2 *(Continued)*

Mapping Population	Molecular Markers	QTL Location	LOD Scores	R^2 Value (%)	References
176 F2 and F3 *S. lycopersicum* Solentos X *S. arcanum* LA2157	P14M60-276P	chr1	4.1	6.8	Chaerani et al. (2007)
–	P14M51-146E	chr2	9	16.2	–
–	P14M51-055P	chr5	6.2	10.5	–
–	P11M48-266E	chr6	6.3	10.8	–
–	P15M62-349P	chr7	8.3	15.2	–
–	P11M60-109P	chr9	8.7	15.5	–
F2, F3, F4 *S. lycopersicum* NCEBR1 X *S. pimpinellifolium* LA2093	–	chr2, chr3, chr4, chr5, chr6, chr7, chr9 and chr12	Combined effect = 44% Individual = 7.6% to13.4%		Foolad et al. (2008)
Selective genotyping of F2 *S. lycopersicum* NCEBR1 X *S. pimpinellifolium* LA2093	–	chr1, chr2, chr3, chr4, chr5, chr6, and chr11	–	–	
172 RIL *S. lycopersicum* NCEBR1 X *S. pimpinellifolium* LA2093	cLEC73K6b, cLEC73I19a, CT205	chr2	3.3	8.0	Ashrafi and Foolad (2015)
–	TG463, cTOFC19J9, CT103	chr2	2.9	5.5	–
–	cLEY-18H8, cTOA24J24, cTOC2J24, cTOC20J21	chr5	5.5	15.3	–

TABLE 5.2 *(Continued)*

Mapping Population	Molecular Markers	QTL Location	LOD Scores	R^2 Value (%)	References
–	TG274, cLEN10H12, TG590, cLEN10H12	chr6	4.4	13.0	–
–	TG343, cLED4N20, TG348, cTOE10J18	chr9	4	11.5	–

Source: Modified from Adhikari et al. (2017).

A total of nine EB resistance major QTLs were located on chromosomes 1, 2, 5, 9, 10, 11, and 12 and two minor QTLs on chromosomes 8 and 11 using the simple interval mapping method in selective BC_1 population of *S. lycopersicum* (NC84173) and *S. habrochaites* (PI126445). The individual effects ranged from 8.4% (chromosome 5) to 21.9% (chromosome 1). The simple interval mapping detected thirteen QTLs in the BC_1S_1 mapping population of the same cross with individual effects varying from 7.9% (chromosome 5) to 25.9% (chromosome 9). The six QTLs from BC_1 population located at chromosomes 1, 5, 9 (two QTLs), 10 and 11 and the QTLs of BC_1S_1 mapping populations located chromosomes 1, 5, 8 (two QTLs), 9, and 10 were proved independent by composite interval mapping (Foolad et al., 2002). The EB resistance in PI 126445 contributed seven QTLs located on chromosomes 3, 4, 5, 6, 8, 10, and 11 in the BC1 mapping populations developed by Zhang, Lin, Nino-Liu, and Foolad. The finding of (Foolad et al., 2002) was parallel for the QTLs located on chromosomes 5, 8, 10, and 11 (Zhang et al., 2003). Meanwhile, (Chaerani et al., 2007) also conducted EB resistance mapping on the F_2 and F_3 population derived from *S. arcanum* (LA2157) and susceptible *S. lycopersicum* (Solentos). Six QTLs located on chromosomes 1, 2, 5, 6, 7, and 9 that explained 7–16% of total phenotypic variance was detected. The QTLs on chromosome 2 and 9 showed effectiveness in both the environment, i.e., field and glasshouse condition where 35% of total phenotypic variance of stem lesions was contributed by QTLs on chromosome 9 alone stating the possibility in EB resistance molecular breeding. The detected QTLs were in concordance with (Foolad et al., 2002; Zhang et al., 2003). The mapping of F_2, F_3 and F_4 populations derived from the cross of *S. pimpinellifolium* LA2093 and tomato breeding line NCEBR-1 yielded 10 QTLs conferring EB resistance. The QTLs were located on chromosomes 2, 3, 4, 5, 6, 7, 9, and 12 with an individual effect of 7.6–13.4% and a combined effect of 44% of total phenotypic variance. The selective genotyping approach in F2 populations of the same crosses detected QTLs on chromosomes 1, 2, 3, 4, 5, 6, and 11 (Foolad et al., 2008). The finding is useful in further EB resistance breeding. A study conducted by (Ashrafi and Foolad, 2015) identified five major QTLs for EB resistance from the RILs, i.e., the F_7, F_8, F_9, and F_{10} generations of the cross *S. pimpinellifolium* LA2093 and tomato breeding line NCEBR-1. The chromosome 2 contributed two QTLs while single QTL from chromosome 5, 6, and 9. The Stable QTLs with largest phenotypic effect for EB resistance were contributed by from chromosomes 5 and 6 with 11–17% and 10–16% phenotypic variation, respectively.

Extensive research has been carried on for identifying the EB resistance QTLs, yet the commercial cultivar is not developed. The reason being the genetic complexity of EB resistance and other linkage drag associated with using the wild species as resistant donors. The study is now going toward transgenic to developed EB-resistant cultivar with desirable horticultural traits. The *rolB* gene of *Agrobacterium rhizogenes* transformed the tomato with EB tolerance with high fruit and nutritional quality (Arshad et al., 2014). The plant growth-promoting rhizobacteria (PGPR) play an important role in EB resistance. Babu et al. (2015) treated tomato plant with PGPR and observed an increase in synthesis of antioxidant peroxidase (POX) and polyphenol oxidase (PPO) enzymes that suppressed the *Aternari solani* a causal agent of EB.

5.5 CONCLUSION

The EB resistance breeding in tomato is a complicated task due to the undesirable linkage drag from the resistant sources. These has resulted in the search for new resistant sources, to validate, utilize, and introduce detected QTLs in cultivated tomato species for the development of early to mid-season maturity with EB and high yielding cultivar. Most of the interspecific crosses of the cultivated species with *S. habrochaites*, *S. pimpinellifolium* and *S. arcanum* yielded QTLs for EB resistance. The QTLs on chromosome 5 and 6 were in concordance with most of the findings and contributed 10–16% (Ashrafi and Foolad, 2015) effect for EB resistance while the largest effect are observed in the QTL from *S. habrochaites* PI 126445 accession chromosome 9, which contributed 25.6% of phenotypic variance (Foolad et al., 2002; Zhang et al., 2003) and the chromosome 9 of the *S. arcanum* LA2157 accession that showed 35% of total phenotypic variance for EB resistance (Chaerani et al., 2007). Unfortunately, few detected QTLs have been validated and have small effects, large marker intervals, i.e., >10 cM that creates the possibility of undesirable linked genes (Foolad et al., 2008). To solve the problems of undesirable linkage, the advent of NGS technique that detects SNPs are employed for fine mapping. The introgression lines (ILs), near-isogenic lines (NILs) and sub-NILs containing the resistance portion from wild species are used for fine mapping. This line breaks undesirable traits prior to introgression in elite cultivars and to identify the physiological basis of QTL effects, pleiotropic effects, and multiple genetic factors controlling the trait and give precise estimation of the effects and chromosomal positions of the detected QTLs (Brouwer and Clair, 2004). The future research can focus on extensive screening of multiple germplasms to identify new resistance sources with strong resistance and minimal undesirable horticultural trait linkage, pyramiding of the reported QTLs into single line, and further fine mapping of the largest QTL effects with recent markers.

KEYWORDS

- **early blight**
- **marker-assisted selection**
- **molecular breeding**
- **next-generation sequencing**
- **plant growth-promoting rhizobacteria**
- **recombinant inbred lines**
- **single nucleotide polymorphisms**

REFERENCES

Adhikari, P., Oh, Y., Panthee, D., (2019). Current status of early blight resistance in tomato: An update. *International Journal of Molecular Sciences, 18*(10). doi: 10.3390/ijms 18102019.

Arshad, W., Ihsan ul, H., Waheed, M. T., Mysore, K. S., & Mirza, B., (2014). Agrobacterium-mediated transformation of tomato with rolB gene results in enhancement of fruit quality and foliar resistance against fungal pathogens. *Plos One, 9*(5), e96979.

Ashrafi, H., & Foolad, M. R., (2015). Characterization of early blight resistance in a recombinant inbred line population of tomato: I. Heritability and trait correlations. *Adv. Stud. Biol., 7*, 131–148.

Ashrafi, H., & Foolad, M. R., (2015). Characterization of early blight resistance in a recombinant inbred line population of tomato: II. Identification of QTLs and their co-localization with candidate resistance genes. *Adv. Stud. Biol., 7*, 149–168.

Babu, A. N., Jogaiah, S., Ito, S., Nagaraj, A. K., & Tran, L. S. P., (2015). Improvement of growth, fruit weight, and early blight disease protection of tomato plants by rhizosphere bacteria is correlated with their beneficial traits and induced biosynthesis of antioxidant peroxidase and polyphenol oxidase. *Plant Sci., 231*, 62–73.

Balkrishna, A. S., Bhushan, B. D., Hussain, K., Panda, S., Meir, S., Rogachev, I., Aharoni, A., et al., (2017). Dynamic metabolic reprogramming of steroidal glycol-alkaloid and phenylpropanoid biosynthesis may impart early blight resistance in wild tomato (*Solanum arcanum* Peralta). *Plant Mol. Biol., 95*(4/5), 411–423.

Barksdale, T. H., (1969). Resistance of tomato seedling to early blight. *Phytopathology, 59*, 443.

Barksdale, T. H., (1971). Field evaluation for tomato early blight resistance. *Plant Dis. Rep., 55*, 807.

Brouwer, D. J., & St Clair, D. A., (2004). Fine mapping of three quantitative trait loci for late blight resistance in tomato using near-isogenic lines (NILs) and sub-NILs. *Theor. Appl. Genet., 108*, 628–638.

Chaerani, R., & Voorrips, R., (2006). Tomato early blight (*Alternaria solani*): The pathogen, genetics, and breeding for resistance. *J. Gen. Plant Pathol., 72*, 335–347.

Chaerani, R., Smulders, M. J. M., Van, D. L. C. G., Vosman, B., Stam, P., & Voorrips, R. E., (2007). QTL identification for early blight resistance (*Alternaria solani*) in a *Solanum lycopersicum* X *S. arcanum* cross. *Theor. Appl. Genet., 114*, 439–450.

Foolad, M. R., & Panthee, D. R., (2012). Marker-assisted selection in tomato breeding. *Crit. Rev. Plant Sci., 31*, 93–123.

Foolad, M. R., & Sharma, A., (2005). molecular markers as selection tools in tomato breeding. *Acta Hort., 695*, 225–240.

Foolad, M. R., (2007). Genome mapping and molecular breeding of tomato. *Int. J. Plant Genom.*, 64358. doi: 10.1155/2007/64358.

Foolad, M. R., Merk, H. L., & Ashrafi, H., (2008). Genetics, genomics and breeding of late blight and early blight resistance in tomato. *Crit. Rev. Plant Sci., 27*, 75–107.

Foolad, M. R., Ntahimpera, N., Christ, B. J., & Lin, G. Y., (2000). Comparison between field, greenhouse, and detached-leaflet evaluations of tomato germplasm for early blight resistance. *Plant Dis., 84*, 967–972.

Foolad, M. R., Zhang, L. P., Khan, A. A., Nino-Liu, D., & Lin, G. Y., (2002). Identification of QTLs for early blight (*Alternaria solani*) resistance in tomato using backcross populations of a *Lycopersicon esculentum* x *L-hirsutum* cross. *Theor. Appl. Genet., 104*, 945–958.

Fulton, T. M., Van, D. H. R., Eannetta, N. T., & Tanksley, S. D., (2002b). Identification, analysis, and utilization of conserved ortholog set markers for comparative genomics in higher plants. *Plant Cell, 14*, 1457–1467.

Gardner, R. G., & Panthee, D., (2012). 'Mountain magic': An early blight and late blight-resistant specialty type F-1 hybrid tomato. *Hortscience, 47*, 299–300.

Gardner, R. G., & Shoemaker, P. B., (1999). 'Mountain supreme' early blight-resistant hybrid tomato and its parents, NC EBR-3 and NC EBR-4. *Hortscience, 34*, 745–746.

Gardner, R. G., (1988). NC-EBR-1 and NC-EBR-2 early blight resistant tomato breeding lines. *Hortscience, 23*, 779–781.

Gardner, R. G., (2000). 'Plum dandy', a hybrid tomato, and its parents, NC EBR-5 and NC EBR-6. *Hortscience, 35*, 962–963.

Johanson, A., & Thurston, H. D., (1990). The effect of cultivar maturity on the resistance of potato to early blight caused by *Alternaria solani. Am. Potato J., 67*, 615–623.

Kalloo, G., & Banarjee, M. K., (1993). Early blight resistance in *Lycopersicon esculentum* Mill. Transferred from *L. pimpinnellifolium* (L.) and *L. hirsutum* f. *glabratum* mull. *Gartenbauwissenschaft, 58*, 238–240.

Locke, S. B., (1949). Resistance to early blight and *Septoria* leaf spot in the genus *Lycopersicon. Phytopathology, 39*, 829–836.

Maiero, M., Ng, T. J., & Barksdale, T. H., (1990). Genetic resistance to early blight in tomato breeding lines. *Hortscience, 25*, 344–346.

Martin, F. W., & Hepperly, P., (1987). Sources of resistance to early blight, *Alternaria-solani*, and transfer to tomato, *Lycopersicon-esculentum. J. Agric. Univ. Puerto Rico, 71*, 85–95.

Nash, A. F., & Gardner, R. G., (1988). Tomato early blight resistance in a breeding line derived from *Lycopersicon esculentum* PI126445. *Plant Dis., 72*, 206–209.

Poysa, V., & Tu, J. C., (1996). Response of cultivars and breeding lines of *Lycopersicon* spp., to *Alternaria solani. Can. Plant Dis. Surv., 76*, 5–8.

Rotem, J., (1994). *The Genus Alternaria: Biology, Epidemiology, and Pathogenicity* (p. 48, 326). The American Phytopathological Society: St. Paul, MN, USA.

Sinden, S. L., Obrien, M. J., & Goth, R. W., (1972). Effect of potato alkaloids on growth of *Alternaria solani. Am. Potato J., 49*, 367.

Thirthamallappa, & Lohithaswa, H. C., (2000). Genetics of resistance to early blight (*Alternaria solani Sorauer*) in tomato (*Lycopersicon esculentum* L.). *Euphytica, 113*, 187–193.

Yamamoto, N., Tsugane, T., Watanabe, M., Yano, K., Maeda, F., Kuwata, C., Torki, M., et al., (2005). Expressed sequence tags from the laboratory grown miniature tomato (*Lycopersicon esculentum*) cultivar micro-tom and mining for single nucleotide polymorphisms and insertion/deletions in tomato cultivars. *Gene, 356*, 127–134.

Yang, W., Bai, X., Eaton, C., Kabelka, E., Eaton, C., Kamoun, S., Van, D. K. E., & Francis, D., (2004). Discovery of single nucleotide polymorphisms in *Lycopersicon esculentum* by computer aided analysis of expressed sequence tags. *Mol. Breeding., 14*, 21–34.

Zhang, L. P., Lin, G. Y., Nino-Liu, D., & Foolad, M. R., (2003). Mapping QTLs conferring early blight (*Alternaria solani)* resistance in a *Lycopersicon esculentum* X *L. hirsutum* cross by selective genotyping. *Mol. Breed., 12*, 3–19.

CHAPTER 6

MOLECULAR APPROACHES FOR THE CONTROL OF SEPTORIA LEAF SPOT IN TOMATO

MD. ABU NAYYER,[1] MD. FEZA AHMAD,[2] MD. SHAMIM,[3] DEEPA LAL,[4] DEEPTI SRIVASTAVA,[1] and V. K. TRIPATHI[5]

[1]*Integral Institute of Agricultural Science and Technology, Integral University, Dasauli, Lucknow – 226021, Uttar Pradesh, India, E-mail: nayyer123@gmail.com (M. A. Nayyer)*

[2]*Department of Horticulture (Fruit and Fruit Technology), Bihar Agricultural University, Sabour, Bhagalpur – 813210, Bihar, India*

[3]*Department of Molecular Biology and Genetic Engineering, Dr. Kalam Agricultural College, Kishanganj, Bihar Agricultural University, Sabour, Bhagalpur – 813210, Bihar, India*

[4]*Department of Horticulture, Babasaheb Bhimrao Ambedkar University, Lucknow – 226025, Uttar Pradesh, India*

[5]*Department of Horticulture, Chandra Shekhar Azad University of Agriculture and Technology, Kanpur, Uttar Pradesh – 208002, India*

ABSTRACT

Tomato is one of the most important multipurpose vegetables around the globe. Amongst the major diseases of tomato, septoria leaf spot (SLS) is one of the most devastating foliar diseases caused by *Septoria lycopersici* Speg. Characterized by multicellular hyaline filiform conidia and dark-brown pycnidia. The major source of inoculum are solanaceous weeds and tomato plant debris, from which conidia are dispersed mainly by rain or overhead irrigation. Various attempts have been made to control disease through chemical, biological, and molecular approaches, so that the disease can be managed and crop loss is prevented. The use of fungicide is efficient to control disease, but they may cause human health hazards and also increase environmental pollution. The use of resistance/tolerance tomato cultivar and species is good for managing SLS, but a high level of resistance is very rare. So, there is a need of development of a resistance-linked marker and induced systemic resistance (ISR) against SLS in tomatoes.

6.1 INTRODUCTION

Tomato (*Solanum lycopersicum* L., 2n = 2x = 24), belongs to the family Solanaceae and originated Peru-Ecuador-Bolivia region of Andes mountains (known as the primary center of origin) and eastern Andes (known as secondary center of origin) from South America. Tomato is an herbaceous spreading plant with a weak woody stem and grows up to the height of 1–3 m produces yellow color flowers; fruits vary in their shape and size from cherry tomatoes to beefsteak tomatoes with 1–2 cm to 10 cm or more in diameter respectively. At the time of ripening color of fruit changes from green to red. Tomato is a climacteric in nature and most commonly used multipurpose vegetables of subtropics and tropics part of world (Govindappa et al., 2013) that provide well-balanced healthy diet as of its high nutritive value, anticancer, and antioxidative properties of lycopene, active carotenoid pigment of tomato. It holds a significant amount of vitamins (mainly B and C), sugars, minerals, phosphorus, iron, necessary amino acids and dietary fibers.

In the world situation, production and productivity of tomato is 211.80 million metric tons and 37.16 metric tons per ha, respectively with an area of more than 5.7 million ha area (Anonymous, 2016). In India, it covers an area of about 0.80 million ha with 19.54 million metric tons production and 21.2 metric tons per ha productivity (Anonymous, 2017). There are many limitations in the cultivation of tomato but diseases play an important role which cause heavier losses in the production of the crop worldwide (Kakati and Nath, 2014). Tomato is acknowledged to susceptible for more than 200 diseases (Shelat et al., 2014). Major fungal diseases which affects the production of tomato are septoria leaf spot, late blight, early blight (EB), fusarium wilt, powdery mildew, corky root rot, and leaf mold (Panthee and Chen, 2010). Amongst the various tomato foliar diseases, septoria leaf spot caused by *Septoria lycopersici* Speg is one of the most overwhelming diseases (Joshi et al., 2015). Controlling of these diseases mainly subjected to the plants treated with fungicides (El-Mougy et al., 2004). Though, application of fungicide cause human health hazards and also helps in increasing the environmental pollution. Thus, the substitution of the eco-friendly approach is desirable for control of different plant diseases (Rojo et al., 2007; Abd-El-Kareem, 2007; Mandal et al., 2009). It has been illustrated that the resistance to septoria leaf spot (SLS) is controlled by a single dominant gene (Barksdale and Stoner, 1978). Whereas, most of the source of resistance lines belongs to different wild species such as, *S. peruvianum, S. pimpinellifolium, S. glandulosum* and *S. habrochaites*. But *S. habrochaites* shows maximum degree of resistance against this disease (Locke, 1949). *S. pimpinellifolium, S. pennelli, S. lycopersicum* var. *cerasiforme,* and *S. chilense* also shows some degree of resistance to this disease. Interspecific crossing of S. *habrochaites* accessions shown higher degree of resistance against this disease. However, these lines show many undesirable horticultural qualities such as indeterminate growth habit, small fruits, late maturity, and low yield are more common. Use molecular markers can mitigate this problem. Molecular markers related to the gene(s) of interest can be used from the selection of plants which is genetically identical to their recurrent parent and possess the required horticultural traits (Joshi et al., 2015).

6.2 SEPTORIA LEAF SPOT (SLS): CAUSE, DAMAGE, AND YIELD LOSSES

Septoria leaf spot (SLS) of tomato caused by *Septoria lycopersici* Speg is a devastating foliar disease that belongs to the phylum Ascomycota. Initially, the infection starts from the lower leaves and progressively proceeds upward under favorable conditions but does not directly infect the fruits (Gleason and Edmunds, 2006). Under favorable environmental conditions, it may cause complete defoliation of plants leads to a higher crop loss, mainly in the humid regions during the time of heavy rainfall, frequent sprinkler or over-head irrigation (Andrus and Reynard, 1945; Delahaut and Stevenson, 2004). This SLS disease was first reported from

Argentina in 1882 and later in the United States in 1896 (Sutton and Waterston, 1966).

Initially, the symptom of SLS disease appears in small, water-soaked spots that soon become circular and about 1/8th inch in diameter with dark edges and grayish-white centers, which is the most distinguishing symptom of this disease. Under favorable environmental conditions, the fruiting body of fungus appear as tiny black specks in the centers of the spots. Severely infected leaves turn yellow, fade, and ultimately fall off (Gleason and Edmunds, 2006). Moderate air temperature and high humidity throughout the cropping season favor the development of disease, which causes premature leaf fall and in significant yield reduction (Sugha and Kumar, 2000). Foliar pathogens lowers the photosynthetic activity in infected leaves by reducing the area of green leaf because of the lesions and also, probably, by stimulating senescence in infected leaves (Christov et al., 2007). Infection may appear at any stage of plant development, but most commonly they appear when plants start fruit setting.

6.3 MANAGEMENT OF SEPTORIA LEAF SPOT (SLS)

6.3.1 CHEMICAL CONTROL OF SEPTORIA LEAF SPOT (SLS)

Many fungicides are used for controlling of SLS in tomato. The commonly used fungicides were reported Mancozeb 75 WP, dithiocarbamates, Copper oxychloride 50 WP (Choulwar and Datar, 1992), Captan (Vakalounakis and Malathrakis, 1988), Copper oxychloride (Maheshwari et al., 1991), and Captafol 80% WP and Ridomil (R) 63.5% WP (Tedla, 1985). Three spray schedule were used viz, spray schedule I-3 sprays of mancozeb 75 WP @ 0.25%; spray schedule II-2 sprays of mancozeb 75 WP @ 0.25% and 1 spray of copper oxychloride 50 WP @ 0.25%, and spray schedule III-1 spray of mancozeb 75 WP @ 25% and 2 sprays of copper oxychloride 50 WP @ 0.25%; each with and without staking along with an unsprayed control at 15 days interval from 40 days after transplanting of 'Roma' cultivar of tomato and they found that leaf spot caused by *S. lycopersici* diseases were less in the spray schedule I and in staked plots (Bhardwaj et al., 1995). Five fungicides (carbendazim, Captan, thiophanate-methyl, blue copper and mancozeb + thiophanate-methyl) were tested in a pot experiment to check their ability to control the leaf spot of tomato caused by *S. lycopersici*. All of the fungicides effectively control the disease but Captan followed by carbendazim + blue copper and mancozeb + thiophanate-methyl give the best control (Ahamad and Ahmad, 2000).

Tedla (1985) worked on spray interval and fungicide spray on tomato and give the score from 0–5 according to disease severity, maximum score means higher disease severity and they found less disease damage were encountered when 0.2% captafol + 0.23% ridomil sprayed at 7 days interval. When spray interval increase from 7 to 21 days, disease severity/infection also increase. Sohi and Sokhi (1970) evaluated the effect of ten fungicides for the control of Septoria leaf spot and they found that weekly sprays of Dithane Z-78 (Zineb) 0.2% or Bordeaux (4:4:50) provide the best control over other treatment, reducing the average defoliation from 95.46% to 53.17% and 48.86%, respectively. Treatment with zineb gave the highest net profit. Govardhan (2001) examined different fungicides against septoria lycopersici under field condition. The fungicides are mancozeb, iprodione, capton, blitox, kavach, bavistin, baynate, and zineb were sprayed in the field. Three spray of each fungicide were given at every 10 days interval. Among the fungicides, bavistin, and mancozeb were found to be highly effective against the disease.

6.3.2 MICROBIAL AND BACTERIAL CONTROL OF SEPTORIA LEAF SPOT (SLS)

It has been reported that *Trichoderma* spp. showed the antagonistic property against various soil-borne plant pathogens and produces many toxic metabolites *in vitro* by *Trichoderma* spp. in the soil (Wright, 1956). A reduction of *S. lycopersici* disease infestation up to 30–40% by *T. viride* (Kashyap and Leukina, 1977). Blum (2000) evaluated the effect of Bacteria and Yeast on severity and incidence reduction of *Septoria lycopersici* on tomato under different experiments and they reported in their first experiment that yeast isolate Y236 (*Cryptococcus laurentii*) reduces the disease incidence and the number of leaf spots significantly ($P \leq 0.05$), further, in second experiment, combined treatment of Y178 (*Candida tenuis*) and Y180 (*C. oleophila*) found most efficient in control of septoria leaf spot, And in their third experiment BTL (*Pseudomonas putida*) and Y182 (*C. oleophila*) found most significant ($P \leq 0.05$) in the control of the disease. A microbial product such as Prestop and Serenade Opti is also used to control SLS in tomato. Biological activity of Prestop is living fungus based, *Gliocladium catenulatum* J1446. This fungus is supposed to work on amalgamation of hyperparasitism and competition, which suppress the activity of pathogen (Mcquilken et al., 2001). Serenade Opti depends on lipopeptides present in the bacterium, Bacillus amyloliquefaciens QST 713. It has been reported that lipopeptides produced by *B. amyloliquefaciens* strains directly overwhelm pathogens by hindering lipid synthesis, membrane integrity/transport and function. In addition, it is also observed that *B. amyloliquefaciens* activates an induced systemic resistance (ISR) process, which reinforces a plant's immune system and thereby decreases susceptibility to pathogen infection. Egel et al. (2019) worked on the efficiency of organic disease control products in field and greenhouse trials on common foliar diseases of tomato and they evaluate Serenade Opti, Regalia, and Sil-Matrix (a silicon product used to manage Septoria leaf spot, EB or bacterial spot) treatments were evaluated separately to each other in the greenhouse as different to alternation treatments in field trials and they found that Serenade Opti was a more effective as compared to Sil-Matrix. They also claim that, this is the first published work to manage Septoria leaf spot by silicon product.

6.4 MOLECULAR APPROACHES FOR THE DEVELOPING RESISTANCE IN SEPTORIA LEAF SPOT (SLS) OF TOMATO

6.4.1 RESISTANCE LINKED MARKER AND QTLS AGAINST SEPTORIA LEAF SPOT (SLS)

Barksdale and Stoner (1978), reported that the resistance to septoria leaf spot is controlled by a single dominant gene named it the Se gene (Andrus and Reynard, 1945). Although most of the source of resistance lines belongs to some wild species comprising *S. peruvianum, S. pimpinellifolium* and *S. glandulosum,* but the maximum level of resistance was found in *S. habrochaites* (Andrus and Reynard, 1945; Locke, 1949). This resistance was found to be linked with late maturity and small fruit size (Poysa and Tu, 1993). Beneficial levels of resistance have also been found in *S. chilense S. pennelli,* and *S. lycopersicum* var. *cerasiforme*. Breeding lines develop from interspecific crossing with *S. habrochaites* accessions have shown a high degree of resistance. However, these interspecific lines have one or many undesirable horticultural traits such as indeterminate growth habit, small fruits, late maturity and low yield (Table 6.1) (Joshi et al., 2015).

TABLE 6.1 Tolerance/Resistance Tomato Accessions and Cultivars Against Septoria Leaf Spot

SL. No.	Cultivated Parent/Lines/Wild Tomato/ Varieties	References
1.	NC 839-2 (2007)-1	Joshi et al. (2015)
2.	Pusa Red Plum	Gupta (1960)
3.	*L. pennellii, L. esculentum var. cerasiforme, L. pimpinellifolium, L. chilense, L. hirsutum*, and *L. peruvianum*	Poysa and Tu (1993)
4.	*Lycopersican pimpinellifolium* line 'PI 422397'	Barksdale and Stoner (1978)
5.	'BWR 1,' 'Marglobe,' 'EC 5889	Sugha and Kumar (1998)
6.	AVTO1173, AVTO1008, AVTO1003, and AVTO1219	Gul et al. (2016)
7.	*S. habrochaites, S. peruvianum* accessions (CNPH-1036, PI-306811, LA-1984, LA-1910, and LA-2744)	Satelis et al. (2010)
8.	S-12 and SH-12	Raina and Razdan (2010)
9.	Tomato Kt-4 a natural cross, EC 4555, EC 2750, EC 7785, EC 6993 and HP2453	Sohi and Sokhi (1969)
	EC 7293 and HP2453	
10.	Rossol and EC 1085	Bedi et al. (1990)

Pandey and Pandey (2002) screened 132 germplasm of tomato which included *Septoria lycopersici, Xanthomonas campestris* pv. *vasicatoria* and *Alternaria solani* for multiple disease effect and their results shows that the among screened lines, only one line LE-415 germplasm was found to be resistant against the disease and six lines were moderately resistant for multiple diseases of tomato. Out of 44 tomato cultivars from different countries has been tested in which 18 varieties from *Solanum lycopersicon* and 5 varieties from subspecies of *Lycopersicon esulentum* for resistance to *S. lycopersici* only two tomato cultivars (one from Denmark and other from) and one tomato variety (*L. esculentum* var. *cerasiforme*) revealed resistance, whereas many wild *Lycopersicon* spp. shows resistance and are considered as promising sources of resistance. *Solanum sisymbrifolium* also shows resistance against *S. lycopersici*, but was not crossable with tomato varieties (Sotirova and Rodeva, 1991).

6.4.2 INDUCED SYSTEMIC RESISTANCE (ISR)

Induced systemic resistance (ISR) is defined as the systemic protection of plants by improvement in their defensive ability against a wide spectrum of pathogens that develops after suitable inducing of infection by a pathogen.

Induction of systemic resistance by plant growth-promoting rhizobacteria (PGPR) and chemicals against different diseases was measured as the most appropriate method in crop protection. The major three differences are found between ISR and other mechanisms. First, the action of ISR is based on the defense mechanisms that are stimulated by inducing agents. Secondly, when ISR once expressed they activate several potential defense mechanisms that include the increased activity of 3-glucanases, chitinases, β-1 and POs (Xue et al., 1998; Schneider and Ullrich, 1994; Maurhofer et al., 1994). Thirdly, an important feature of ISR is the wide spectrum of pathogens that can be controlled by a single inducing agent (Wei et al., 1996; Hoffland et al., 1996). Thus ISR seems to be the result of various mechanisms which together are more effective against a wide range of bacterial, fungal, and viral pathogens. Induced resistance and induced responses have been well-documented and have been found in several plant taxa (Karban and Baldwin, 1997). Systemic resistance mechanisms are instigated in crop plants by treatment with chemical inducers, such as benzothiadiazole (BTH), isonicotinic acid (INA), salicylic acid (SA), and probenazole (Gorlach et al., 1996; Pieterse et al., 1998; De Meyer et al., 1999; Sakamoto et al., 1999).

Anand et al. (2007) worked on plant defense enzyme activities with systemic resistance to leaf spot leaf and EB induced in tomato plants by Pseudomonas fluorescens and azoxystrobin at three different levels of concentrations, viz., 31.25, 62.50, and 125 g a.i./ha (mancozeb (1.0/kg ha) and Pseudomonas fluorescens (10.0 kg/ha) respectively and evaluate the activity of defense enzymes polyphenol oxidase (PPO), peroxidase (POX), phenylalanine ammonia-lyase (PAL), β-1, catalase, 3-glucanase, chitinase, and total phenols. This study shows that azoxystrobin remarkable changes in PO, PAL, PPO, β-1,3-glucanase, chitinase, catalase, and total phenols. β-1,3-glucanase and PO are related to cross-linking of cell wall components, suberin monomers, polymerization of lignin and subsequent resistance to pathogen in several host-pathogen interactions (Reuveni et al., 1995). Induction of systemic resistance by *P. fluorescens* in plants gave several defense related genes, and it is known that the defense genes are sleeping genes, and appropriate stimuli or signals are required to activate them. Inducing the plants particular defense mechanisms by earlier application of biological inducer is supposed to be a unique plant protection strategy.

Anand et al. (2010) evaluate the effect of azoxystrobin (Amistar 25 SC) against leaf spot diseases of tomato and their results show that the higher doses of azoxystrobin (125 and 62.50 g a.i. ha^{-1}) shown 100% reduction in the incidence of leaf spot followed by its lower dose, In the second season of an experiment, all the doses of azoxystrobin were found effective against the SLS without any phytotoxic effect with no detectable level of residues are found in the harvested fruits.

Silva et al. (2004) studied the induction of systemic resistance by *Bacillus cereus* against foliar diseases of tomato, and they reported that no symptom of septoria leaf spot were observed during the beginning of the experiment, but control plant shows highest degree of infestation if once infection was established. Treatment T_1 (*B. cereus* + 20 spray of Chlorothalonil at 2 g/l a.i.) controls the plants satisfactorily against Septoria leaf spot as compare to other treatments, approving the benefits of integrating control techniques. They also reported that septoria infection is more severe at cooler temperature. At the beginning of the field experiment, ambient temperature was quite high, which may results in the nonappearance of infection during the first weeks of the study. Benedict (1971) evaluate the effect of quality and intensity of light on POX activity related with Septoria leaf spot of tomato and they reported that the scans of POX isozymes through paper electrophoresis exhibited a visible increase in the bimodal pattern of staining intensity reflects a high level of activity of major enzymatic components in the leaves after 4 days of inoculation with the fungus. Minimal values of activity were obtained at 6 days, enhanced isozyme activity happened at later stages with the maturation of pycnidia and pycnospores and progression of the disease. The effect of light intensity is more after 10 days of inoculation. They also reported that the light intensity is a vital factor that evaluates the symptom expression and also helpful in evaluates the enzyme activity in this disease.

6.4.3 INTEGRATED MANAGEMENT OF DISEASE

The most favorable condition for the development of plant diseases are presence of extra moisture on the leaves and a higher concentration of moisture in the soil. Shortage, and specifically excess water content, increase the severity of various diseases and pests of tomatoes. Drip and furrow system of irrigation, which does not allow to wet the aerial part of the plants, reduces the incidence of many diseases (Marouelli et al., 2005). Cabral et al. (2013) worked on diverse irrigation systems and water management strategies for management of Septoria leaf spot in tomato and their results proved that the conventional system of sprinkler irrigation favors more SLS epidemics, although the furrow and drip systems were seen to be efficient in obstructing the dissemination of the pathogen. The size of the splashed drops of water on the plants had a strong effect on the severity of the disease, which developed a more severe infection as the size of the drop increased. Finally, increasing the time interval between sprinkler irrigations, and the adoption of soil organic mulch hindered the onset of disease may act synergistically to further reduction in disease severity level. Raina and Razdan (2010) experimented on the effect of weather factors on the seasonal occurrence of *Septoria lycopersici* shows that rainy days/percentage of total rainfall coupled with relative humidity causes a maximum spread of disease. They also screened 41 cultivars for disease resistance against Septoria leaf spot in which only two varieties (SH-12 and S-12) showed resistance at each tested location. Further, they reported that spraying the tomato crop with Dithane M-45 or Dithane Z-78 at 0.2% each, proved highly effective in limiting the spread of disease under field conditions and thereby alleviating the yield returns. Nahak and Sahu (2015) reported that the application of *Azadirachta indica* aqueous extract reduces the leaf spot disease from 40th day onwards, and maximum reduction was observed on 80th day in comparison to control through a metabolic change in plants includes induction of antioxidant defensive enzymes and accumulation of phenol. Anwar et al. (2017) worked on integrated management of major fungal diseases of tomato in Kashmir valley, India with different management practices viz., organic amendment, cultural practices, seedling root-dip in bioagents, fungicide, and 2 foliar sprays of chemicals 20 days after the onset of disease and they found that the plants treated with Staking + Removal of basal leaves + Mulching (mustard pod straw @ 4 kg/m +Sprays of mancozeb 75WP @

2 g/L water followed by carbendazim 50 WP @ 0.5 g/L water proved best in controlling diseases and revealed a minimum disease severity of 4.4% with highest disease of control 80.85% in case of Septoria leaf spot as compared to other treatments. Keerthi et al. (2019) worked on detection of tomato leaf disease based on SVM classifier, and they suggest that SLS can be controlled by taking actions such as watering done at the base of the plant during morning time in place of evening to ensure that the leaves are wet for a minimum period of time, use of disease-resistant plants and crop rotation, etc.

Pattnaik et al. (2012) worked on effect of ten different medicinal plants extract in different concentrations such as, *Pongamia piñata, Azadirachta indica, Aegle Marmelos, Piper nigrum, Brassica campestris, Euphorbia tirucalli, Ageratum conyzoides, Tagetes Patula, Vitex Negundu*, and *Ziziphus jujube* and they found that *P. nigrum* showed maximum inhibition 31.06±0.04% of *Septoria lycopersici* mycelial growth and *A. indica* showed low inhibition of mycelial growth. The remaining plants show were less reduction of mycelium growth.

6.5 CONCLUSION

Tomato is a widely used crop and affected by various biotic and abiotic stresses. Among biotic stress Septoria leaf spot disease (SLS) is a most devasting disease of tomato. Therefore, there is an urgent need to overcome this disease. Although various management tools and practices such as chemicals, biocontrol methods are used for the management of this disease but their effectiveness on a large scale is not up to the mark and not durable. Durable resistance is influenced by environmental factors. Therefore, there is an urgent need to use other means of disease management. The use of resistant tomato cultivars can minimize the effect of environmentally destructive pesticides. Recent advances in tomato genomics can provide additional information about the resistant genes that could be sustainable and environment favorable. Besides this, molecular breeding can also be used effectively once the resistance source is identified. The present chapter outlines the different traditional methods and some integrated methods that can be used for the management of this disease. Since, till date, no resistant source is found, so the use of molecular methods could open a new door in the identification of new resistant sources.

KEYWORDS

- **disease inoculum**
- **disease management**
- **molecular markers**
- **phenylalanine ammonia-lyase**
- **plant growth-promoting rhizobacteria**
- **resistance mechanism**
- ***Septoria lycopersici***

REFERENCES

Abd-El-Kareem, F., (2007). Induced resistance in bean plants against root rot and *Alternaria* leaf spot diseases using biotic and abiotic inducers under field conditions. *Research Journal of Agriculture and Biological Sciences, 3*(6), 767–774.

Ahamad, S., & Ahmad, N., (2000). Evaluation of systemic and non-systemic fungicides against *Septoria* leaf spot of tomato. *Annals of Plant Protection Sciences, 8*, 117–119.

Anand, T., Chandrasekaran, A., Kuttalam, S., & Samiyappan, R., (2010). Evaluation of azoxystrobin (Amistar 25 SC) against early leaf blight and leaf spot diseases of tomato. *Journal of Agricultural Technology, 6*, 469–485.

Anand, T., Chandrasekaran, A., Kuttalam, S., Raguchander, T., Prakasam, V., & Samiyappan, R., (2007). Association of some plant defense enzyme activities with systemic resistance to early leaf blight and leaf spot induced in tomato plants by azoxystrobin and *Pseudomonas fluorescens*. *Journal of Plant Interactions, 2*, 233–244.

Andrus, C. F., & Reynard, G. B., (1945). Resistance to septoria leaf spot and its inheritance in tomatoes. *Phytopathology, 35*, 16–24.

Anonymous, (2017). *Area and Production of Horticultural Crops-All India, 2016–17 (Third Advance est.)*. Government of India, Ministry of Agriculture and Farmers Welfare, Department of Agriculture, Cooperation and Farmers Welfare (Horticulture Statistics Division).

Anwar, A., Bhat, M., Mughal, M. N., Mir, G. H., & Ambardar, V. K., (2017). Integrated management of major fungal diseases of tomato in Kashmir Valley, India. *International Journal of Current Microbiology and Applied Sciences, 6*(8), 2454–2458.

Barksdale, T. H., & Stoner, A. K., (1978). Resistance in tomato to *Septoria lycopersici*. *Plant Disease Reporter, 62*, 844–847.

Bedi, J. S., Sokhi, S. S., Munshi, G. D., & Mohan, C., (1990). Identification of resistance to some diseases of tomato lines. *Pl. Dis. Res., 5*, 188–190.

Benedict, W. G., (1971). Effect of intensity and quality of light on peroxidase activity associated with *Septoria* leaf spot of tomato. *Canadian Journal of Botany., 49*, 1721–1726.

Bhardwaj, C. L., Thakur, D. R., & Jamwal, R. S., (1995). Effect of fungicide spray and staking on diseases and disorders of tomato (*Lycopersicon esculentum*). *Indian J. of Agril. Sc., 65*, 148–151.

Cabral, R. N., Marouelli, W. A., Lage, D. A., & Café-Filho, A. C., (2013). *Septoria* leaf spot in organic tomatoes under diverse irrigation systems and water management strategies. *Horticultura. Brasileira, 31*, 392–400.

Choulwar, A. B., & Datar, V. V., (1992). Management of tomato early blight with chemicals. *J. Maharastra Agric Uni., 17*, 214–216.

Christou, P., (2018). Foreword. In: *Biotechnologies of Crop Improvement, Volume 2: Transgenic Approaches* (Vol. 2).

Christov, I., Stefanov, D., Velinov, T., Goltsev, V., Georgieva, K., Abracheva, P., Genova, Y., & Christov, N., (2007). The symptomless leaf infection with grapevine leafroll associated virus 3 in grown *in vitro* plants as a simple model system for investigation of viral effects on photosynthesis. *J. Plant Physiol., 164*, 1124–1133.

De Meyer, G., Capieau, K., Audenaert, K., Buchala, A., Metraux, J. P., & Hotfite, M., (1999). Nanogram amounts of salicylic acid produced by the rhizobacterium *Pseudomonas aeruginosa* 7NSK 2 activate the systemic acquired resistance pathway in bean. *Molec. Plant-Microbe Interact., 12*, 450–458.

Delahaut, K., & Stevenson, W., (2004). *Tomato Disorders: Early Blight and Septoria Leaf Spot.* The University of Wisconsin, Madison, WI.

Egel, D. S., Hoagland, L., Davis, J., Marchino, C., & Bloomquist, M., (2019). Efficacy of organic disease control products on common foliar diseases of tomato in field and greenhouse trials. *Crop Protection*, (Accepted manuscript).

El-Mougy, N. S., Abd-El-Karem, F., El-Gamal, N. G., & Fotouh, Y. O., (2004). Application of fungicides alternatives for controlling cowpea root rot diseases under greenhouse and field conditions. *Egypt. J. Phytopathol., 32*, 23–35.

Gleason, M. L., & Edmunds, B. A., (2006). *Tomato Diseases and Disorders.* University Extension PM 1266, Iowa State University, Ames.

Gorlach, J., Volrath, S., Knauf- Beiter, G., Hengy, O., Beckhove, U., Kogel, K. H., Oostendorp, M., et al., (1996). Benzothiadiazole, a novel class of inducers of systemic acquired resistance, activates gene expression and disease resistance in wheat. *Plant Cell, 8*, 629–643.

Govardhan, V. P., (2001). *Studies on Septoria Leaf Spot of Tomato (Lycopersicon esculentum Mill.) Caused by Septoria lycopersici Speg.* MSc (Agri.) Thesis, Univ. Agric. Sci., Bangalore, Karnataka (India).

Gul, Z., Ahmed, M., Ullah, K. Z., Khan, B., & Iqbal, M., (2016). Evaluation of tomato lines against *Septoria* leaf spot under field conditions and its effect on fruit yield. *Agricultural Sciences, 07*, 181–186.

Gupta, S. C., (1960). Susceptibility of 11 varieties of tomato plants to *Septoria* leaf spot disease. *Proc. Natl. Acad. Sc.l, India, Sect. B., 30*, 98–100.

Hoffland, E., Hakulinem, J., & Van, P. J. A., (1996). Comparison of systemic resistance induced by avirulent and nonpathogenic *Pseudomonas* species. *Phytopathology, 86*, 757–762.

Joshi, B. K., Louws, F. J., Yenco, G. C., Sosinski, B. R., Arellano, C., & Panthee, D. R., (2015). Molecular markers for *Septoria* leaf spot (*Septoria lycopersicii* Speg.) resistance in tomato (*Solanum Lycopersicum* L.). *Nepal Journal of Biotechnology, 3*(1), 40–47.

Kakati, N., & Nath, P. D., (2014). Sustainable management of tomato leaf curl virus disease and its vector, *Bemisia tabaci* through integration of physical barrier with biopesticides. *International J. of Innovative Res. and Devel., 3*(2), 132–140.

Karban, R., & Baldwin, I. T., (1997). *Induced Responses to Herbivory* (pp. 56–63). Chicago: Chicago Press.

Kashyap, U., & Leukina, L. M., (1977). Effect of microorganisms isolated from tomato leaves on the mycelial growth of some pathogenic fungi. *Biologiya., 1*, 65–69.

Keerthi, J., Maloji, S., & Krishna, P. G., (2019). An approach of tomato leaf disease detection based on SVM classifier. *International Journal of Recent Technology and Engineering, 7*, 697–704.

Lahlali, R., Peng, G., Gossen, B. D., McGregor, L., Yu, F. Q., Hynes, R. K., Hwang, S. F., et al., (2013). Evidence that the biofungicide Serenade® (*Bacillus subtilis*) suppresses clubroot on canola via antibiosis and induced host resistance. *Phytopathology, 103*, 245–254.

Locke, S. B., (1949). Resistance to early blight and *Septoria* leaf spot in the genus *Lycopersicon*. *Phytopathology, 39*, 829–836.

Maheshwari, S. K., Gupta, P. C., & Gandhi, S. K., (1991). Evaluation of various fungi toxicants against early blight of tomato (*Lycopersicon esculentum* Mill.). *Agric. Sci. Digest., 11*, 201, 202.

Mandal, S., Mallicka, N., & Mitraa, A., (2009). Salicylic acid-induced resistance to *Fusarium oxysporum f. sp. lycopersici* in tomato. *Plant Physiology and Biochem., 47*(7), 642–649.

Marouelli, W. A., Medeiros, M. A., Souza, R. F., & Resende, F. V., (2011). Production of organic tomatoes irrigated by sprinkler and drip systems, as single crop and intercropped with coriander. *Horticultura Brasileira, 29*(3), 429–434.

Maurhofer, M., Hase, C., Meuwly, P., Metraux, J. P., & Defago, G., (1994). Induction of systemic resistance of tobacco-to-tobacco necrosis virus by the root-colonizing *Pseudomonas fluorescens* strain CHAO: Influence of the gacA gene and of pyoverdine production. *Phytopathology, 84*, 139–146.

Mcquilken, M. P., Gemmel, J., & Lahdenpera, M. L., (2001). *Gliocladium catenulatum* as a potential biological control agent of damping-off in bedding plants. *J. Phytopathol., 149*, 171–178.

Nahak, G., & Sahu, R. K., (2015). Biopesticidal effect of leaf extract of neem (*Azadirachta indica* A. Juss) on growth parameters and diseases of tomato. *J. of Appl. and Natural Sc., 7*(1), 482–488.

Pandey, P. K., & Pandey, K. K., (2002). Field screening of different tomato germplasm lines against *Septoria, Alternaria* and bacterial disease complex at seedling stage. *J. Mycol. Pl. Pathol., 32*, 234, 235.

Panthee, D. R., & Chen, F., (2010). Genomics of fungal disease resistance in tomato. *Curr. Genomics, 11*, 30–39.

Pattnaik, M. M., Kar, M., & Sahu, R. K., (2012). Bioefficacy of some plant extracts on growth parameters and control of diseases in *Lycopersicum esculentum. Asian Journal of Plant Science and Research, 2*, 129–142.

Pieterse, C. M., Ewws, S. C., Van, P. J. A., Knoester, M., Laan, K., Gerrits, H., Weisbeek, P. J., & Van, L. L. C., (1998). A novel signaling pathway controlling induced systemic resistance in Arabidopsis. *Plant Cell, 10*, 1571–1580.

Poysa, V., & Tu, J. C., (1993). Response of cultivars and breeding lines of *Lycopersicon* spp. to *Septoria lycopersici*. *Canadian Plant Disease Survey, 73*, 9–13.

Raina, P. K., & Razdan, V. K., (2010). Influence of weather factors and management of *Septoria* leaf spot of tomato in sub-tropics. *Indian Phytopathology, 63*, 26–29.

Reuveni, M., Agapov, V., & Reuveni, R., (1995). Induced systemic protection to powdery mildew in cucumber plants by phosphate and potassium salts: Effect of inoculum concentration and post-inoculation treatment. *Can J. Plant Pathol., 17*, 247–251.

Rojo, F. G., Reynoso, M. M., Sofia, M. F., & Torres, A. M., (2007). Biological control by *Trichoderma* species of *Fusarium solani* causing peanut brown root rot under field conditions. *Crop Protection, 26*, 549–555.

Sakamoto, K., Tada, Y., Yokozeki, Y., Akari, H., Hayashi, N., Fujimura, T., & Ichikawa, N., (1999). Chemical induction of disease resistance in rice correlated with the expression of a gene encoding a nucleotide-binding site and leucine-rich repeats. *Plant Molec. Biol., 40*, 847–855.

Satelis, J. F., Boiteux, L. S., & Reis, A., (2010). Resistance to *Septoria lycopersici* in *Solanum* (section *Lycopersicon*) species and in progenies of *S. lycopersicum* × *S. peruvianum*. *Sci. Agric. (Piracicaba, Braz.)., 67*, 334–341.

Schneider, S., & Ullrich, W. R., (1994). Differential induction of resistance and enhanced enzyme activities in cucumber and tobacco-caused by treatment with various abiotic and biotic inducers. *Physiol. Molec. Plant Pathol., 45*, 291–304.

Shelat, M., Murari, S., Sharma, M. C., Subramanian, R. B., Jummanah, J., & Jarullah, B., (2014). Prevalence and distribution of Tomato leaf curl virus in major agroclimatic zones of Gujarat. *Advances in Biosci. and Biotech.*, 5, 1–3.

Silva, H. S. A., Romeiro, R. S., Carrer, F. R., Pereira, J. L. A., Mizubuti, E. S. G., & Mounteer, A., (2004). Induction of systemic resistance by *Bacillus cereus* against tomato foliar diseases under field conditions. *Journal of Phytopathology, 152*, 371–375.

Sohi, H. S., & Sokhi, B. S., (1969). Relative resistance and susceptibility of different varieties of tomato to *Septaria lycopersici* Speg. *Indian J. Hort., 26*, 83–86.

Sohi, H. S., & Sokhi, S. S., (1970). Chemical control of defoliation disease of tomato caused by *Septoria lycopersici* speg. *Indian J. of Horticulture, 27*, 201–204.

Sotirova, V., & Rodeva, R., (1991). Sources of resistance in tomato to *Septoria lycopersici* speg. *Archio fur Phytopathology and Pflazenschutz., 26*, 469–471.

Sugha, S. K., & Kumar, S., (1998). Reaction of tomato (*Lycopersicon esculentum*) genotypes to *Septoria* leaf spot (*Septoria lycopersici*). *Indian J. of Agril. Sc., 68*, 709, 710.

Sugha, S. K., & Kumar, S., (2000). Factors affecting the development of *Septoria* leaf spot of tomato. *Indian Phytopathology, 53*, 178–180.

Sutton, B. C., & Waterston, J. M., (1996). *Septoria lycopersici* (p. 1–2). Kew: Commonwealth mycological Institute. N.89: Descriptions of pathogenic fungi and bacteria.

Tedla, T., (1985). Effect of captafol and ridomil (R) mz in the control of late blight (*Phytophthora infestance*) and *Septoria* leaf spot (*Septoria lycopersici*) on tomato. *Acta Horticulturae, 158*, 389–399.

Vakalounakis, D. J., & Malathrakis, N. E., (1988). A cucumber disease caused by *Alternaria* alternate and its control. *J. Phytopathol., 12*, 325–326.

Wei, L., Kloepper, J. W., & Tuzun, S., (1996). Induced systemic resistance to cucumber diseases and increased plant growth-promoting rhizobacteria under field conditions. *Phytopathology, 86*, 221–224.

Wright, J. M., (1956). The production of antibiotics in soil and production of gliotoxin wheat straw buried in soil. *Ami. Appl. Biol., 44*, 461–465.

Xue, L., Charest, P. M., & Jabaji-Hare, S. H., (1998). Systemic induction of peroxidases, b-1, 3-glucanases, chitinases and resistance in bean plants by binucleate *Rhizoctonia* species. *Phytopathology, 88*, 359–365.

CHAPTER 7

MOLECULAR APPROACHES FOR THE CONTROL OF CERCOSPORA LEAF SPOT IN TOMATO

SANTOSH KUMAR,[1] MEHI LAL,[2] TRIBHUWAN KUMAR,[3] and MAHESH KUMAR[3]

[1]*Department of Plant Pathology, Bihar Agricultural University, Sabour – 813 210, Bihar, India, E-mail: santosh35433@gmail.com (Santosh Kumar)*

[2]*Plant Protection Section, ICAR-Central Potato Research Institute, Regional Station, Modipuram, Meerut – 250 110, Uttar Pradesh, India*

[3]*Department of Molecular Biology and Genetic Engineering, Bihar Agricultural University, Sabour – 813 210, Bihar, India*

ABSTRACT

Tomato, an important vegetable crop, is well suited to biotic stress, but their production is challenged by several kinds of pathogen such as fungi, bacteria, and viruses. Cercospora leaf spot caused by *Pseudocercospora fuligena*, also known as Cercospora leaf mold is the very minor fungal disease and found in almost all continents of the globe. The wide host range is due to the production of a photoactivated toxin, cercosporin. Integrated disease management (IDM), including host plant resistance (HPR), agronomic practices, use of fungicides and bio-pesticides applied in complementation can be used as a way of management to control this disease. In addition, both breeding and biotechnological approaches can be used to develop resistant lines against this disease. The process of heterosis, one of the breeding approaches can be used to increase productivity and disease resistance. As far as biotechnological intervention in the control of this disease is concerned, hardly any report is available. Transformation with cercosporin auto resistance genes taken from the fungus, and transformation with designed constructs to silence the production of cercosporin may be used like tobacco in tomato to develop Cercospora leaf spot tolerant lines. Moreover, marker-assisted selection (MAS) of Cercospora resistant QTL and their introgression in the high yielding line of tomato may be an effective way to get rid of this disease. This book chapter incorporates information on Cercospora leaf spot of tomato and various way of management to control this disease, including biotechnological intervention.

7.1 INTRODUCTION

Tomato (*Solanum lycopersicum* L.) is an important vegetable crop of Solanaceaethe family. It is native of South America and is cultivated across the globe in outdoor fields, greenhouses, and net houses. Tropical to temperate climate are the ideal ones for the production of this crop. China, India, USA, Turkey, Egypt, Iran, Italy, Spain, and Brazil are the leading country which are involved in the production of this crop. Among these countries, the rank of India lies next to China in terms of both areas used for the cultivation of tomatoes and production. Tomato is mostly grown as a summer crop in India, but it also can be grown throughout the year. India has about 0.79 million hectare production area with a production of 19.76 million tons (Horticulture Statistics at a Glance, 2018). Tomato contains vitamins like 'A' and 'C' and antioxidant in enough quantity. That is why it has more demand and is one of the most valuable ingredients of dishes. Tomatoes are cooked as vegetables and processed as pickles, chutneys, soups, ketchup, sauces, etc., in addition to fresh and raw fruits.

7.2 CERCOSPORA LEAF SPOT

Tomatoes are also infected by various members of biotic and mesobiotic pathogens rendering various disease. Cercospora leaf mold, also known as Cercospora leaf spot is a very minor disease and found in almost all continents of the earth. It may change its nomenclature as a major disease, due to change in climate conditions over the years. Tomatoes grown in tropical and sub-tropical areas too are affected by this disease beside India. Fruit number and weight gets reduced up to 20% and 7%, respectively due to infestation of this disease (Hartman and Wang, 1992). Defoliation results in crop losses in between 15 and 20%. As far as Cercospora leaf spot of tomato is concerned, very little information is available regarding various approaches to manage this disease.

7.3 CAUSAL ORGANISM

Cercospora belongs to the largest genera of hyphomycetes. Species of Cercospora cause leaf spot disease in various cultivated and non-cultivated plants resulting in considerable losses. This disease was first documented in the Philippines in 1938 and the pathogen was described as *Cercospora fuligena* (Hartman and Wang, 1992). Later, Deighton (1976) declared that this fungus belongs to the genus *Pseudocercospora fuligena*. Conidiophore of this fungus is brown to dark brown, 16–70 × 3.5–5 μm, with slightly thickened conidial scars. While *Conidia* is 15–120 × 3.5–5 μm in size, pale straw-colored, 1–9-septate, smooth or minutely verrucose, subcylindrical, straight or slightly curved, tip broadly rounded, basal cell narrowing abruptly to a truncate base, with an indistinct, narrow scar.

7.4 HOST RANGE

Fungi belonging to genus *Cercospora* cause huge crop losses infecting wide range of crop plants at global level including sugar beet, beans, soybean, cowpea, corn, and groundnut, as well as many vegetables and ornamental species (Lartey et al., 2010). The wide host range and lack of disease resistance in many host species are due to the production of a photoactivated perylene quinone toxin, cercosporin (Daube and Ehrenshaft, 2000; Daub et al., 2005, 2013). Alternative hosts of this pathogen extend to the reservoir in plants which cause infections in the next growing season.

7.5 SYMPTOMATOLOGY

Foliage is the part of the plant on which the symptoms first appear, but they may also be seen on petioles, stems, and fruit peduncles. Small yellow imprecise spots with no margins in the leaf are the initial symptom of this disease (Figure 7.1). Small brown lesions start to form in the yellow discolored area also. Later on, these small yellow spots turn to larger necrotic spots with a marked zone on the upper and underside of the leaves. As the disease develops, the small lesion coalesces to form large patches and blighting in tomatoes (Figure 7.2). The affected leaves wilt, dry with the passage of time, and remain hanging on the plant in a dark soot-covered fashion. When the disease is severe, defoliation of the leaves will restrict fruit development and result in sunburn and tanning of fruit.

FIGURE 7.1 Initial symptom in the leaf as small yellow indefinite spots without margins.

FIGURE 7.2 Blighting and necrosis of the leaf.

7.6 DISEASE CYCLE AND FAVORABLE WEATHER CONDITIONS

The fungus survives on infected plant debris in the soil and alternative hosts like black nightshade over winter. Survival of pathogens on seeds and plant debris in soil is considered as the primary source of inoculum, especially the infected cotyledons attached with the seedlings or lying in the soil surface (Shree Kumar, 1974). During favorable condition, conidia is probably disseminated by wind, splashing water from overhead irrigation and rain as well workers, farm equipment and tools into new susceptible host. When conidia are blown to a wet, warm leaves, it causes secondary infection within two to three days. High humidity, warm, and wet weather conditions (temperature 27°C) favor the development of this disease and sporulation. Leaf moisture from rains or dew is an added factor liable to disease severity. Generally, on the underside of the leaves, high humid conditions favor rapid production of conidia.

7.7 MANAGEMENT STRATEGIES

The management strategies of this disease involve cultural, botanical, host plant resistant, molecular (biotechnological), and chemicals.

7.7.1 CULTURAL PRACTICES

Cultural practices mostly affect the environmental conditions favorable for the growth, building up of inoculums, survival, and their spread. Manipulation in sowing date with proper row spacing may be used to cope with the losses incurred due to leaf spot of Cercospora. The efficacy of this manipulation would differ from place to place, as it is the temperature and relative humidity in an area that determine reduction or increase in plant mortality during the crop season. Cultural practices like turning plant debris in the soil and removal and rouging of infected plant reduce the primary inoculum. Burying or destroying the remains of a tomato harvest reduces the amount of fungus inoculums, which may be able to infect new crops. The fungus can survive several years on solanaceous weeds, so removal of nearby volunteer plants and solanaceous weeds are compulsory. Pruning and spacing plants for adequate air movement in the field helps to reduce the favorable condition of the fungus. It is wise worthy to avoid planting of new tomato plant while diseased field remains nearby. Overhead irrigation in the tomato field should also be avoided. Intercropping by planting alternate rows of tomato and another suitable non-solanaceous crop, such as maize or sorghum, limits the spread of the disease within a field but not eliminate it.

7.7.2 BOTANICAL MANAGEMENT

The use of plant extracts with antifungal activity offers an alternative method for the management of leaf spot disease of groundnut because it is highly economical, environmentally safe, and easily available resource to poor farmers (Rahman and Hossain, 1996). Many research workers have tried to find out safe and economical control of plant diseases using extracts of different plant parts (Hasan et al., 2005; Bdliya and Alkali, 2008). Aage et al. (2003) reported that most of the botanicals have found to be effective against Cercospora leaf spot disease of groundnut.

7.7.3 HOST PLANT RESISTANT

Progress in breeding for Cercospora leaf spot resistance in tomato is limited due to its minor occurrence. Breeding resistant cultivars is one of the best means of reducing crop yield losses from the infection of this pathogen. For evolvement of Cercospora leaf spot resistant varieties, there is a need of screening of existing lines/germplasm/cultivars against Cercospora spp. Varietal screening has been reported in the other crops against Cercospora leaf spot. There is also a need to breed resistant cultivars in developed countries to reduce farmers' dependence on chemicals.

7.7.4 FUNGICIDAL MANAGEMENT

Fungicidal management is one of the important strategies in fungal disease management because of fact that it is easily applicable, economical with regards to production, quick action and effectiveness. The use of chemicals for the control of Cercospora leaf spot of many crops has been practiced for a long time give better results (Backman et al., 1977). Spray of fungicide mixture (carbendazim 0.05% + mancozeb 0.2%) effectively controls the disease. Some other effective systemic fungicides are benomyl, bavistin, brestanol may also give better performance. Spray of Sulphur and copper fungicides, gives good control of leaf spots. Some experts recommend seed treatment with carbendazim @ 1.5 g/kg seed as a preventative measure. Weekly spray of benomyl, after three weeks of planting, gave the best control of the Cercospora leaf spot diseases in cowpea (Amadi, 1995). Johanson et al. (1998) reported that fungicidal mixture (mancozeb 0.2% and carbendazim 0.1%) effectively controls the Cercospora leaf spot disease of groundnut. The above fungicides mentioned for chemical management, these are only for research findings, now many of them have been banned or used under restriction.

7.7.5 MOLECULAR APPROACHES

Cercospora renders a reduction in crop yield at global level on many crop species. The wide host range and lack of disease resistance in many host plant species is due to the fungus' production of a photo-activated perylenequinone toxin, cercosporin (Thomas et al., 2020). Cercosporin is a compound that absorbs light energy to generate singlet oxygen and other reactive oxygen species (ROS) (Daub, 1982; Daub and Hangarter, 1983; Dobrowolski and Foote, 1983; Yamazaki et al., 1975). The cercosporin-generated ROS cause peroxidation of the host cell membrane lipids, leading to membrane breakdown, death of host cells and leakage of nutrients required by the fungus for tissue colonization (Daub, 1982; Daub and Briggs, 1983). Few biotechnological initiatives have been taken to give rise resistant line against Cercospora in tobacco Plant. Two strategies have been used in the Tobacco plant against *Cercospora nicotianae*: (a) transformation with cercosporin auto resistance genes taken out from the fungus; and (b) transformation with designed constructs to silence the production of cercosporin during disease development. Thereafter it was found that both expression of fungal cercosporin auto resistance genes and silencing of the cercosporin pathway were effective for engineering resistance to Cercospora diseases where cercosporin plays a pivotal role. However, this sort of report is not available in the tomato plant. These strategies may be applied in tomato as well to develop resistant lines. The process of heterosis can be used for increasing productivity and disease resistance also. In this regard, the hybrids, W × P and wild × insulata may be chosen for least leaf spot disease severity and high fruit yield (Amaefula et al., 2014). Moreover, marker-assisted selection (MAS) of Cercospora resistant quantitative trait loci (QTL) and their introgression in the background of high yielding cultivar of tomato may be an effective and sustainable approach in the direction of development of Cercospora resistant high yielding cultivar.

7.8 CONCLUSION

Tomato is an important component of Indian and other tropical cuisines. Diseases no doubt are one of the major constraints to their efficient production. Managing the yield and grain quality reducing disease involves a number of strategies. Major emphasis has been given on economic importance, etiology, symptomatology, disease cycle, epidemiology, and management strategies of Cercospora leaf spot of tomato in this chapter. Holistic approach is effective in this case since no one strategy for sustainable control of this disease is grossly effective at all times and in every condition. Host plant resistance (HPR), fungicides, botanicals, and agronomic practices such as crop rotation, inter/mixed cropping, change in date of sowing should be potential viable parts for integrated management. HPR, fungicides, natural plant products, biofungicides, botanicals, and agronomic practices are the potentially novel options for IDM. Besides eco-disruption, use of standard fungicides may not be economically feasible in many low-input farming systems of India. Therefore, tomato health management for sustainable fruit production in such farming systems should also involve the latest advancements in the identification of biocontrol agents and biotechnological approaches such as marker-assisted selection, genetic engineering, and hybridization to develop cultivars with resistance to a wide range of diseases as a control measures. The application of information technology in the modeling of disease, development of decision support systems, and utilization of remote sensing to refine, broaden, and disseminate IDM technologies should also be taken under consideration.

KEYWORDS

- **Cercospora leaf spot**
- **host plant resistance**
- **management strategies**
- **molecular approaches**
- **quantitative trait loci**
- **reactive oxygen species**
- ***Solanum lycopersicum* Linnaeus**

REFERENCES

Aage, V. E., Gaikwad, S. J., Behere, G. T., & Tajane, V. S., (2003). Efficacy of extracts of certain indigenous medicinal plants against *Cercospora* leaf spot of groundnut. *Journal of Oils and Crops, 13*, 140–144.

Amadi, J. E., (1995). Chemical control of *Cercospora* leaf spot disease of cowpea (*Vigna unguiculata* (L.) Walp. *AgResearch, 1*, 101–107.

Amaefula, C., Agbo, C. U., Echezona, B. C., & Nwofia, G. E., (2014). Field reactions of interspecific hybrids of tomato lines of leaf spot *Lycopersicon* Mill.) Lines to leaf spot disease. *Agro-Science Journal of Tropical Agriculture, Food, Environment and Extension, 13*, 15–23.

Anonymous, (2018). *Horticulture Statistics at a Glance-2018*. Horticulture statistics division, DAC and Farmers Welfare, Ministry of agriculture and farmer's welfare, Government of India.

Bdliya, B. S., & Aikali, G., (2008). Efficacy of some plant extracts in the management of *Cercospora* leaf spot of groundnut in the Sudan savanna of Nigeria. *Journal of Phytopathology Plant Protection, 32*, 154–163.

Daub, M. E., & Briggs, S. P., (1983). Changes in tobacco cell membrane composition and structure caused by the fungal toxin, cercosporin. *Plant Physiol., 71*, 763–766.

Daub, M. E., & Ehrenshaft, M., (2000). The photoactivated *Cercospora* toxin cercosporin: Contributions to plant disease and fundamental biology. *Ann Rev Phytopath., 38*, 461–490.

Daub, M. E., & Hangarter, R. P., (1983). Production of singlet oxygen and superoxide by the fungal toxin, cercosporin. *Plant Physiol., 73*, 855–857.

Daub, M. E., (1982). Cercosporin, a photosensitizing toxin from *Cercospora* species. *Phytopathology, 72*, 370–374.

Daub, M. E., (1982). Peroxidation of tobacco membrane lipids by the photosensitizing toxin, cercosporin. *Plant Physiol., 69*, 1361–1364.

Daub, M. E., Herrero, S., & Chung, K. R., (2005). Photoactivated perylenequinone toxins in fungal pathogenesis of plants. *FEMS Microbiol. Lett., 252*, 197–206.

Daub, M. E., Herrero, S., & Chung, K. R., (2013). Reactive oxygen species in plant pathogenesis: The role of perylenequinone photosensitizers. *Antiox Redox Signaling, 19*, 970–989.

Deighton, F. C., (1976). Studies on *Cercospora* and allied genera. VI. *Pseudocer-cospora* Speg., *Pantospora* Cif. and *Cercoseptoria* Petr. *Mycological Papers, 140*, 1–168.

Dobrowolski, D. C., & Foote, C. S., (1983). Chemistry of singlet oxygen 46. Quantum yield of cercosporin-sensitized singlet oxygen formation. *Angewandte Chemie., 95*, 729–730.

Ganie, S. A., Ghani, M. Y., Anjum, Q., Nissar, Q., Rehman, S. U., & Dar, W. A., (2013). Integrated management of early blight of potato under Kashmir valley conditions. *African Journal of Agriculture Research, 8*, 4318–4325.

Hartman, G. L., & Wang, T. C., (1992). Black leaf mold development and its effect on tomato yield. *Plant Disease, 76*, 462–465.

Hasan, M. M., Chowdhury, S. P., Alam, S., Hossain, B., & Alam, M. S., (2005). Antifungal effect of plant extracts on seed borne fungi of wheat seeds regarding seed germination, seedling health and vigor index. *Pakistan Journal Biological Science, 8*, 1284–1289.

FAO, (2012). *Production of Tomato by Countries*. http://www.fao.org/faostat/en/#data/QC/visualize.

Johnson, M., Rao, M. M., & Meenakumari, K. V. S., (1998). Chemical control of groundnut late leaf spot in Anantapur district of Andhra Pradesh. *Indian Phytopathol., 51*, 382–384.

Lartey, R. T., Weiland, J., Panella, L., Crous, P., & Windels, C. E., (2010). *Cercospora Leaf Spot of Sugar Beet and Related Species*. St. Paul, MN: APS Press.

Pande, S., Sharma, M., Kumari, S., Gaur, P. M., Chen, W., Kaur, L., MacLeod, W., et al., (2009). Integrated foliar diseases management of legumes. *International Conference on Grain Legumes: Quality Improvement, Value Addition and Trade.* Indian Society of Pulses Research and Development, Indian Institute of Pulses Research, Kanpur, India.

Rahman, M. A., & Hossain, I., (1996). Controlling *Cercospora* leaf spot of okra with plant extracts. *Bangladesh Horticulture, 24*, 147–149.

Shree, K. K., (1974). *Studies on Foliar Fungal Diseases of Urid Leading to Their Control* (p. 84). MSc Ag. Thesis JNKVV, Jabalpur, M.P., India.

Thomas, E., Herrero, S., Eng, H., Gomaa, N., Gillikin, J., Noar, R., et al., (2020). Engineering *Cercospora* disease resistance via expression of *Cercospora nicotianae* cercosporin-resistance genes and silencing of cercosporin production in tobacco. *PLoS One, 15*, 1–19.

Yamazaki, S., Okubo, A., Akiyama, Y., & Fuwa, K., (1975). Cercosporin, a novel photodynamic pigment isolated from *Cercospora kikuchii. Agric Biol Chem., 39*, 287–288.

CHAPTER 8

MOLECULAR APPROACHES FOR THE CONTROL TOMATO LEAF CURL VIRUS (TLCV)

RAMESH KUMAR SINGH, NAGENDRA RAI, and MAJOR SINGH

ICAR-Indian Institute of Vegetable Research (IIVR), Jakhini (Shahanshahpur), Varanasi – 221305, Uttar Pradesh, India, E-mail: nrai1964@gmail.com (N. Rai)

ABSTRACT

Tomato leaf curl virus (TLCV) is a universal issue of tomatoes. This Geminivirus is transmitted by whitefly (*Bemisia tabaci*) and a circular single-stranded DNA. Although, tomato leaf curl disease (ToLCD) in the Varanasi region is spread by TLCGV. Most of the resistant genes have been reported from wild species. Many molecular markers are used to identify and develop the TLCV resistant genotypes. This is a fact that the molecular advances have been provided many prospects of linked markers and loci on chromosomes. However, the alteration of resistant genes using genetic engineering is the time-honored method for developing TLCV resistant sources. The modern techniques amalgamated with conventional ones that offer great opportunities for improvement of the TLCV resistant tomato genotypes and will pave the future approaches of resistance breeding in tomato.

8.1 INTRODUCTION

Till date, limited cultivars have been reported for resistance to TLCV, but probable resistant sources are available in wild spp., e.g., *Solanum chilense, Solanum Chmielewski, Solanum habrochaites, Solanum pennellii, Solanum peruvianum,* and *Solanum pimpinellifolium* (Singh et al., 2014, 2015a, b). Since, uses of chemicals for management of TLCV are fruitless and risky for the atmosphere and fitness of living things. Therefore, the search of resistant sources by whitefly vaccination methods and their introgression with susceptible lines for developing resistant varieties to TLCV are the secure and permanent substitute to solve these constraints (Konate et al., 2008; Singh et al., 2014, 2015a, b; Naveed et al., 2015; Arooj et al., 2017). The ratio of total sugar and phenol, chlorophyll substances in plant leaves and reduction in leaf area supply as signs of resistance and susceptibility of the germplasm (Marco, 1975; Gevorkyan et al., 1976; Banerjee and Kalloo, 1989a–c; Singh et al., 2015b). Total phenol content is responsible for resistant capacity and obtained less susceptible lines (Singh et al., 2010, 2015b). During the past two decades, concentration and focused effort have been directed to TLCV resistance breeding in India and for more efficient approaches are the mandates of the day to know the genetics of resistant (Banerjee and Kalloo, 1987a; Lapidot and Friedmann, 2002; Maruthi et al., 2003). Many genetic maps have been consisted by molecular markers and developed in recent years for many crop plants (Paran et al., 1995). Marker-assisted selection (MAS) in breeding programs had been confirmed as a competent approach to defeat many diseases (Foolad, 2007), and gene tagging and gene mapping using simple sequence repeat (SSR) markers have been successfully used for TLCV resistance in tomato (Foolad, 2007; Benor et al., 2008). Presently, five tomato yellow leaf curl virus (TYLCV) resistances genes, e.g., *Ty-1*, *Ty-2*, *Ty-3*, *Ty-4, Ty-5*, and *Ty-6* have been reported by using a number of molecular markers (Zamir et al., 1994; Hanson et al., 2006; Ji et al., 2007b, 2008; Anbinder et al., 2009; Hutton and Scott, 2013, 2015) but few studies has been published using SSR markers in tomato for Indian TLCV resistance (Frary et al., 2005; Singh et al., 2014, 2015c). Genetic engineered (transgenic) plants of tomato were found resistant to viruses under *In Vivo* and *In Viro* conditions and this brought to therefore two strategies *viz.,* coat protein (CP) and satellite RNA refereed as resistance to TLCV by gene alteration (Raj et al., 2005; Paduchuri et al., 2010). In the present endeavor, an attempt was made to search the available resources of resistance along with the predictable and modern approaches to develop resistance against TLCV.

8.2 BACKGROUND

Tomato (*Solanum lycopersicum* L., 2n = 2x = 24), an herbaceous plant, is grown globally (Peralta et al., 2005; Singh et al., 2015e). The tomato crop a sufferer of many problems like, viruses, fungi, bacteria, and nematodes and they hampered the tomato yield (Banerjee and Kalloo, 1987a, b). TLCV, a whitefly (*Bemisia tabaci*) transmitted Geminivirus is a major problem for tomato production (Kalloo and Banerjee, 1990; Singh et al., 2010, 2015a, d, e). In many temperate, tropical, and subtropical areas, TLCV has caused heavy yield losses ranging between 50 to 100% (Singh et al., 2010). Yield reduction is depending on the cropping seasons or growth stage at which the plant becomes infected. Reduction in yield of tomato because of the severity of TLCV and yield loss recorded country and region-wise, e.g., Nigeria, Lebanon, the Mediterranean region, Sudan, Malawi, Kenya, Tanzania, and India with 23%, 40%, 50%, 63%, 75%, 80%, 99–100%, respectively (Pico et al., 1998; Maruthi et al., 2003; Singh et al., 2015a). The symptom of TLCV disease on tomato plants has characterized by upward and inwards leaf curling, curved mode, twisting of leaf lamina between the layer, bump on leaf surface, reduction in leaf area, reduced the internodes, stunted plants, deformed growing tips, low colorific in leaflets, dropping the flower and reduction in fruit size (Banerjee and Kalloo, 1987b; Singh et al., 2015a, b). First-time TLCV was recorded by Cowland in 1932 from Sudan and in India it was first time reported by Vasudeva and Samraj (1948). First case of TLCV identified in the eastern Mediterranean on tomato crops (Cohen and Antignus, 1994). Singh (2013) has been explored that the family *Geminiviridae* is composed of four genera, based on the number of genomic components, types of insect vector, host range and phylogenic relationship, mastervirus, curtovirus, begomovirus, and topocuvirus (Figure 8.1). Tomato crop is infected by at least 35 viruses. TLCV is belonging to the genus *Begomovirus* and family *Geminiviridae* with circular and single-stranded DNA (Fauquet et al., 2003). This disease is transmitted by a vector whitefly *Bemisia tabaci* (Pico et al., 1998; Anbinder et al., 2009; Singh et al., 2010, 2014, 2015a–c). There is no evidence for the genera *Alfamovirus, Potexvirus,* or *Closterovirus* but these genera have been reported proves of their occurrence on tomato in the European region (Fauquet and Mayo, 1999; Brunt et al., 1990; Mayo and Brunt, 2005). Tomato leaf curl virus (TLCV) is categorized on behalf of regions and symptoms worldwide including the states of India (Singh et al., 2019). In the Indian subcontinent, TLCV has famed by different names, *viz.,* TLCGV, TLCNDV, TLCBV, TLCKV, and TLCPV (Table 8.1). Earlier, according to some reports, ToLCV in the Varanasi region has been detected as TLCGV with the same symptom and strain (Chakraborty et al., 2003; Sharma and Prasad, 2017; Singh et al., 2019).

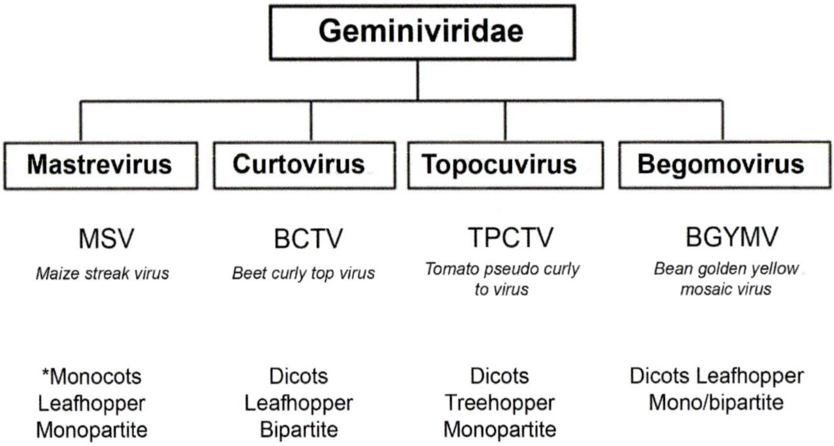

FIGURE 8.1 Classification of the family *Geminiviridae.*

The adult whiteflies (*Bemisia tabaci*) maintain on brinjal, pointed gourd, and on other host plants in both field and glasshouse conditions. Earlier, colonies of adult whiteflies were maintained on eggplants (brinjal) under controlled conditions (glasshouse) for virus spreading in the Varanasi region (Chakraborty et al., 2003; Singh et al., 2010, 2015a). Pico et al. (1998) has been reported a ratio of 1:10 of healthy seedlings and infected plants and should be changed at regular intervals of 72 hours for providing a regular source of inoculums. Two whitefly inoculation methods, e.g., natural and artificial (mass and cage) were used for screening of tomato genotypes against TLCV disease. In the case of large breeding populations, the open field screening is more authentic. The screening under field conditions can be done when the environmental conditions are conducive for whitefly perpetuation (Banerjee and Kalloo, 1987a; Pico et al., 1998; Singh et al., 2010, 2015a–c, e,

2018). Major fruit set reduction has observed after both artificial and natural inoculations, but infection was severe in mass and cage inoculations (Vidavski et al., 2008; Singh et al., 2015a, b, 2018). The relationship of disease severity and environmental (rainfall, temperature, and humidity) variations indicates a positive relation. In some studies, it was seen, although the rainfall, temperature, and humidity was more in rainy than winter seasons (Figure 8.2) then disease incidence (CI%) of TLCV was more in winter than rainy season (Singh et al., 2010, 2015a, e). Probably, less humidity was more responsible for the increase in the number of whitefly in winter season (Singh et al., 2015a, c; Naveed et al., 2015). It may be due to vector (whiteflies) could not survive in high temperature and they favor winter season for infection (Singh et al., 2015a, e). The infection of TLCV can spread through the path of plant sap. The infection may reach in meristematic tissues via xylem and phloem (Gafni, 2003). But the symptom of TLCV can be identified by the visual appearance. This study confirms the facts that TLCV susceptible varieties were less chlorophyll content due to chlorosis in leaves which is in conformation with the study of Marco (1975), who stated that the after disease in infected plants, interveinal chlorosis may occurs.

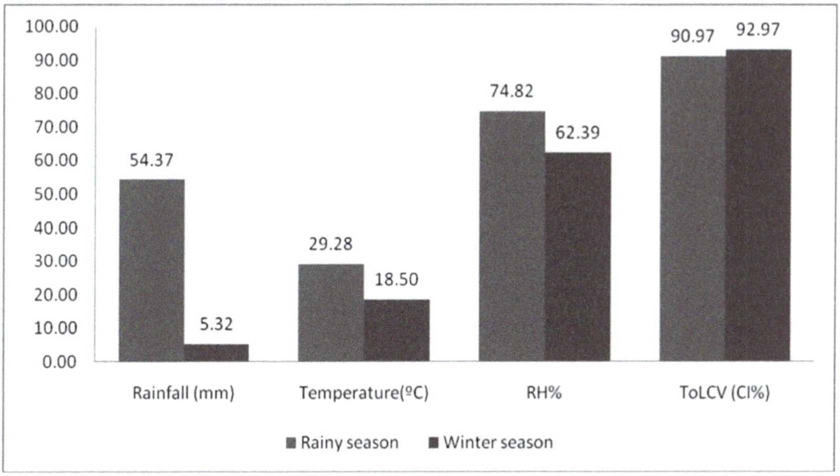

FIGURE 8.2 Response of TLCV disease is seasonal (rainy and winter) variation during study period.

TABLE 8.1 Different Name of Tomato Leaf Curl Viruses and Their Strain Within India

Name of Virus	Monopartite/ Bipartite[a]	Suggestion(s)
Indian tomato leaf curl virus (ITmLCV)	Bipartite	Kalloo and Banerjee (1990)
Tomato leaf curl virus (TLCV)	Bipartite	Banerjee and Kalloo (1987a, b); Singh et al. (2015a, 2019)
Tomato leaf curl virus (ToLCV)	Bipartite	Singh et al. (2010, 2014, 2015a, b, 2019)
Tomato leaf curl disease (ToLCD/TLCD)	Monopartite	Muniyappa et al. (2002)
Tomato leaf curl Gujarat virus	Bipartite	Chakraborty et al. (2003)
Tomato leaf curl virus New Delhi	Bipartite	Sharma and Prasad (2017)
Tomato leaf curl Bangalore virus	Monopartite	Srivastava et al. (1995)
Tomato leaf curl Karnataka virus	Monopartite	Chakraborty et al. (2003)
Tomato leaf curl Patna virus	Bipartite	Kumari et al. (2010)

[a]Monopartite and bipartite are classical nuclear location sequences (NLSs) pattern which makes probable the express into the core.

8.3 RESISTANCE SOURCES AND GENETICS

Limited genotypes have been reported as resistance source against TLCV among the *S. lycopersicum* genotypes (Muniyappa et al., 2002; Singh et al., 2015a–c, 2018, 2019). Most of the tomato cultivars have been found susceptible to TLCV and utilized with wild accessions for breeding approaches (Singh et al., 2015a). Earlier, a wild species *L. peruvianum* exhibited low incidence and number of inbreds, i.e., H-2, H-11, H-17, H-23, H-24, and H-36 developed by crossing *L. hirsutum f. glabaratum* into *Lycopersicon esculentum* (Kalloo and Banerjee, 1990) for TLCV resistant (Table 8.2) for TLCV. A numbers of resistance lines developed from *L. pimpinellifolium*, *LA121* (Pilowski and Cohen, 1974). While, *L. pimpinellifolium* and *L. hirsutum* reported as partial resistance, but *L. chilense* showed high resistance levels (Zakay et al., 1991; Maruthi et al., 2003) against TLCV (Table 8.2). However, another wild accessions *L. chilense* 1969 and *L. hirsutum* LA 1777 were also resistant to TLCV (Kheyr-Pour et al., 1994).

TABLE 8.2 Resistant Sources for TLCV in Tomatoes (Cultivars, Wild, and Wild Derivatives)

Resistant Sources	Species	References
Kashi Aman, Kashi Adarsh, Kashi Chayan and Arka Rakshak; TLB-111, TLB-119 and TLB-183	*S. lycopersicum*	Breeder's group of ICAR-IIVR, Varanasi, and ICAR-IIHR, Bangalore (2013–2016); Muniyappa et al. (2002)
H-88-78-1, H-88-78-2, and H-88-78-4; H-2, H-11, H-17, H-23, H-24, H-36 and H-86; *S. hirsutum*	*S. lycopersicum* derivative of *S. habrochaites f. glaboratum*	Rai et al. (2010); Singh et al. (2015a, d); Sharma and Prasad (2017); Banerjee and Kalloo (1989a); Kalloo and Banerjee (1990); Hanson et al. (2000)
Avinash-2, Rashmi, Arka Vikas, Arka Rakshak and Arka Samrat	*S. lycopersicum* derived by hybrids varieties	Syngenta, India; Indo-American hybrid seeds, India; ICAR-IIHR, Hesaraghatta, Bangalore, India
EC-520058, EC-520060 and EC-520061; 901-1, 901-2, 902, 906-7, 908, 910, 913; P1390658 and P1390659; LA 1777	*S. habrochaites syn. S. hirsutum*	Singh et al. (2014, 2015a–c); Vidavsky and Czosnek (1998); Muniyappa et al. (1991); Kheyr-Pour et al. (1994)
EC-520070, EC-520071, EC-520077, EC-520079, EC-520065, EC-520074, EC-521078, EC-521080; Pim2; LA 121	*S. pimpinellifolium*	Singh et al. (2015a); Vidavsky et al. (1998); Pilowski and Cohen (1974)
WIR-5032; Chil1; Ty-50, Ty-52; TY0-2, TY0-5, TY0-15, TY0-23; FA9, FA24, FA38, FA491, FA709, FA710, FA712, FA716, FA736, FA907; 3750, 3761; 1969	*S. chilense*	Singh et al. (2014, 2015a–c, 2018b); Vidavsky et al. (1998); Zamir et al. (1994); Hebrew University of Jerusalem; Zeraim Gedera Seed Co., Israel; A. B. Seeds, Israel; Kheyr-Pour et al. (1994)
WIR-3957, WIR-4360, WIR-4361; Per1; TY-70; P1127830 and P1127831	*S. Arcanum* syn. *S. peruvianum*	Singh et al. (2014, 2015a–c, 2018b); Vidavsky et al. (1998); Muniyappa et al. (1991)
EC-520049	*S. chmielewskii*	Singh et al. (2014, 2015a, b, 2018b)

Source: Maruthi et al. (2003); Singh et al. (2018, 2019).

Resistant breeding is a solid approach for developing resistant gene to biotic and abiotic stresses. In tomato, it has been difficult to coalesce the resistance gene for numerous diseases and other important yield characters and quality traits like lycopene, ascorbic acid, TSS, etc. Although, the less chance for developing resistant varieties having better fruit shape, fruit size and quality by use of wild species. While, it has been possible that the resistant tomato with good fruit quality by using *S. lycopersicum* (Singh et al., 2015a, d). Previously, two cultivars Punjab Chhuhara and H-88-78-4 has been utilized for developing TLCV resistance and high yielder variety using F_2 and back cross progenies (Singh et al., 2008). They obtained the segregation ratio 1:2:1 as resistant: intermediate: susceptible for TLCV (Table 8.3). They found when the backcrosses used with the susceptible parents segregated into 2:1 ratio of intermediate: susceptible type in almost, while they used when backcrosses with resistant parents the segregation ratio was 1:2 indicated resistant and intermediate, respectively (Singh et al., 2008). However, in many cases for obtaining resistant varieties with good fruit quality, the crosses developed between *S. lycopersicum* and wild species (Singh et al., 2014). Previously, it has been reported that the resistance of TLCV is governed by a single gene but sometimes controlled by multiple genes (Lapidot and Friedmann, 2002; Sharma and Prasad, 2017). The inter-specific crosses between susceptible cultivars, namely ACC99, AC142, Collection No.2, Kalyanpur Angurlata, HS101, HS102, HS110, Pusa Ruby, and Punjab Chhuhara along with resistant wild accession *L. hirsutum F. glabratum* (B6013), *L. hirsutum F. typicum* (A1904), *L. pimpinellifolium* (A1921), and *L. peruvianum* (LA385) and these resistant lines were monogenic and incompletely dominant over susceptibility (Banerjee and Kalloo, 1987a, b; Ji et al., 2007a; Vidavski et al., 2008). They also observed, these crosses were good combiners for high yield, fruit quality, and high level of resistance and also exhibited additive and dominant gene action (Singh et al., 2008). Similarly, in another study crosses between *S. lycopersicum* and *S. habrochaites* (EC-520061) express the inhibitory (13:3), monogenic dominant (3:1) and intermediate dominant (1:2:1) genetic ratios (Singh et al., 2015d). However, a cross between *S. lycopersicum* and H-88-78-1 (*S. lycopersicum,* a derivative of *S. hirsutum f. glabratum*) exhibited 3:1 genetic ratio and dominant epistatic gene action (Table 8.3) for TLCV in each population (Singh et al., 2015a, d). The phenotypical dominance was exhibited up to 80% coverage of H-88-78-1 in the F_2 population which may be due to the *S. habrochaites f. glabratum* background known for possessing tight gene linkage for TLCV resistance (Singh et al., 2015c). In support of this study, a cross *S. habrochaites* × *L. esculentum* (E. 6206) exhibited more than 85% coverage of *S. habrochaites* (LA-1777) genome (Momotaz et al., 2007).

TABLE 8.3 Responsible Inheritance for TLCV between Interspecific and Intraspecific Crosses

Cross	Mendelian Ratio	References
TLBR-3 × H-88-78-1, PKM-1 × H-88-78-1, FLA-7421 × H-88-78-1, Vaibhaw × H-88-78-1, Punjab Chhuhara × H-88-78-1, Arka Vikash × H-88-78-1, CO-3 × H-88-78-1, DVRT-2 × H-88-78-1, H-86 × EC-521080, H-24 × EC-521080, Punjab Chhuhara × EC-520049, H-86 × EC-520049, H-24 × EC-520049, DVRT-2 × EC-520049, Punjab Chhuhara × EC-528372, Punjab Chhuhara × WIR-3957, DVRT-2 × WIR-3957, Punjab Chhuhara × WIR-5032	3:1	Singh et al. (2015d, 2018)
HS101 × B6013, HS102 × B6013, HS110 × B6013, Pusa Ruby × B6013, Punjab Chhuhara × B6013, Punjab Chhuhara × EC-520061, H-24 × EC-520061, H-86 × EC-520061, DVRT-2 × EC-520061	13:3	Banerjee and Kalloo (1989a); Singh et al. (2015c)
Punjab Chhuhara × EC-521080, DVRT-2 × EC-521080, DVRT-2 × EC-528372, H-86 × WIR-5032, H-24 × WIR-5032, DVRT-2 × WIR-5032	1:2:1	Singh et al. (2018)
H-86 × EC-528372, H-24 × EC-528372, H-86 × WIR-3957, H-24 × WIR-3957	1:3	Singh et al. (2018)

Source: Singh et al. (2019); H-88-78-1 (*S. lycopersicum* derivative of *S. habrochaites f. glaboratum*); EC-520061 (*S. habrochaites*); EC-521080 (*S. pimpinellifolium*); EC-520049 (*S. chmielewskii*); EC-521080 (*S. chmielewskii*); EC-528372 (*S. ceraseforme*); WIR-5032 (*S. chilense*); WIR-3957 (*S. Arcanum* syn. *S. peruvianum*).

8.4 MOLECULAR MARKERS AND THEIR APPLICATION

Molecular markers indicated genetic differences between two individuals that can't represent the target gene only can display the phenotypic variation. Genetic markers that are tightly linked may be referred to as 'gene tag' and these markers do not affect the phenotype of the gene of interest (Collard et al., 2005). The genetic markers displayed specific genomic positions within chromosomes referred to as 'loci.' Among the molecular markers are the isoenzyme and DNA markers popularly termed as molecular markers (Tanksley et al., 1982; Edwards et al., 1987). A technique used for mapping of gene sequences is also known as molecular mapping or linkage map. In other words, the linkage map would be measured as a 'path map' of the chromosome's derivative of two different parents or a population studied for molecular mapping, commonly called mapping population (Paterson, 1996). Tagging and mapping of single genes traits with molecular/biochemical marker in tomato including morphological, physiological, and disease resistance traits (Rick, 1974). The latest published classical linkage map of the 12 tomato chromosomes like physiological, isozyme, and disease-resistant genes (Tanksley, 1993). In 1986, the first linkage map of tomato has been published 18 isozyme and 98 DNA markers of cDNA clones (Bernatzky and Tanksley, 1986). However, the first high density linkage map of tomato using 1030 markers, published in 1992 (Tanksley et al., 1992). Marker-based selection, a very useful tool to maximize the utilization of existent resources. Controlling of different agronomic traits can be quickly brought together in an existing variety. Furthermore, the genes responsible for resistance to TLCV can also be pyramided. Molecular mapping has been successfully applied in several crop-breeding programs. Molecular mapping programs based on introducing more effective multiple genes from highly resistance varieties, and this is the best approach to control the disease.

A variety of approaches have been used to achieve Geminivirus resistance using classical breeding and biotechnological plans (Lapidot and Friedmann, 2002). Many molecular markers were developed based on polymorphism in wide crosses and are not informative when used within closely related germplasm, whereas a saturated linkage map of tomato is available when applied across *S. lycopersicum* × *S. pinnelli* (Fulton et al., 2002). The populations derived from crosses with closely related species and crosses within *S. lycopersicum,* require additional efforts to construct linkage maps and to associate a marker with phenotype. The molecular markers consist of a specific molecule which shows easily detectable differences among different strains of similar or different species during gene map of various species. In the case of plants, two types of molecular markers used for the preparation of genome maps, restriction fragments length polymorphism (RFLP) and Random amplified Polymorphic DNA (RAPD).

8.4.1 RANDOMLY AMPLIFIED POLYMORPHIC DNA (RAPD)

The RAPD is molecular markers that behave as dominant alleles. With RAPD short primers were also used to target homologous sites in the genome (Williams et al., 2004). RAPD are not as potent polymorphic as

SSRs. Polymorphism in RAPD is revealed by different amplified fragments due to dissimilarities in the frequency of target DNA sequences. RAPD have been used in studies of *Solanum* species to select somatic hybrids (Baird et al., 1992) and to detect variation among anther-derived monoploids (Singsit and Ozias-Akins, 1993). It is pronounced as rapids. RAPDs are produced by repeated cycle of DNA denaturation-renaturation-DNA replication in PCR (polymerase chain reaction) equipment. RAPDs are generated by using random sequence 9–12 base oligonucleotides. The RAPD amplification reaction is performed on genomic DNA with an arbitrary oligonucleotide primer. It results in amplification of several discrete DNA products. Thus the RAPDs markers can be regarded in the same manner as RFLP markers and similarly used for preparation of RAPD amps. RAPDs maybe generate an information in 4 weeks which would have taken about 2 years to obtain using RFLPs maps are being prepared for several crop species like maize, tomato, soybean, rice, sugarcane, sunflower, etc. RFLPs has been used to develop first high-density linkage map in tomato (Barnatzky and Tanksley, 1986; Tanksley et al., 1992).

Klein-Lankhorst et al. (1991), studied a new DNA polymorphism on the base of amplification by PCR. The genomic DNA molecule of an Israeli isolate of TYLCV was amplified from total DNA extracts of TYLCV-infected plants (*L. esculentum* 'M-82') by the use of PCR (Navot et al., 1992). They used "RAPD markers," for the construction of genetic maps. They found "fingerprints" of *L. esculentum, L. pennellii,* and the *L. esculentum* chromosomes 6 substitution line LA 1641, which carries chromosomes 6 from *L. pennellii*, 3 chromosomes-6-specific RAPD markers could be directly identified in the DNA fragments. Zamir et al. (1994), also studied that TYLCV Geminivirus in wild resistant species *L. chilence* and a cultivated tomato *L. esculentum* through RAPD. *Ty-1*, was mapped to chromosomes-6, two modifier genes were mapped to chromosomes 3 and 7. Ji and Scott (2006) have been studied that two resistant genes *Ty1* and *Ty3* may be linked, and both genes in a single genome may provide superior recombinant.

8.4.2 RESTRICTION FRAGMENTS LENGTH POLYMORPHIC (RFLPS) MARKER

The sequence of RFLP markers can recover, primer design, alleles amplify and sequencing performed to identify the nature of available polymorphism.

In the RFLP markers, single or low copy genomic fragments which are suitable for marker development of closely related species, were corresponded (Miller and Tanksley, 1990; Frary et al., 2005). These markers are based on specific restriction enzymes and probes used to detect alteration in different strains/species in the position of recognition sites of the given enzyme in the chromosome's region identified by the probe. RFLP denotes that solitary restriction enzymes make fragments of diverse length from the same stretch of genomic DNA from different strains of a species or from different related species. In the RFLP analysis, DNA into smaller fragments are separated by the gel electrophoresis. In tomato, resistance gene *Ty-1* and *Ty-3* against TYLCV has already been identified (Zamir et al., 1994; Ji et al., 2007a).

Hanson et al. (2000) also suggested that mapping of DNA fragment introgressed into cultivated tomato presumably from the wild species *L. hirsutum* Humb. and Bopnl. and found to associated with TYLCV resistance. Further, they observed that the presence of one wild tomato introgression each on chromosomes 8 and 11 and control tomato line Ty-52 (homologous for Ty-1 allele for TYLCV tolerance) were exposed to viruliferous whiteflies (*Bemisia tabaci* Gen.) in greenhouse. The number of mapped genes in the form of cDNAs has increased considerably with the introduction of RFLP markers. The current tomato RFLP map was constructed using an F_2 population of the interspecific cross *L. esculentum* × *L. pennellii* and contains more than 1030 markers, which were distributed over 1276 cM. (Tanksley et al., 1992). Praveen et al. (2005) studied the transgenic tomato resistant to tomato leaf curl disease (ToLCD) using replicare (rep) gene sequence.

8.4.3 MICROSATELLITES OR SIMPLE SEQUENCE REPEATS (SSRS), MARKER

Microsatellites, or SSRs, is a class of repetitive sequences which are widely-distributed in all eukaryotic genomes. SSRs, highly variable regions of genome in animal and plant, that consist of short DNA sequences where frequent reallocates result in differences in the number of repeats. SSRs, also called microsatellites, are highly polymorphic, are distributed throughout the genomes of plants and animals (Tautz and Renz, 1984), and occur intragenically and intergenically (Weber, 1990). Microsatellites are highly accepted markers because of their co-dominant heritage, high wealth, huge extent of allelic range, and SSR size variation. The regions surrounding the repeats are highly conserved (Smulders et al., 1997), and can be used to design primers that will amplify across the repeat during a PCR. For SSR and idles, great polymorphisms can be detected among 20 bp to 1 kb using gel electrophoresis. Fragment size differences from 2 to 20 bp are classically detected on polyacrylamide gels and stained with ethidium bromide or silver. According to Cato et al. (2001), one disadvantage to developing SSR or SNP markers in transcribed sequences is their low polymorphism between costly related tomato species and within cultivated germplasm. Subsequently using different search parameters, putative SSRs were detected in the ESTs and genomic sequences (He et al., 2002; Frary et al., 2005). Earlier, Singh et al. (2015c) has been used SSRs markers in F_2 segregating population of tomato and identified two linked markers $SSR218_{170-145}$ and $SSR304_{158-186}$ on 15 cM and 35 cM distance for TLCV resistance.

8.4.4 SINGLE NUCLEOTIDE POLYMORPHISM (SNPS)

Polymorphisms corresponding to differences at mono-nucleotide position occur about every 1.3 kb (Cooper et al., 1985) and single nucleotide dissimilarity in genome sequence of an individual among a population, known as SNPs. They constitute the most abundant molecular markers in the genome and are widely distributed throughout genomes, although their occurrence and distribution varies among species. An important subset corresponds to mutations in genes that are associated with diseases or other phenotypes. Positional cloning based on SNPs may accelerate the identification of disease traits and a range of biologically informative mutations (Wang et al., 1998). The SNPs are usually more prevalent in the non-coding regions of the genome. Within the coding regions, an SNP is either non-synonymous and results in an amino acid sequence change (Sunyaev et al., 1999), or it is synonymous and does not alter the amino acid sequence. Improvements in sequencing skill and accessibility of a rising number of EST chains has completed direct analysis of genetic difference at the sequence level of DNA (Buetow et al., 1999; Soleimani et al., 2003). The majority of SNP genotyping assays are based on one or two of the following molecular mechanisms: allele specific hybridization, primer extension, oligonucleotide ligation, and invasive cleavage (Sobrino et al., 2005).

The DNA markers have been developed based on polymorphism in various crosses, but they did not give a response when used closely related germplasm to construct linkage map as associated a marker with a phenotype (Tanksley et al., 1992; Fulton et al., 2002). RFLPs used for looking genetic diversity with several crop species and their wild relatives. A high density RFLP and AFLP maps in tomato F_2 populations *L. esculentum* × *L. pennellii* had been constructed (Frary et al., 2005). This map developed on 1482 cM using 67 RFLP and 1175 AFLP markers. The genomic DNA molecule of an Israeli isolate of TYLCV infected plants of tomato (*L. esculentum* 'M-82') amplified by the use of RAPD markers (Navot et al., 1992), and some *Ty1* and *Ty3* genes were also mapped on chromosomes 6, 3, and 7 for TYLCV by using RAPD and RFLP markers (Zamir et al., 1994; Ji et al., 2007a, b; Ji and Scott, 2006). Subsequently using different search parameters, putative SSRs were detected in the ESTs database and genomic sequences (Frary et al., 2005; Singh et al., 2014). However, disadvantage of developing SSR or SNPs markers in transcribed sequences are their low polymorphism among tomato (Benor et al., 2008; Singh et al., 2014, 2015e; Sharma and Prasad, 2017).

8.5 GENE TAGGING AND MAPPING

The chromosomes derived from two diverse genotypes and population used for gene tagging and mapping, referred to as mapping population (Paterson, 1996). It can also be defined as the population used for gene mapping is commonly called a mapping population, and this mapping populations are developed from the crosses. In a study, a saturated linkage map of tomato developed by *S. lycopersicum* × *S. pinnelli* for TYLCV resistance (Fulton et al., 2002). Gene mapping is typically used when phenotyping in a mapping population are easier and/cheaper than genotyping using molecular markers and disadvantages of this method, it is not efficient in determining the effects of QTLs (Tanksley, 1993). A population of *L. esculentum* × *L. cheesmanii* of 97 recombinant inbred lines (RILs) was used to create a genetic map (Paran et al., 1995) and recorded 1:1 ratio between two homozygous classes. In many reports, breeders used wild genetic background to build lines with high levels of resistance, e.g.,

S. peruvianum (Friedmann et al., 1998; Vidavsky and Czosnek, 1998), *S. chilense* (Zamir et al., 1994), *S. pimpinellifolium* (Vidavsky et al., 1998), *S. pinnelli* (Frary et al., 2005) and *S. habrochaites* (Hanson et al., 2000; Singh et al., 2015d).

Tagging or mapping of single genes traits include many morphological, physiological, and disease resistance traits using with molecular and biochemical marker in tomato (Tanksley, 1993). The gene tagging, a short route of gene mapping and can be used to identify markers for QTLs mapping by two 'short-cut' methods *viz.*, bulked segregant analysis (BSA) and selective genotyping (Singh et al., 2014, 2015e). BSA, used to tag the markers located in a specific region of chromosome (Michelmore et al., 1991). Briefly, pools or bulks of DNA samples are combined from '10–20' individual plants from a segregating population for a trait of interest (Singh et al., 2014, 2015e). BSA technique may also be used to identify additional markers linked to specific chromosomal regions (Michelmore et al., 1991). Many marker such as RAPD or AFLP, are generally preferred for BSA to identify linkage between genes of interest (Michelmore et al., 1991; Singh et al., 2015c). Two putatively linked markers $SSR218_{170-145}$ and $SSR304_{158-186}$ at 15 cM and 35 cM distance (Table 8.4) has been identified for TLCV resistance using BSA methods through SSRs markers in a F_2 segregating population of *S. lycopersicum* × *S. habrochaites* (Singh et al., 2014, 2015c, 2019).

Molecular mapping has been applied in several tomato breeding programs and produced many varieties/lines with good yield attributes and resistant capacity. In 1986, the first molecular linkage map of tomato has constructed using 18 isozymes and 98 DNA markers (Bernatzky and Tanksley, 1986). In 1992, the linkage map of tomato was developed comprising of 1030 markers (Tanksley et al., 1992). The current scenario of the tomato map, considered in the molecular and classical markers can be mapped on 12 linkage group. Recommended relationship between genetic and physical distance is about 750 kb per cM in tomato. Earlier, >1000 molecular markers have been mapped in tomato and covered 1276 map units (Tanksley et al., 1992).

The genetic map was constructed between *L. pennellii* and *L. esculentum* on chromosome 6 which can be directly identified among the set of amplified DNA fragments in an Israeli isolate of TYLCV (Navot et al., 1992). Whereas, a genetic map was built by using 132 markers in a RIL population of *L. esculentum* × *L. cheesmanii* (Paran et al., 1995). Earlier,

5 QTLs viz., *Ty-1, Ty-2, Ty-3, Ty-4, Ty-5*, and *Ty-6* recognized through physical map (Figure 8.3), for TYLCV resistance (Anbinder et al., 2009; Hutton and Scott, 2013, 2015; Caro et al., 2015), among them *Ty-1* and *Ty-3* were mapped using CAPS (cleaved amplified polymorphic sequence), SCAR (allelic specific sequence characterized amplified region), RFLP, and RAPD markers on the chromosome number 6 of *S. chilense* that was closely linked with Mi (QTL responsible for root-knot nematode resistance gene) (Zamir et al., 1994; Hanson et al., 2000, 2006; Ji and Scott, 2006; Ji et al., 2007b). A minor QTL *Ty-4* was also mapped in advance breeding lines of *S. chilense* derivatives on the long arm of chromosome 3 between the PCR based markers C2_At4g17300 and C2_At5g60610 (Ji et al., 2008, 2009). The *Ty-2* gene derived from *S. habrochaites* was mapped to the chromosome 11 using RFLP markers (Hanson et al., 2006). A major QTL *Ty-5* was identified and mapped on chromosome 4 (Table 8.4) in an advance breeding line Ty-172 derived from *S. peruvianum* using DNA markers (Anbinder et al., 2009). A limited segregating population and RILs were used for gene mapping against Indian TLCV and limited markers of *TY*-n gene had been validated against TLCV (Singh et al., 2014, 2015e). Recently another *Ty-6* gene has been identified and mapped on chromosomes 10 (Figure 8.3) and developed by Fla. 8383 and *S. chilense* LA2779 using SNPs markers (Hutton and Scott, 2013, 2015).

MAS was employed to select *Ty-3* lines from different intraspecific crosses. Large F_2 populations consisting of 200 plants were grown for three crosses that were developed using superior parents. Combination of MAS and pedigree selection was used to select 22 F_2 plants and F_3 families derived from each of the selected F_2 plants were grown (Ji and Scott, 2006; Ji et al., 2007b). A total five lines carrying *Ty-3* allele were selected following within-family selection. Pedigree selection was performed based on the visual selection criteria for horticultural traits. A total of 30 hybrids were developed based on a combination of *Ty-2* and *Ty-3* lines, and these will be evaluated during both early and main tomato growing seasons. The parental lines used in the generation of these hybrids included previously developed *Ty-2* and *Ty-3* carrying lines. Marker assays were performed on both the seed and pollen parent plants using the *Ty-2* and *Ty-3* diagnostic markers. The plants that were positive for these diagnostic markers were selected for hybridization (Zamir et al., 1994; Hanson et al., 2000, 2006; Ji and Scott, 2006; Ji et al., 2007b).

TABLE 8.4 Acknowledged Gene for *Leaf Curl Virus* Resistant in Tomato

Name of Gene	Background	Chromosome Number	Molecular Markers	Flanking Markers of Both Side	References
TY-1	*S. chilense*	6	RAPD, SCAR, RFLP, and CAPS	C2_At5g61510 and C2At3g10920/T1456	Zamir et al. (1994)
TY-2	H-24 (*S. habrochaites*)	11	RFLP	C2_AT1g07960 and T0302	Hanson et al. (2000)
TY-3	*S. chilense*	6	RAPD, SCAR, RFLP, and CAPS	C2_At5g05690 and C2_At5g41480 To507 and T0693	Ji et al. (2007a)
TY-4	*S. chilense*	3	RAPD	C2_AT4g17300, T1320 and C2_AT5g60160	Ji et al. (2008, 2009)
TY-5	*S. peruvianum*	4	RAPD, SSRs	SSR43 and TG182	Anbinder et al. (2009)
TY-6	*S. chilense*	10	SNPs	–	Hutton and Scott (2013)
SSR218170–145 and *SSR304158–186*	*S. habrochaites* (EC-520061)	10 and 7	SSRs	–	Singh et al. (2015e)

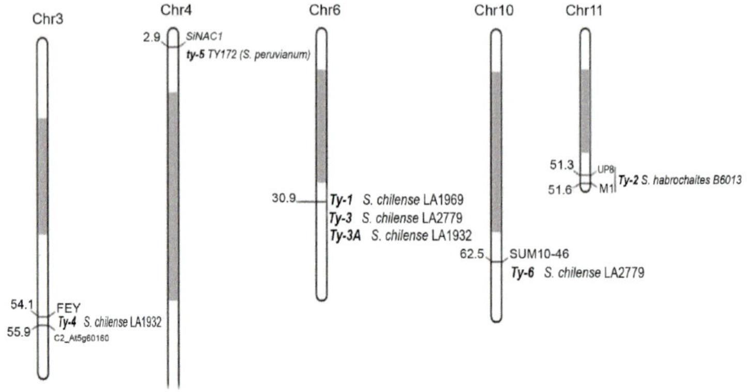

FIGURE 8.3 Chromosome's location of *Ty-1 to Ty-6* gene and derivative tomato species responsible for resistance to TYLCV
Source: Reprinted with permission from Singh et al., 2019. © Taylor & Francis.

The microarray experiment revealed a drastic down-regulation of proline rich protein (PRP) under drought stress condition. PRPs are important cell wall proteins having different roles in diverse plant species. PRPs are involved in plant growth and germination of seeds to cell death and play the lead role in both biotic and abiotic stresses. Its expression is up-regulate in cold, salt, and heat stress, while interestingly it is down-regulated in drought stress. This provoked us to further investigate tissue specific expression of the PRP gene. Water stress has showed to one month old plant withholding irrigation till the appearance of stress symptoms in plants. For PRP gene in both control and stressed plants, the quantitative real time PCR (Rt PCR) has used. The uppermost down-regulation of PRP has detected in stem, leaf, and root tissues (Figure 8.4).

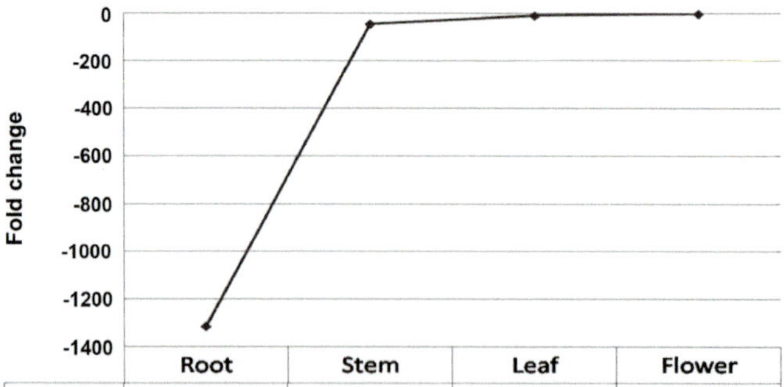

	Root	Stem	Leaf	Flower
Series1	-1314.228	-45.56961	-10.12605	-2.114036

FIGURE 8.4 Expression of qPCR for StPRP in a range of tissues of tomato.

8.6 GENE TRANSFORMATION FOR TRANSGENIC RESISTANCE

The transgenic tomatoes have gained popularity during the last decade. The transformation system in tomato has been appreciate using *Agrobacterium tumifaciens* (Padchuri et al., 2010). A number of resistant varieties of tomato for TLCV, have been used through conventional breeding, but they could not produce a stable variety for TLCV resistant (Raj et al., 2005). However, several workers attempted method of *Agrobacterium using nucleo*-capsid protein (CP) of TYLCV, cucumber mosaic virus (CMV) and *Tobacco virus* (TMV) for genetic transformation to produce transgenic tomato plants (Fuchs et al., 1996; Zhuk and Rassokha, 1992). According to Raj et al. (2005), these transgenic tomato plants were found to be resistant to the respective viruses either under greenhouse or field conditions, and the given two components, CP and satellite RNA have been used to obtain transgenic plants resistant to TLCV (Paduchuri et al., 2010). The transformation of an inbred line of tomato for TLCV nucleoprotein gene cassette produced high levels of resistance to the virus in cultivars derivative of wild species by using breeding programs (Ultzen et al., 1995). The CP of TYLCV, as a single-stranded (ss) DNA binding protein and isolate of *spotted wilt tospovirus* (TSWV-BL) were transferred into TMV resistant tomato line *via Agrobacterium* transformation (Palanichelvam et al., 1998). Whereas, tomato yellow leaf curl Sardinia virus (TYLCSV) shortened replication-associated protein (Rep) gene was used to transform genotype of tomato plants and prepared transgenic plants were agro-inoculated with Tunisian infectious strain of TYLCSV (Ben et al., 2009). Disgracefully, the CRISPR/Cas9 tools has essential in field conditions and would be applicable to establish the disease resistance against ToLCNDV (Sharma and Prasad, 2017). The limited resistance genes have been identified against begomoviruses, and the most important of them being the *Ty* series of genes available in wild tomato (*Solanum chilense*) against TYLCV and its alleles have been used into commercial tomato. However, suitable molecular tools for using this technique are not available (Verlaan et al., 2011). In a study, the fine-mapping was applied with 12,000 RILs derived by *Ty-1* and *Ty-3* genes for TYLCV resistant (Verlaan et al., 2013). Recently, six TLCV genes expressed individually in tomato by inducing different levels (Gorovits et al., 2017). They explained in the current study, individual TYLCV gene is capable of suppressing the HSP90-dependent death and HSFA2 deactivation resulting C_2 caused a decrease in the severity of death phenotypes and the expression of V_2 strengthened cell death. They also found that the C_2 or V_2 affected to stress under viral infection.

8.7 GENOME EDITING AND EMBRYO RESCUE IN TOMATO

Reverse genetics methods has been used man time for molecular understanding of genes and their role. These have need of the development of mutant plants of a desired gene, is a complex assignment. The tool of genome editing CRISPR/Cas9 has accelerated this practice with easy steps. The work on consistency of genome editing procedure in tomato for basic studies has been initiated. For this purpose, CRISPR/Cas9 nuclease and a gRNA for targeting the desired gene will be used. Presently, two genes are selected as target genes to develop mutant plants (Pudchuri et al., 2010). The accession LA2157 of *S. arcanum* crossed with tomato cultivar Kashi Amrit and hybrid was successfully isolated using embryo rescue. The embryo rescued plants were tested for hybridity using morphological and DNA marker assay. The differences in the morphology of leaf, inflorescence, and fruits confirmed the hybridity of the plant. Similarly, the marker assay performed using CAPS marker in tomato which was located on Chr. 1 for confirmation of the hybridity (Figure 8.5).

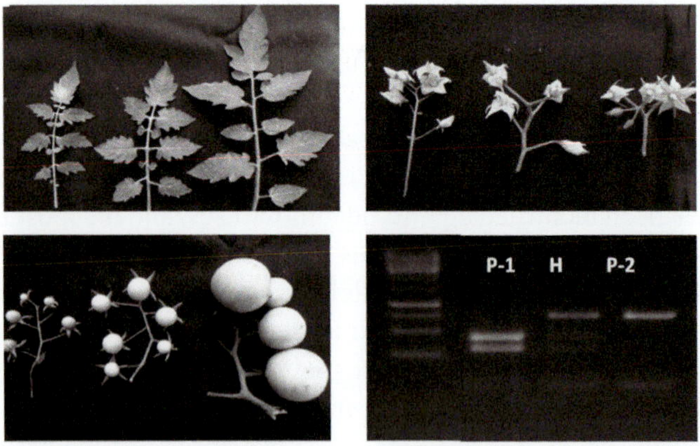

FIGURE 8.5 Variations in morphology of leaf and inflorescence and also differences in marker analysis (CAPS marker: C2At3g12685 digested by Msp-I) between the parents and interspecific hybrid of embryo rescued. The leaf, inflorescence, fruit of hybrids and DNA samples are in the middle of the panel.

8.8 CONCLUSION

Identification of resistance gene for TLCV has created a major issue around the worldwide. Generally, artificial screening is applying as an appropriate technique to identify the resistant genotypes but during the hybridization use of wild sources will more potent as donor parents. The wild sources are controlled by dominant gene and promising material for genetic improvement of resistance tomato. Biochemical markers and physiological markers can exploit and escape the barriers of interspecific hybridization, as well as to expedite and shorten the resistance genes in tomato breeding programs. So, in tomatoes a number of molecular markers have been used *viz.,* RAPD, RFLP, SCAR, and AFLP to identify QTLs for resistance genes *Ty-1, Ty-2, Ty-3, Ty-4, Ty-5*, and *Ty-6* to leaf curl virus but lacking the use of SSRs markers for *leaf curl virus* resistance gene except *Ty5*. Molecular breeding for resistant program is sustainable, and this could be an efficient method to found a new resistant source. Only little resistant varieties of tomato have been developed by conventional breeding for TLCV disease. So, a protocol developed for tomato transformation has a wide scope in the generation of transgenic plants to develop TLCV resistant varieties.

KEYWORDS

- **bulked segregant analysis**
- **mapping population**
- **molecular breeding**
- **nuclear location sequences**
- **polymerase chain reaction**
- **resistant to TLCV**
- **transgenic approaches**

REFERENCES

Anbinder, I., Reuveni, M., Azari, R., Ilan, P., Nahon, S., Shlomo, H., Chen, L., et al., (2009). Molecular dissection of tomato leaf curl virus resistance in tomato line *Ty172* derived from *Solanum peruvianum. Theor. Appl. Genet., 119*(3), 519–530.

Arooj, S., Iftikhar, Y., Kamran, M., Ullah, M. I., Mubeen, M., Shakeel, Q., Zeerak, N., & Bilqees, I., (2017). Management of tomato leaf curl virus through non-chemicals in relation to environmental factors. *Pak. J. Phytopathol., 29*(1), 41–46.

Baird, E., Cooper-Bland, S., Waugh, R., De Maine, M., & Powell, W., (1992). Molecular characterization of inter- and intra-specific somatic hybrids of potato using randomly amplified polymorphic DNA (RAPD) markers. *Mol. Gen. Genet., 233*, 469–475.

Banerjee, M. K., & Kalloo, G., (1987a). Sources and inheritance of resistant to leaf curl virus in *Lycopersicon. Theor. Appl. Genet., 73*, 707–710.

Banerjee, M. K., & Kalloo, G., (1987b). Inheritance of tomato leaf curls virus resistance in *Lycopersicon hirsutum f. glabratum. Euphyt., 36*, 581–584.

Banerjee, M. K., & Kalloo, G., (1989a). The inheritance of earliness and fruit weight in crosses between cultivated tomatoes and two wild species of *Lycopersicon. Plant Breed., 102*, 148–152.

Banerjee, M. K., & Kalloo, G., (1989b). Effect of tomato leaf curls virus on biochemical attributes in resistant/susceptible plants of tomato (*Lycopersicon esculentum* Mill.). *Veg. Sci., 16*(1), 21–31.

Banerjee, M. K., & Kalloo, G., (1989c). Role of phenols in resistance to tomato leaf curl virus, fusarium wilt and fruit borer in *Lycopersicon. Curr. Sci., 58*(10), 575.

Ben, T. H., Gharsallah, C. S., Lengliz, R., Maxwell, D. P., Marrakchi, M., Fakhfakh, H., & Gorsane, F. (2009). Use of tomato leaf curl virus (TYLCV) truncated Rep gene sequence to engineer TYLCV resistance in tomato plants. *Acta Virol., 53*(2), 99–104.

Benor, S., Zhang, M., Wang, Z., & Zhang, H., (2008). Assessment of genetic variation in tomato (*Solanum lycopersicum* L.) inbred lines using SSR molecular markers. *J. Gen. Genom., 35*, 373–379.

Bernatzky, R., & Tanksley, S. D., (1986). Towards a saturated linkage map in tomato-based on isozymes and random cDNA sequences. *Genet., 112*, 887–898.

Brunt, A., Grabtree, K., & Gibbs, A., (1990). *Viruses of Tropical Plants* (p. 707). CAB International, UK.

Buetow, K. H., Edmonson, M. N., & Cassidy, A. B., (1999). Reliable identification of large numbers of candidate SNPs from public EST data. *Nat. Genet., 21*, 323–325.

Caro, M., Verlaan, M. G., Julian, O., Finkers, R., Wolters, A. M. A., Hutton, S. F., Scott, J. W., et al., (2015). Assessing the genetic variation of Ty-1 and Ty-3 alleles conferring resistance to tomato yellow leaf curl virus in a broad tomato germplasm. *Mol. Breed., 35*, 132.

Chakraborty, S., Pandey, P. K., Banerjee, M. K., Kalloo, G., & Fauquet, C. M., (2003). Tomato leaf curl Gujarat virus, a new *Begomovirus* species causing a severe leaf curl disease of tomato in Varanasi, India. *Phytopathol., 93*, 1485–1495.

Cohen, H., & Antignus, Y., (1994). Tomato yellow leaf curl virus, a whitefly-borne Geminivirus of Tomatoes. *Adv. Dis. Vect. Res., 10*, 259–288.

Collard, B. C. Y., Jahufer, M. Z. Z., Brouwer, J. B., & Pang, E. C. K., (2005). An introduction to markers, quantitative trait loci (QTL) mapping and marker-assisted selection for crop improvement: The basic concepts. *Euphyt., 142*, 169–196. doi: 10.1007/s10681-005-1681-5.

Cooper, D. N., Smith, B. A., Cooke, H. J., Niemann, S., & Schmidtke, J., (1985). An estimate of unique DNA sequence heterozygosity in the human genome. *Hum. Genet., 69*, 201–205.

Edwards, M. D., Stuber, C. W., & Wendel, J. F., (1987). Molecular-marker facilitated investigation of quantitative-trait loci in maize: I. Numbers, genomic distribution and types of gene action. *Genet., 116*, 113–125.

Fauquet, C. M., Bisaro, D. M., Briddon, R. W., Brown, J. K., Harrison, B. D., Rybicki, E. P., Stenger, D. C., & Stanley, J., (2003). Revision of taxonomic criteria for species demarcation in the family *Geminiviredae*, and an updated list of *Begomovirus* species. *Arch. Virol., 148*, 405–421.

Fauquet, M. C., & Mayo, M. A., (1999). Abbreviations for plant virus names. *Arch. Virol., 144*, 6.

Foolad, M. R., (2007). Genome mapping and molecular breeding of tomato. *Int. J. Plant Gen., 2007*(2), 64358. doi: 10.1155/2007/64358.

Frary, A., Xu, Y., Liu, J., Mitchell, S., Tedeschi, E., & Tanksley, S. D., (2005). Development a set of PCR- based anchor markers encompassing the tomato genome and evaluation of their usefulness for genetics and breeding experiments. *Theor. Appl. Genet., 111*, 291–312.

Friedmann, M., Lapidot, M., Cohen, S., & Pilowski, M., (1998). A novel source of resistance to tomato yellow leaf curl virus exhibiting a symptomless reaction to viral infection. *Journal of American Society for Hort. Sci., 123,* 1004–1007.

Fuchs, M., Provvidenti, R., Slighton, J. L., & Gonsalves, D., (1996). Evaluation of transgenic tomato plants expressing the coat protein gene of cucumber mosaic virus strain W. L. under field conditions. *Plant Dis., 80*, 270–275.

Fulton, T. M., Vander, H. R., Eanetta, N. T., & Tanksley, S. D., (2002). Identification, analysis, and utilization of conserved ortholog set markers for comparative genomics in higher plants. *Plant Cell, 14*, 1457–1467.

Gafni, Y., (2003). Tomato yellow leaf curl virus, the intracellular dynamics of a plant DNA virus. *Mol. Plant Pathol., 4*, 9–15.

Gevorkyan, Z. G., Adhzhemyam, L. A., & Gasparyan, P. V., (1976). Tomato leaf curl disease, its pathomorphology and effect on physiological and biochemical properties of plants (Russian summary). *IZUS-Kh Nauk., 5*, 38–40.

Gorovits, R., Moshe, A., Amrani, L., Kleinberger, R., Anfoka, G., & Czosnek, H., (2017). The six-tomato yellow leaf curl virus genes expressed individually in tomato induce different levels of plant stress response attenuation. *Cell Stress Chapter., 22*(3), 345–355.

Hanson, P., Bernacchi, M., Green, S., Tanksley, S. D., Muniyappa, V., & Padmaja, A. S., (2000). Mapping a wild tomato introgression associated with tomato yellow leaf curl virus resistance in a cultivated tomato line. *J. Am. Soc. Hort. Sci., 15*, 15–20.

Hanson, P., Green, S. K., & Kuo, G., (2006). *TY-2*, a gene on chromosomes 11 conditioning *Geminivirus* resistance in tomato. *Rep. Tomato Genet. Coop., 56*, 17–18.

He, C., Poysa, V., & Yu, K., (2002). Development and characterization of simple sequence repeat (SSR) markers and their use in determining relationships among *L. esculentum* cultivars. *Theor. Appl. Genet., 106*, 363–373.

Hutton, S. F., & Scott, J. W., (2013). Fine-mapping and cloning of *Ty-1* and *Ty-3*; and mapping of a new TYLCV resistance locus, "*Ty-6*". In: *Tomato Breeders Round Table Proceedings*. Chiang Mai, Thailand.

Hutton, S. F., & Scott, J. W., (2015). *Ty-6*, a major *Begomovirus* resistance gene located on chromosome 10. *Rep. Tomato Genet. Coop., 64,* 14–18.

Ji, Y., & Scott, J. W., (2006). *Ty-3*, a *Begomovirus* resistance locus linked to *Ty-1* on chromosome 6. *Rep. Tomato Genet. Coop., 56,* 22–25.

Ji, Y., Schuster, D. J., & Scott, J. W., (2007b). *Ty-3*, a *begomovirus* resistance locus near the tomato yellow leaf curl virus resistance locus *Ty-1* on chromosome 6 of tomato. *Mol. Breed., 20,* 271–284.

Ji, Y., Scott, J. W., & Maxwell, D. P., (2009). Molecular mapping of *Ty-4*, a new tomato yellow leaf curl virus resistance locus on chromosome 3 of tomato. *J. Am. Soc. Hort. Sci., 134,* 281–288.

Ji, Y., Scott, J. W., Hanson, P., Graham, E., & Maxwell, D. P., (2007a). Sources of resistance, inheritance, and location of genetic loci conferring resistance to members of the tomato-infecting *Begomoviruses*. In: Czosnek, H., (ed), *Tomato Yellow Leaf Curl Virus Disease* (pp. 343–362). Spr. Netherl.

Ji, Y., Scott, J. W., Maxwell, D. P., & Schuster, D. J., (2008). *Ty-4*, a tomato yellow leaf curl virus resistance gene on chromosome 3 of tomato. *Tom. Genet. Coop. Rep., 58,* 29–31.

Kalloo, G., & Banerjee, M. K., (1990). Transfer of tomato leaf curl virus resistance from *Lycopersicon hirsutum f. glabraum to L. esculetum. Plant Breed., 105*(2), 156–159.

Kheyr-Pour, A., Gronenborn, B., & Czosnek, H., (1994). Agroinoculation of tomato yellow leaf curl virus (TYLCV) the virus resistance of wild *Lycopersicon species*. *Plant Breed., 112*, 228–233.

Konate, G., Barro, N., Fargette, D., Swanson, M. M., & Harrison, B. D., (2008). Occurrence of whitefly transmitted Gemini viruses in crops in Burkina Fasso and their serological detection and differentiation. *Annals of Applied Biology, 126*(1), 121–129.

Kumari, P., Chattopadhyay, B., Singh, A. K., & Chakraborty, S., (2010). A new *Begomovirus* species causing tomato leaf curl disease in Patna, India. *Plant Dis., 93,* 545.

Lapidot, M., & Friedmann, M., (2002). Breeding for resistance to whitefly-transmitted Gemini viruses. *Ann. Appl. Biol., 140*, 109–127.

Marco, S., (1975). Chlorophyll content of tomato yellow leaf curl virus-infected Tomatoes in relation to virus resistance. *Phytoparasitica, 3*(2), 141–144.

Maruthi, M. N., Czosnek, H., Vidavski, F., Tarba, S. Y., Milo, J., Leviatov, S., Venkatesh, H. M., et al., (2003). Comparison of resistance to tomato leaf curl virus (India) and tomato yellow leaf curl virus (Israel) among *Lycopersicon* wild species, breeding lines and hybrids. *Europ. J. Plant Pathol., 109*, 1–11.

Mayo, M. A., & Brunt, A. A., (2005). *Plant Virus Taxonomy: Families, Genera, and Type Species of Plant Viruses and Viroids* (p. 1). ICTV.

Michelmore, R., Paran, I., & Kesseli, R., (1991). Identification of markers linked to disease-resistance genes by bulked segregant analysis: A rapid method to detect markers in specific genomic regions by using segregating populations. *Proc. Nat. Acad. Sci., USA., 88*, 9828–9832.

Miller, J. C., & Tanksley, S. D., (1990). RFLP analysis of phylogenetic relationship and genetic variation in gene *Lycopersicon. Theor. Appl. Genet., 80*, 437–438.

Momotaz, A., Scott, J. W., & Schuster, D. J., (2007). *Solanum habrochaites* accession LA 1777 recombinant inbred lines are not resistant to tomato yellow leaf curl virus or tomato mottle virus. *HortScience, 42*(5), 1149–1152.

Muniyappa, V., Padmaja, A. S., Venkatesh, H. M., Sharma, A., Chandrasekhar, S., Kulkarni, R. S., Hanson, P. M., et al., (2002). Tomato leaf curl virus resistance tomato lines TLB111, TLB130, and TLB182. *HortScience., 37*(3), 603–606.

Naveed, K., Imran, M., Riaz, A., Azeem, M., & Tahir, M. I., (2015). Impact of environmental factors on tomato leaf curl virus and its management through plant extracts. *Int. J. Afric. Asian Stud., 11,* 44–52.

Navot, N., Zeidan, M., Pichersky, E., Zamir, D., & Czosnek, H., (1992). Use of polymerase chain reaction to amplify tomato yellow leaf curl virus DNA from infected plants and viruliferous whiteflies. *Phytopathol., 82,* 1199–1202.

Paduchuri, P., Gohokar, S., Thamke, B., & Subhas, M., (2010). Transgenic tomatoes: A review. *Int. J. Adv. Biot. Res., 1*(2), 69–72.

Palanichelvam, K., Kunik, T., Citovsky, V., & Gafni, Y., (1998). The capsid protein of tomato yellow leaf curl virus binds cooperatively to single-stranded DNA. *J. Gen. Virol., 79,* 2829–2833.

Paran, I., Goldmann, I., Tanksley, S. D., & Zamir, D., (1995). Recombinant inbred lines for genetic mapping in tomato. *Theor. Appl. Genet., 90*(3, 4), 542–548.

Paterson, A. H., (1996). Making genetic maps. In: Paterson, A. H., (ed.), *Genome Mapping in Plants* (pp. 23–39). R.G. Landies company San Diego, California: Acad. Press, Aust. Tex.

Peralta, I. E., Knapp, S., & Spooner, D. M., (2005). New species of wild tomatoes (*Solanum* section *lycopersicon*: Solanaceae) from Northern Peru. *Syst. Bot., 30*(2), 424–434.

Pico, B., Diez, M. J., & Nuez, F., (1998). Evaluation of whitefly mediated inoculation techniques to screen *Lycopersicon esculentum* and wild relatives for resistance to TYLCV. *Euphyt., 101*(3), 259–271.

Pilowski, M., & Cohen, S., (1974). Inheritance of resistance to tomato yellow leaf curl virus in tomatoes. *Phytopath., 64,* 632–635.

Praveen, S., Mishra, A. K., & Dasgupta, A., (2005). Antisense suppression of replicase gene expression recovers tomato plants from leaf curl infection. *Plant Sci., 168,* 1011–1014.

Raj, S. K., Singh, R., Pandey, S. K., & Singh, B. P., (2005). *Agrobacterium*-mediated tomato transformation and regeneration of transgenic lines expressing *Tomato leaf curl virus* coat protein gene for resistance against *TLCV* infection. *Curr. Sci., 88*(10), 1674–1679.

Rick, C. M., (1974). High soluble-solids content in large-fruited tomato lines derived from a wild green-fruited species. *Hilgard., 42,* 493–510.

Sharma, N., & Prasad, M., (2017). An insight into plant-tomato leaf curl New Delhi virus interaction. *Nucl., 60,* 335–348.

Singh, A. K., Rai, G. K., Singh, M., Singh, S. K., & Singh, S., (2008). Inheritance of resistance to tomato leaf curl virus in tomato (*L. esculentum* Mill.). *Veg. Sci., 35*(2), 194–196.

Singh, R. K., Rai, N., & Singh, S. N., (2010). Response of tomato genotypes to tomato leaf curl virus. *Ind. J. Agri. Sci., 80*(8), 755–758.

Singh, R. K., Rai, N., Kumar, P., & Singh, A. K., (2015d). Inheritance study in tomato for tomato leaf curl virus (TLCV) resistance. *Ind. J. Agric. Sci., 85*(7), 896–901.

Singh, R. K., Rai, N., Lima, J. M., Singh, M., Singh, S. N., & Kumar, S., (2015c). Genetic and molecular characterizations of *Tomato leaf curl virus* resistance in tomato. *The J. Hort. Sci. Biotech., Eng., 90*(5), 503–510.

Singh, R. K., Rai, N., Singh, A. K., Kumar, P., & Singh, B., (2019). A critical review on tomato leaf curl virus resistance in tomato. *Int. J. Veg. Sci., 25*(4), 373–393. doi: 10.1080/19315260.2018.1520379.

Singh, R. K., Rai, N., Singh, A. K., Kumar, P., Chaubey, T., Singh, B., & Singh, S. N., (2018). Elucidation of diversity among F1 hybrids to examine heterosis and genetic inheritance for horticultural traits and ToLCV resistance in tomato. *J. Gen., 97*(1), 67–78.

Singh, R. K., Rai, N., Singh, M., Saha, S., & Singh, S. N., (2015a). Detection of tomato leaf curl virus resistance and inheritance in tomato (*S. lycopersicum* L.). *J. Agric. Sci. Camb., 153*(1), 78–89. doi: 10.1017/S0021859613000932.

Singh, R. K., Rai, N., Singh, M., Singh, R., & Kumar, P., (2015e). Effect of climate change on tomato leaf curl virus (TLCV) disease in tomatoes. *Ind. J. Agric. Sci., 85*(2), 290–292.

Singh, R. K., Rai, N., Singh, M., Singh, S. N., & Srivastava, K., (2014). Genetic analysis to identify good combiners for TLCV resistance and yield components in tomato using interspecific hybridization. *J. Genet., 93,* 623–629.

Singh, R. K., Rai, N., Singh, M., Singh, S. N., & Srivastava, K., (2015b). Selection of tomato genotypes resistant to tomato leaf curls virus disease using biochemical and physiological markers. *J. Agric. Sci., Camb., 153*(4), 646–655.

Singsit, C., & Ozias-Akins, P., (1993). Genetic variation in monoploids of diploid potatoes and detection of clone-specific random amplified polymorphic DNA markers. *Plant Cell Rep., 12,* 144–148. https://doi.org/10.1007/BF00239095.

Smulders, M. J. M., Bredemeijer, G., Rus- Kortkaas, W., Arens, P., & Vosman, B., (1997). Use of short microsatellite from database sequence to generate polymorphism among *L. esculetum* cultivars and accessions of other *Lycopersicon* species. *Theor. Appl. Genet., 97,* 264–272.

Sobrino, B., Brion, M., & Carracedo, A., (2005). SNPs in forensic genetics: A review on SNP typing methodologies. *For. Sci. Int., 154*(2/3), 181–494.

Soleimani, V. D., Bernard, B., & Johnson, D. A., (2003). Efficient validation of single nucleotide polymorphisms in plants by allele-specific PCR, with an example from barley. *Plant Mol. Biol. Rep., 21*(3), 281–288.

Srivastava, K. M., Hallan, V., Raizada, R. K., Chandra, G., Singh, B. P., & Sane, P. V., (1995). Molecular cloning of Indian tomato leaf curl virus genome following a simple method of concentrating the supercoiled replicative form of viral DNA. *J. Virol. Meth., 52,* 297–304.

Sunyaev, S. R., Eisenhaber, F., Rodchenkov, I. V., Eisenhaber, B., Tumanyan, V. G., & Kuznetsov, E. N., (1999). PSIC: Profile extraction from sequence alignments with position-specific counts of independent observations. *Prot. Eng., 12*, 387–394.

Tanksley, S. D., (1993). Linkage map of tomato (*Lycopersicon esculentum* Mill.) (2n=24). In: O' Brien, S., (ed), *Genetic Map: Locus Map of Complex Genomes* (pp. 6, 39, 36, 60). Cold Spring Harbor Lab Press, USA.

Tanksley, S. D., Ganal, M. W., Prince, J. P., De Vicente, M. C., Bainerbale, M. W., Broun, P., Fulton, T. M., et al., (1992). High density molecular linkage maps of tomato and potato genomes. *Biolog. Inf. Pract. Appl. Genet., 123*, 1141–1160.

Tautz, D., & Renz, M., (1984). Simple sequences are ubiquitous repetitive components of eukaryotic genomes. *Nucl. Aci. Res., 12*(10), 4127–4138.

Ultzen, T., Gielen, J., Venema, F., Westerbroek, A., Haan, P., De Tan, M., et al., (1995). Resistance to tomato spotted wilt virus in transgenic tomato hybrids. *Euphyt., 85*, 159–168.

Vasudeva, R. S., & Samraj, J., (1948). A leaf curls disease in tomato. *Phytopath., 38*, 364–369.

Verlaan, M. G., Hutton, S. F., Ibrahem, R. M., Kormelink, R., Visser, R. G. F., Scott, J. W., Edwards, J. D., & Bai, Y., (2013). The tomato yellow leaf curl virus resistance genes Ty-1 and Ty-3 are allelic and code for DFDGD-Class RNA-dependent RNA polymerases. *PLoS Gen., 9*(3), 1003399. doi: 10.1371/journal.pgen.1003399.

Verlaan, M. G., Szinay, D., Hutton, S. F., De Jong, H., Kormelink, R., Visser, R. G. F., Scott, J. W., & Bai, Y., (2011). Chromosomal rearrangements between tomato and *Solanum chilense* hamper mapping and breeding of the TYLCV resistance gene *Ty-1*. *Plant J., 68*, 1093–1103.

Vidavski, F., Czosnek, H., Gazit, S., Levy, D., & Lapidot, M., (2008). Pyramiding of genes conferring resistance to tomato yellow leaf curl virus from different wild tomato species. *Plant Breed, 127*(6), 625–631.

Vidavsky, F., & Czosnek, H., (1998). Tomato breeding lines resistant and tolerant to tomato yellow leaf curl virus issued from *Lycopersicon hirsutum*. *Phytopath., 77*(10), 910–914.

Vidavsky, F., Leviatove, S., Milo, J., Rabinowitch, H. D., Kedar, N., & Czosnek, H., (1998). Response of tolerant breeding line of tomato, *Lycopersicon esculentum*, originating from three different sources *(L. pruvianum, L. pimpinellifolium*, and *L. chilense)* to early controlled inoculation by tomato yellow leaf curl virus (TYLCV). *Plant Breed., 117*(2), 165–169.

Wang, D. G., Fan, J. B., Siao, C. J., Berno, A., Young, P., Sapolsky, R., Ghandour, G., Perkins, N., Winchester, E., Spencer, J., et al., (1998). Large-scale identification, mapping, and genotyping of single-nucleotide polymorphisms in the human genome. *Sci., 280*, 1077–1082.

Weber, J. L., (1990). Informativeness of human (dC-dA)n. (dG-dT)n polymorphisms. *Genom., 7*, 524–530.

Williams, K., Willsmore, K., Hoppo, S., Eckermann, P., & Zwer, P., (2004). Mapping of quantitative trait loci for yield, quality and disease resistance. In: Peltonen-Saino, P., & Topi-Hulmi, M., (eds.), *Proc. 7th Int. Oat Conf.* (p. 71). Helsinki, Final. www.mtt.fi/met/pdf/met51.pdf (accessed on 1 March 2021).

Zakay, Y., Novot, N., Zeidan, M., Kedar, N., Rabinowitch, H. D., Czosnek, H., & Zamir, D., (1991). Screening *Lycopersicon* accession for resistance to tomato leaf curl virus: Presence of viral DNA and symptom developments. *Plant-dis. St. Paul, Minn.: Am. Phytopath. J., 75*(3), 279–281.

Zamir, D., Michelson, E. I., Zakay, Y., Novot, N., Zeidan, M., Sarfatti, M., Eshed, Y., et al., (1994). Mapping and interrogation of a tomato yellow leaf curl virus tolerant gene, *Ty-1*. *Theor. Appl. Genet., 88*(2), 141–146.

Zhuk, I. P., & Rassokha, S. N., (1992). Regeneration and selection of somatic clones of tomato for resistance to TMV. *Russian Acad. Agricultural Sciences,* pp. 11, 12, 18–21.

CHAPTER 9

MOLECULAR ADVANCES OF THE TOBACCO MOSAIC VIRUS INFECTING TOMATO

MAHESH KUMAR,[1] MD. SHAMIM,[1] V. B. JHA,[1] TUSHAR RANJAN,[2] SANTOSH KUMAR,[3] HARI OM,[1] RAVI RANJAN KUMAR,[2] VINOD KUMAR,[2] RAVI KESHRI,[2] and M. S. NIMMY[4]

[1]*Department of Molecular Biology and Genetic Engineering, Dr. Kalam Agricultural College (Bihar Agricultural University, Sabour), Kishanganj, Bihar – 855107, India,*
E-mail: maheshkumara2z@gmail.com (M. Kumar)

[2]*Department of Molecular Biology and Genetic Engineering, Bihar Agricultural University, Sabour, Bhagalpur – 813210, Bihar, India*

[3]*Department of Plant Pathology, Bihar Agricultural University, Sabour, Bhagalpur – 813210, Bihar, India*

[4]*NRC on Plant Biotechnology, IARI, Pusa Campus, New Delhi, India*

ABSTRACT

Viruses are the obligate organisms; use machineries of the host cell for accumulation and multiplication. Tobacco mosaic virus (TMV) is a most limiting factor for the tomato production. The quality of the tomato (*Solanum lycopersicum* L.) is reduced by virus after hijacking the most sophisticated host defense system and causing various types of symptoms. To establish the disease in their host plant requires infection cycle. This infection cycle includes; entry of the virus mechanical injure part, replication, and multiplication in infected cell, movement of virion particle or genetic material from infected cell to healthy cell through plasmodesmata and finally systemic movement through vasculature to cause disease in whole plant. So, in this chapter, we described present information of TMV taxonomy, genome organization morphology and structure. Subsequently, the replication mechanism of the TMV and its life science were briefly elaborated. In the last part of this chapter, possible control and measure of TMV was discussed.

9.1 INTRODUCTION

Tobacco mosaic virus (TMV) is considered as a unique place in the past. The synonyms of the TMV common strains are wild type, vulgare U1 (Siegel and Wildman, 1954) OM, Japanese common strain (Nozu and Okada, 1968) Korean common strain (Koh et al., 1992). TMV was first described by Adolf Mayer in 1886 and concluded that it is similar to bacterial infections and transferred between plants. The first perfect report on the virus, i.e., a non-bacterial infectious agent, was given by Dmitri Ivanovsky in 1892 after showing the infectious nature of the sap even after filtering through the finest Chamberland filters. M. Beijerinck was repeated the Ivanovsky experiment in 1898; concluded same and checked its infectious nature on tobacco plant. In host cell of the tobacco plant showed infection. He termed as this infectious agent as a "virus" to reveal mosaic disease of tobacco is non-bacterial nature. W. M. Stanley was crystallized first virus that is TMV in 1935; for his contribution, he was awarded with Nobel Prize in Chemistry in 1946. TMV RNA was purified by the H. Fraenkel-Conrad and R. Williams in 1955 and experimentally proved that coat protein (CP) and purified TMV RNA assemble. Rosalind Franklin designed and built a model of TMV is hollow, not solid, and hypothesized that genetic material of the TMV RNA, i.e., single-stranded in the 1958 World's Fair at Brussels.

TMV has been reported from worldwide. It has a wide host range, infecting more than 200 species of dicotyledonous and monocotyledonous families of plant (Shew and Lucas, 1991).

Tobacco mosaic disease investigations lead to its advancement for the discovery of several essential basic concepts in current biology. These include; TMV virion properties, mechanisms of Virion assembly, self-assembly of viruses and understanding of TMV-host interaction.

TMV is a first plant viral vector that is used for the expression of the heterologous protein in the plant. After discovery and utilization of the

Green Fluorescent Protein has revolutionized the understanding in the field of plant-virus communications and cell biology.

In the present book chapter, we are focusing on our current knowledge of TMV genome, structure, replication mechanism, transmission, control, and management.

9.2 TAXONOMY OF TMV

According to recent ICTV classification, TMV is placed under the family *Virgaviridae* which comes under a Martellivirales order. The *Virgaviridae* family includes seven genus: *Goravirus Furovirus, Hordeivirus, Peclu-virus, Pomovirus, Tobamovirus,* and *Tobravirus*. Among them, *tobamovirus* is the largest genus. TMV is the representing type species. Characteristics properties of genera in the family *Virgaviridae* is mentioned in Table 9.1 and distinguishing feature of the family mentioned in Table 9.2.

TABLE 9.1 Distinguishing Feature of the Family *Virgaviridae*

Characteristic	Description		
Taxonomy	*Virgaviridae* includes seven genera and approximately sixty species		
Transmission	The transmission of the different genera of the *virgaviridae* was given below:		
	SL. No.	**Genera**	**Transmission**
	1.	Pomoviruses	Plasmodiophorids
	2.	Furoviruses	
	3.	Peculviruses	
	4.	Tobraviruses	Nematodes
	5.	Goraviruses	Pollen and/or seed
	6.	Hordeiviruses	
	7.	Tobamoviruses	No known vectors
Morphology	Usually, virion particle is rod-shaped particles, non-enveloped. The diameter of the virion is about 20 nanometers length up to three hundred nanometers. Exception: *Tobamovirus* genus members.		
Genetic material and genome	Genetic material of the *Virgaviridae* is single stranded positive RNA. Genome size varies from 6.3 to 13 kb.		
Replication	Replication occurs in the cytoplasm of the host cell.		
Translation	From RNAs of subgenomic or genomic.		

TABLE 9.2 Characteristics Features for Genera of *Virgaviridae* Family

Genus	RNAs	RdRP	MP	CP	3′ Structure	Type Species
Goravirus	Two	RT	Triple Gene Block	22 KDa	t-RNA	*Gentian ovary ringspot virus* (GORV)
Furovirus	Two	RT	30 KDa	19K+RT	t-RNA^{Val}	Wheat mosaic virus
Hordeivirus	Three	Separate	Triple Gene Block	22 KDa	t-RNA^{Tyr}	Barley stripe mosaic virus
Pecluvirus	Two	RT	Triple Gene Block	23 KDa	t-RNA^{Val}	Indian peanut clump virus, peanut clump virus
Pomovirus	Three	RT	Triple Gene Block	20 KDa+RT	t-RNA^{Val}	Potato mop-top virus
Tobamovirus	One	RT	30 KDa	17 to 18 KDa	t-RNA^{His}	Tobacco mosaic virus
Tobravirus	Two	RT	30 KDa	22 to 24 KDa	t-RNA^{-}	Tobacco rattle virus.

9.3 MORPHOLOGY AND STRUCTURE OF TMV

In general, viruses are much smaller than the bacteria, and their diameter varies from 15 nm to 300 nm. Como- and nanoviruses are comes under the small plant virus's category; sizes vary from 17 to 30 nm. Shape of the bigger viruses can be rod-shaped and its length varies between 65–350 nm and width between 15–25 nm. Bacilliform particles have a range of 30–500 nm in length and width of 3–8 nm. The largest filamentous plant viruses are known to have a virus particle which measures up to 1000 nm (*Citrus tristeza clasterovirus*) and a width in between 3–20 nm. The naked viral RNA has coiled structure with a diameter of around 10 nm (Citovsky et al., 1992). Mimivirus was largest characterized virus, with a capsid diameter of 400 nm. *Megavirus chilensis* was known to be the largest virus (Arslan et al., 2011) before the discovery of pandoravirus (Nadège et al., 2013) and *Pithovirus sibericum* (Matthieu et al., 2014). The icosasedral capsid diameter of the *Megavirus chilensis* in native condition was measured 520 nm (Arslan et al., 2011). Recently, *Pithovirus sibericum* virus was discovered from Siberian region (Matthieu et al., 2014). This virus infects amoeba and its size approximately 1500 nm in length and 500 nm in diameter, making it the largest virus yet found. The size of this virus is 50% larger than the previous largest known viruses, i.e., pandoraviruses (Nadège et al., 2013).

Viruses display a wide range of sizes and shapes. In general, there are four main morphological virus types: Helical, Icosahedral or near-spherical, prolate, and enveloped. TMV falls under simple rod-shaped helical virus. The virions are characterized as a centrally located single-stranded RNA (5.6%) enveloped by a CP (94.4%). The length and diameter of the virion particle is about 3,000 Å and 180 Å. Protein CP is made up of the capsomeres; about 2,130 capsomeres is required the protein coat. The central core of the rod is about 40 Å in diameter. Each capsomere is a grape like structure containing about 158 amino acids and having a molecular weight of 17,000 Dalton as determined by Knight.

9.4 GENOME, GENOME ORGANIZATION, REPLICATION MECHANISM AND LIFE CYCLE

9.4.1 GENOME

TMV comes under (+) sense ss-RNA genome. Apart from TMV, about 80% of viruses have (+) sense ss-RNA genome which affects both plants and animals (Mandahar, 2006). Genomes of virus play a vital role in the during the infection cycle. It mimics as the host mRNAs and used for the specific viral protein synthesis from their genes, templates for transcription of negative-sense RNA copies, genome replication. After this, assembling of the all the virus component leads to the formation of progeny virions (Buck, 1996; Dreher, 1999).

The (+) sense RNA virus genome size may varies between 3.5 to 30 kb (Koonin and Dolja, 1993). The largest genome size was reported from the *Pandoravirus* and its size varies from 1.9 to 2.5 megabase pair (Nadège et al., 2013). The other largest viruses are *Mimivirus*, *Pithovirus*, and *Megavirus*, and its genome size varies from 1.0 to 2.3 megabase pair (Matthieu et al., 2014). The size of the TMV genome is 6.4 kb.

9.4.2 GENOME ORGANIZATION

Genome of the viruses having both coding and non-coding regions. The noncoding region contains message related to the virus as a *cis*-acting element sequence that manages synthesis for full-length positive (+) and negative (-) strand RNAs, translation of the viral protein, transcription of subgenomic RNA. Viral genome may be monopartite-when all the genetic information is contained in a single RNA, or multipartite when genetic information contains more than two RNAs. Each nucleic acid has two ends; 3'end/terminus/untranslated region or (3'UTR) and 5'end/terminus/ untranslated region or (5'UTR). These two ends are maintained properly by viruses for their fitness. The coding region is called open reading frame (ORF) and its codes for the protein. Generally, plant viruses have compressed genome to bear extensively overlapping ORF's. For example, grapevine virus has a monopartite RNA genome of 8000 nt, which contains five ORFs, out of which 3 ORFs are overlapping and cover 1000 nt.

The total size of tobamoviral genome varies from 6.3 to 6.4 kb. The genome of this virus is linear, monopartite, (+) sense ss-RNA in nature. There is disagreement of the certain TMV gene sequences. Generally, TMV RNA encoded by at least four proteins. Encoded proteins; viz; 126 kDa, 183-kDa, movement protein (MP) and CP are mentioned in Figure 9.1.

The 5′-proximal ORF of the TMV is directly translated the encoded 126 kDa, 183-kDa proteins genomic RNA. Larger 183-kDa protein is synthesized by read-through termination codon of the 126-kDa protein sequence (Skuzeski et al., 1991). Sometimes, special t-RNAs (termed suppressor t-RNAs) mis-read a termination codon as a codon leads for synthesis of larger protein in TMV. Generally, a UAG codon is suppressed by certain tyrosine-specific t-RNAs (Beier et al., 1984). Viral replication complex mainly include 126 kDa and 183-kDa proteins.

MP (30-kDa) and CP (17.5 kDa) are encoded by the downstream of the 126/183-kDa ORF (Figure 9.1). Full-length genomic RNA couldn't synthesize MP and CP directly. Generally, in eukaryotic protein synthesis machinery is not able to re-initiate protein synthesis following termination. Only ORF closest to the 5′ cap is normally translated in a eukaryotic system of polycistronic mRNAs (Kozak, 2002). The synthesis and translations of TMV CP and MP occur only after the synthesis of two subgenomic mRNAs, which are generated during viral infection (Hunter et al., 1976; Sulzinski et al., 1985).

I1 RNA is another type of the subgenomic RNA has been reported from the TMV infected tobacco tissue (Beachy and Zaitlin, 1977; Sulzinski et al., 1985). The 54-kDa protein is encoded by I1 subgenomic RNA formed from read-through portion of the 183-kDa replicase protein (Sulzinski et al., 1985).

Atabekov and associates were identified p4 is another gene product of TMV, i.e., 4-kDa. It is encoded by an ORF (ORF X). According to Morozov et al. (1993); p4 coding region overlaps sequence of 3′ end MP ORF and 5′ end for the CP gene. Later on, p4 was reported from TMV (Fedorkin et al., 1995; Dorokhov et al., 1994).

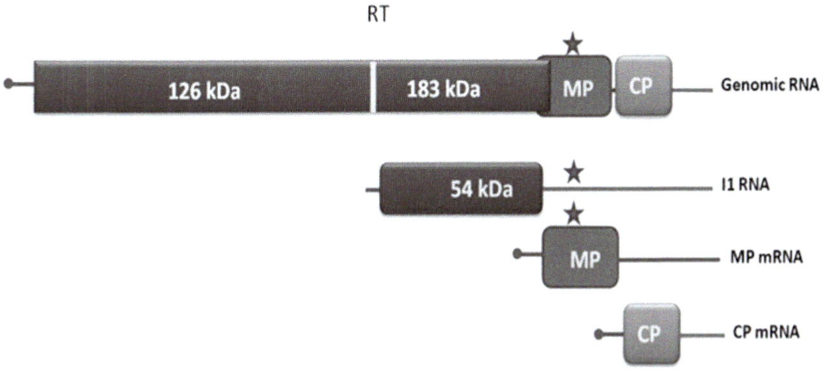

FIGURE 9.1 Genome organization of the tobamovirus. The pictorial representation of the genome organization of the subgroup 1a tobamovirus. Here common strain (U1 strain of TMV) has been shown in the figure. Rectangular boxes indicate Coding region or ORFs of the TMV, and horizontal lines represent RNA. 183 kDa and 126 kDa replicase protein of ORF is shown by demarcation of the by RT. MP and CP indicates movement protein and coat protein gene sequences of the TMV. Rectangular box having 54 kDa indicates 54 kDa protein translated by the Subgenomic RNA (I1). 5′-Cap structures are represented by small round red color circle (o). Origin of assembly sequence (OAS) is represented by a star.

9.4.3 *REPLICATION MECHANISM OF THE TMV*

Once the virus comes in contact with the plant cell, it has to enter into the cell to develop infection. For establishing the infection; TMV RNA has to be launched virus particle into the host cell. The disassembly of the virus is determined by cell environment of the host cell and host factors. Compared to the extracellular environment; inside the cell, pH, and concentrations of the Ca^{2+} is lower compare to outside; due to these reasons, the concentration of the hydrogen and calcium ions reduced and this leads to the un-stabilization of RNA-CP interaction (Culver, 2002; Shaw, 1999).

After uncoating, viral genomic RNA release into the cytoplasm of the host cell. Once the RNA molecule enters into single-cell and viral RNA act as a functional analog of a poly(A)-bound protein (PABP), after the initiation of the recruitment of 40S ribosomal subunits and/or by enhancing stability of the viral RNA. It interacts with the host translation factor, eIF4F (contains three subunits eIF4A, eIF4G, and eIF4E) and 126 and 183 kDa is translated for RNA synthesis.

Wilson was described in 1984 regarding the uncoating of the viral RNA and co-translational disassembly as the host CP subunits replaced ribosomes during translation of the 126 KDa and 183 kDa replicase proteins. Co-translational disassembly was first time discovered by Wilson *in vitro* translation study.

The methyltransferase and helicase domains are coded by 126 kDa and 183 kDa proteins. According to the Wu et al. (1994), RDRP is encoded by the c-terminal domain of the 183 kDa protein. About 126 and 183 kDa proteins together are known as the replication proteins. The interaction of 5′ cap with 4G initiation factor stimulates translation of the TMV RNA to formation of the replicase proteins 126 and 183 kDa.

Replication complex of the TMV and ToMV is formed from with the help of MP, vRNA, and host proteins on the membrane (Heinlein et al., 1998; Hagiwara et al., 2003; Más and Beachy, 1999; de Castro et al., 2012; Laliberté and Sanfaçon, 2010). Initially, 126 kDa recruits viral RNAs from translation machinery then 126 and 183 kDa and viral RNAs to membrane structures.

Replication complex combined with genomic RNA at 3′ UTR region, and RNA dependent RNA polymerase (RdRp) utilize the viral RNA full length molecule as a template for synthesis of the negative strand of the RNA molecules that is complementary in nature. Unwinding of the resulted dsRNA molecule by the helicase activity of the RdRp, releases template and the single-stranded minus-sense RNA (Buck, 1999).

Initiation of positive strands formation starts with the sub-genomic promoters on the negative strand of RNA were identified by replicase proteins to form the sub-genomic RNA (plus-sense RNA). The amount of plus-sense RNA manufactured for the duration of a TMV infection is greater than the quantity of minus-strand (-) RNA formed (Aoki and Takebe, 1975; Buck, 1999). After that helicase enzyme remove the synthesized genomic and subgenomic RNAs from negative-sense templates. Afterward, the methyltransferase domain of the replicase complex caps new molecules of the genomic RNA, the subgenomic MP mRNA, (Grdzelishvili et al., 2000) and probably subgenomic CP mRNA the (Keith and Frankel-Conrat, 1975; Zimmern, 1975). The positive strands of progeny RNA molecules were formed. The sequences, in 5'-UTR of RNAs, which are required for RNA replication *in vivo* or for synthesis of positive-strand RNA *in vitro*, have been identified. The natural negative-strand RNA acts as template for positive-strand RNA synthesis *in vivo* (Grdzelishvili et al., 2000). Translation of all the viral proteins is translated from the synthesized RNA molecules. The proteins forms will be 126- and 183-kDa TMV replicase proteins, CP and MP. Finally, assembly of complete virus particle. All the progeny RNA assemble into functional virion particles and capable of moving from one cell to another cell with the help of MP.

9.4.4 *LIFE CYCLE OF TOBACCO MOSAIC VIRUS (TMV)*

TMV penetrates and enters into the host cells and completes its replication within infected host cells as mentioned in replication mechanism as mentioned above section. After uncoating of the virion particle; genetic material, i.e., RNA released in cytoplasm. In the cytoplasm; TMV utilize the host machineries for their replication, translation, and the formation of the new virion particles. The new viral nucleic acid is considered to organize the protein subunit around it, resulting in the formation of complete virus particle, the virion. The life cycle of the TMV is mentioned in simplified manner in Figure 9.2.

9.5 TRANSMISSION OF THE VIRUS

Viral disease of plant transmitted and spread by the insect vector and debris containing virus. It can be spread either through mechanical means due to rubbing manually, wind, animals, and implements or by various other biological means like insects, nematodes, fungi, grafting, and through seed. Broadly, transmission of the viruses divided into two categories; first 'vertical transmission' and secondly horizontal transmission. In vertical transmission; the viruses are transmitted from one generation to another generation through seed. But in case of "horizontal transmission" spreads of the virus by the means of the agents like insects, wind, man, and the water. About 88% of plant virus species uses an arthropod vector for their maintenance and survival (Andret-Link and Fuchs, 2005). The remaining 12% vector-transmitted plant viruses use fungi, plasmodiophorids, and nematodes. Genera of the *Virgaviridae* family such as, pomoviruses peculviruses and furoviruses have spread via plasmodiophorids, whereas nematodes are the transmitting agent for the tobraviruses. But in the case of hordeiviruses and goraviruses; they are transmitting or spreading via pollen grain and/or seed.

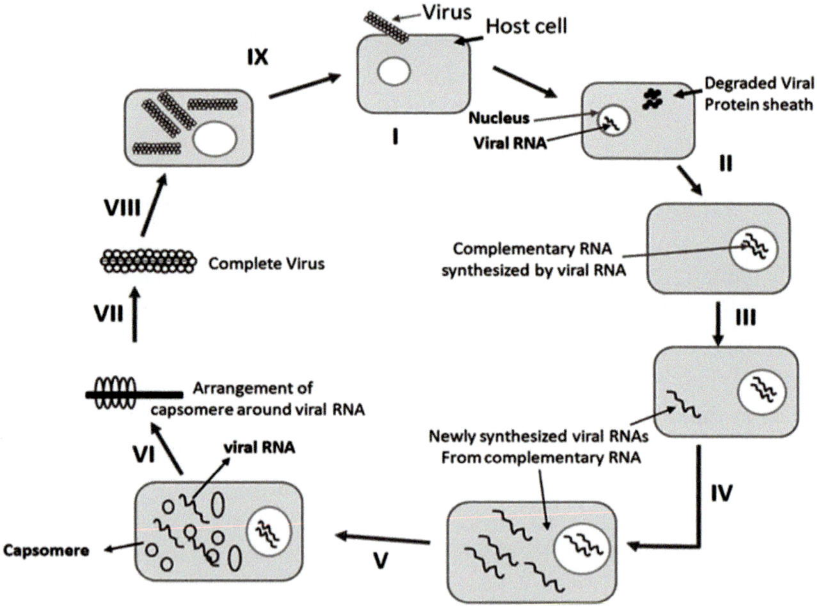

FIGURE 9.2 Life cycle of tobacco mosaic virus: I. virus particle entering to the host cell through mechanically injured tissue; II and III. Viral RNA enters inside the nucleus and synthesized a complimentary copy of RNA from viral RNA; IV. and V. Newly synthesized complementary RNA comes out from the nucleus to cytoplasm and synthesizes different polypeptide chain or proteins responsible for the viral replication, movement, and assembly; VI and VII. Arrangement of the capsomeres around viral-RNA; VIII. Complete virus particle; IX. Host cell containing many viruses.

9.5.1 TRANSMISSION BY VECTORS

TMV is known to transmit mechanically. Harris and Bradley (1973) reported that minor spread through chewing insects and mechanical wound. Broadbent and Fletcher (1963) reported fragments of infected tissue may survive in the soil and act as the inoculums for the infection via roots. TMV is a very persistent nature and survive on cloth and glasshouse structures.

9.5.2 TRANSMISSION THROUGH SEED

Generally, TMV does not spread through seed or pollen. As a matter of fact, viral inoculum is efficiently transmitted when it enters the embryo, and viruses intact to the seed coat may not survive germination when seed coat separates from the seedlings.

9.6 SYMPTOMATOLOGY AND HOST

9.6.1 SYMPTOMATOLOGY

Symptoms may be classified into either local or systemic. When symptoms are confined only to a particular part that develops near the site of entry on leaves, they are called local infections. On the other hand, when the virus causes disease in the whole plant, it is called systemic infection. In contrast, many viruses can also cause infection without showing any visible symptoms. These hosts are called symptomatic, and the viruses are called latent viruses.

Usually, local symptoms appear on the inoculated leaves only. They can be discrete isometric lesions. Infected cells lose chlorophyll and other pigments, which results in symptoms such as chlorotic spots and chlorotic rings. Necrotic spots and necrotic lesions are observed when the infected cells die. In contrast to local infection, systemic symptoms appear on the uninoculated leaves also. They are the most important symptoms because they can affect any part of the plant such as flower, fruit, and petiole. The most important symptom in a systemic infection is mosaic, the appearance of irregular and unfixed pattern of green and chlorotic areas of leaf. Yellowing and necrosis are other patterns of systemic infection (Hull, 2002).

TMV symptom initially starts a translucent greenish color between the veins of young leaves; subsequently developed into "mosaic" or spotted outline of light and dark green areas in the leaves. Initially, symptoms start from the younger leaves and spread to the whole plan. This doesn't result in the death of the plant, however, the plant showed student growth if infection occurs early in the season. Lower leaves are subjected to "mosaic burn," especially during periods of hot and dry weather. The developments of the symptoms are depending on the environmental conditions, genetic makeup of the host plant and the virus strain. TMV too contaminates tomato and leads to the stunted growth, fruit distortion, discoloration of fruit, and ultimately reduction production and productivity.

The intensity of the symptom depends on the environmental factors and plant species. Environmental factors sometimes promote the development of the symptoms while other conditions mask or hide symptoms; for example, TMV infects crops like grapes and apples. Symptoms associated with TMV infections are:

- Mosaic pattern of light and dark green (or yellow and green) on the leaves;
- Yellow spotting on leaves;
- Yellow streaking of leaves (especially monocots);
- Distinct yellowing only of veins;
- Stunting;
- Malformation of leaves or growing points.

Sometimes of the above symptoms can also be caused by excesses of mineral, insect feeding, high temperature, herbicides, growth regulators and mineral deficiencies. Symptoms are one of the parameters for the identification of the TMV diseases but need to be authenticated by molecular methods also.

9.6.2 HOST

9.6.2.1 DIAGNOSTIC HOST

Nicotiana tabacum cvs. Samsun, Burley, White Burley and Xanthi symptoms appear as vein clearing in young leaves appear first later on mosaic, distortion, and blistering. Virus moves on the upper leaves after 3–4 days post-inoculation. There is no symptoms shown by the inoculated leaves except faint chlorotic lesions and that is overcome by the application of the nitrogenous fertilizer.

Chenopodium amaranticolor, C. quinoa, and *Phaseolus vulgaris* cv. Pinto, developed necrotic lesions at inoculated and infection sites. It doesn't show any systemic symptoms below the temperature 28°C. *N. clevelandii* and *N. benthamiana* showed initially a local necrosis subsequently development of the systemic necrosis lesion; ultimately death of the plant.

9.6.2.2 PROPAGATION HOST

TMV is propagated in the *N. tabacum*, cvs. Turkish, Turkish Samsun, Samsun or Xanthi (nn).

9.6.2.3 ASSAY HOST

Local lesion assays are most frequently performed with *N. glutinosa*, *N. tabacum* cvs. Xanthi nc, Xanthi NN, Samsun NN, *Phaseolus vulgaris* cv Pinto, *Chenopodium amaranticolor* or *C. quinoa*.

9.7 MANAGEMENT OF TMV

Management of the TMV infecting the crops may be managed by adopting different practices.

9.7.1 CULTURAL PRACTICES

One of the universal methods to control TMV is sanitation that includes removal of all weeds as this is a reservoir of TMV, crop debris, infected plants. Apart from these tools are disinfects with the disinfectants. To avoid the virus infection; propagation of the plants via seed rather than vegetative and use of the virus free seed and planting materials. To restrict the virus spread; crop rotation must be practiced to avoid infected soil/ seedbeds for a minimum of 2 years. In the recent past, genetic engineering approaches revolutionized the controlling of the TMV application that is elaborated in subsequent heading.

9.7.2 RESISTANCE GENE TO TOBAMOVIRUSES

9.7.2.1 THE N RESISTANCE GENE

A number of genes have been known in a variety of hosts for conferring resistance to tobamoviruses. According to Okada (1999); some of these identified resistance gene have been proven important in crop protection. Among the resistance genes, the *N* gene is the more vigorously studied and introduced into numerous cultivars of tobacco from *N. glutinosa* (Dunigan et al., 1987). Identified dominant resistance genes are having capacity to counter to the almost the entire tobamoviruses with a hypersensitive response (HR). HR is protection of the healthy from infected cell through the mechanism of the programmed cell death. HR is responsible for preventing the virus spread from a place of infection to the remaining part of the plant (Weststeijn, 1981), and it was monitored by TMV. GFP fusion to infect *NN* genotype tobacco plants (Murphy et al., 2001; Wright et al., 2000).

Ross (1961) reported that initiation of the HR by TMV as the result of the induction of SAR; an improved level of resistance for a broad range of infectious agents. For the initiation of SAR; SA plays a vital role (Malamy et al., 1990; Métraux et al., 1990) for regulation of the signal transduction pathway; results into induction of pathogenesis-related (PR) proteins, and finally it add to resistant to bacterial and fungal pathogens (Carr and Klessig, 1989). SA too elicits tolerant to viruses by the altering movement and replication of viruses may be due to rising the levels of host RdRp (Gilliland et al., 2003; Murphy and Carr, 2002).

Whitham et al. (1994) were isolated the *N* gene from tobacco using transposon tagging. The *N* gene having a full-length protein with leucine-rich repeat (LRR) domains, toll interleukin receptor-like (TIR) and nucleotide-binding site (NBS). The NBS and TIR domains engaged in signaling whereas LRR domain participated in ligand binding (Erickson et al., 1999a). After the alternative splicing of the *N* gene; resulted in full length N protein (132 kDa) or a truncated protein of 75 kDa, called Ntr. The Ntr contains TIR and NBS domains and only a portion of the LRR domain. Marathe et al. (2002) reported that both gene products, i.e., N protein and Ntr protein are responsible for the resistance to TMV.

N gene-mediated HR response triggered through 50-kDa protein product from helicase domain of the 126-kDa replicase protein (Erickson et al., 1999b). On the other hand, it is measured not likely that of Ntr or N proteins interact directly with the 126-kDa protein (Marathe et al., 2002).

9.7.2.2 RESISTANCE GENES THAT ARE ELICITED BY THE TOBAMOVIRAL CP

Nicotiana sylvestris contains *N'* gene; a dominant resistance gene; responsible for triggering the HR when it is infected with some TMV strains (Culver and Dawson, 1989; Culver et al., 1994). Site-directed mutagenesis of CP revealed the importance of the interaction of the CP residue and host factors (Taraporewala and Culver, 1996, 1997). Culver in 2002; concluded his finding; CP monomer or small aggregate act as the elicitor for HR. For example, CP gene of the tobamovirus elicits for HR in pepper; containing the *L* resistance genes, a few of that may be counteracting by a number of strains of the tobamovirus pepper mild mottle virus with appropriate mutations in their *CP* genes (Dardick et al., 1999; Berzal-Herranz et al., 1995; de la Cruz et al., 1997). CP too governs the elicitation of the tobamovirus induced HR in brinjal (*Solanum melongena*) (Dardick et al., 1999; Dardick and Culver, 1997).

9.7.2.3 RESISTANCE GENE IN TOMATO

Resistance's genes (*Tm-1*, *Tm-2*, and *Tm-22)* present in the tomatoes are utilized to protect from TMV in commercial cultivars of tomato (Fraser, 1985; Hull, 2002). *Tm-2* and *Tm-22* are promotes to express a HR not in favor of tomato mosaic virus (ToMV). The *Tm-1* gene supports a scanty or no virus multiplication in homozygous plant.

The study of a *Tm-1* resistance breaking mutant, ToMV Lta1, Meshi et al. (1988) showed that mutations mapped to the 126-kDa replicase protein sequence. The results suggest that either the *Tm-1* gene encodes an inhibitor of replication or it encodes a defective host factor needed for 126-kDa protein function. Weber and Pfitzner (1998) reported that movement protein (MP) can act as the elicitor for both the *Tm-2* and the *Tm-22* genes.

9.7.2.4 GENETICALLY ENGINEERED RESISTANCE

Cross protection is the protection of a plant from a severe virus strain by using mild strain of the closely related virus, for example, protection of severe strain of tobamovirus infection cross protected from mild strain tobamovirus by interference with uncoating (Fraser, 1998).

Whitham et al. (1996) were isolated natural resistance *N* gene and successfully transferred into tomato. Similarly, Bendahmane et al. (1999) and Liu et al. (2002) were isolated natural resistance *N* gene and successfully transferred *N. benthamiana.* The response to TMV infection in these transgenic plants is either slower (in tomato: Whitham et al., 1996), or in the case of some of the *N*-transgenic *N. benthamiana* lines the resistance does not work well against wild-type TMV (Peart et al., 2002).

9.8 CONCLUSION

Green fluorescent gene becomes one of the most important reporter proteins to study the cell biological studies. The understanding of TMV replication, accumulation, and spread were widening in the recent past. Advancement of the technologies in the field of labeling of the various components of the viruses, hardware, and software of the imaging microscope will help in the uncovering the various steps of the involved infection process of the virus.

KEYWORDS

- **coat protein**
- **genome organization**
- **movement protein**
- **nucleotide-binding site**
- **open reading frame**
- **replications**
- **tobacco mosaic virus**

REFERENCES

Andret-Link, P., & Fuchs, M., (2005). Transmission specificity of plant viruses by vectors. Journal of Plant Pathology, 87, 3.

Aoki, S., & Takebe, I., (1975). Replication of tobacco mosaic virus RNA in tobacco mesophyll protoplasts *in vitro*. *Virology*, *65*, 343–345.

Arslan, D., Legendre, M., Seltzer, V., Abergel, C., & Claverie, J. M., (2011). Distant *Mimivirus* relative with a larger genome highlights the fundamental features of Megaviridae. *Proc. Natl. Acad. Sci. USA, 108*(42), 17486–1749.

Beachy, R. N., & Zaitlin, M., (1977). Characterization and *in vitro* translation of the RNAs from less than- full-length, virus-related, nucleoprotein rods present in tobacco mosaic virus preparations. *Virology, 81*, 160–169.

Beachy, R. N., (1999). Coat-protein-mediated resistance to tobacco mosaic virus: Discovery mechanisms and exploitation. *Phil. Trans. R. Soc. Lond., 354*, 659–664.

Beier, H., Barciszewska, M., Krupp, G., Mitnacht, R., & Gross, H. J., (1984). UAG readthrough during TMV RNA translation-isolation and sequence of two transfer RNAsTYR with suppressor activity from tobacco plants. *EMBO J.*, *3*, 351–356.

Bendahmane, A., Kanyuka, K., & Baulcombe, D. C., (1999). The *Rx* gene from potato controls separate virus resistance and cell death responses. *Plant Cell, 11*, 781–791.

Berzal-Herranz, A., De La Cruz, A., Tenllado, F., Diaz-Ruiz, J. R., Lopez, L., Sanz, A. I., Vaquero, C., et al., (1995). The *Capsicum* L3 gene-mediated resistance against the tobamoviruses is elicited by the coat protein. *Virology, 209*, 498–505.

Broadbent, L., & Fletcher, (1963). The epidemiology of tomato mosaic. *Annals of Applied Biology, 52,* 233.

Broadbent, L., (1965). The epidemiology of tomato mosaic. VIII. Virus infection through tomato roots. *Annals of Applied Biology, 55*, 57.

Buck, K. W., (1996). Comparison of the replication of positive-stranded RNA viruses of plants and animals. *Adv. Virus Res., 47*, 159–251.

Buck, K. W., (1999). Replication of tobacco mosaic virus RNA. *Phil. Trans. R. Soc. Lond., 354*, 613–627.

Carr, J. P., & Klessig, D. F., (1989). The pathogenesis-related proteins of plants. In: Setlow, J. K., (ed.), *Genetic Engineering: Principles and Methods* (Vol. 11, pp. 65–100).

Citovsky, V., Wong, M. L., Shaw, A., Prasad, B. V. V., & Zambryski, P., (1992). Visualization and characterization of tobacco mosaic virus movement protein binding to single-stranded nucleic acids. *Plant Cell, 4,* 397–411.

Culver, J. N., & Dawson, W. O., (1989). Tobacco mosaic virus coat protein: An elicitor of the hypersensitive reaction but not required for the development of mosaic symptoms in *Nicotiana sylvestris*. *Virology, 173*, 755–758.

Culver, J. N., (2002). Tobacco mosaic virus assembly and disassembly: Determinants in pathogenicity and resistance. *Annu. Rev. of Phytopathol., 40*, 287–308.

Culver, J. N., Stubbs, G., & Dawson, W. O., (1994). Structure-function relationship between tobacco mosaic virus coat protein and hypersensitivity in *Nicotiana sylvestris*. *J. Mol. Biol., 242*, 130–138.

Dardick, C. D., & Culver, J. N., (1997). Tobamovirus coat proteins: Elicitors of the hypersensitive response in *Solanum melongena* (eggplant). *Mol. Plant-Microbe Interact., 10*, 776–778.

Dardick, C. D., Taraporewala, Z. F., Lu, B., & Culver, J. N., (1999). Comparison of tobamovirus coat protein structural features that affect elicitor activity in pepper, eggplant and tobacco. *Mol. Plant-Microbe. Interact., 12*, 247–251.

De Castro, I. F., Volonté, L., & Risco, C., (2012). Virus factories: Biogenesis and structural design. *Cell Microbiol., 15*, 24–34.

De La Cruz, A., Lopez, L., Tenllado, F., Diaz, R. J. R., Sanz, A. I., Vaquero, C., Serra, M. T., & Garcia-Luque, I., (1997). The coat protein is required for the elicitation of the Capsicum L2 gene-mediated resistance against the tobamoviruses. *Mol. Plant Microbe. Interact., 10*, 107–113.

Dorokhov, Y. L., Ivanov, P. A., Novikov, V. K., Agranovsky, A. A., Morozov, S. Y., Efimov, V. A., Casper, R., & Atabekov, J. G., (1994). Complete nucleotide sequence and genome organization of a tobamovirus infecting *Cruciferae* plants. *FEBS Lett., 350*, 5–8.

Dreher, T. W., (1999). Function of 3' untranslated regions of positive-stranded RNA viral genomes. *Annu. Rev. Phytopathol., 37*, 151–174.

Dunigan, D. D., Golemboski, D. B., & Zaitlin, M., (1987). Analysis of the *N*-gene of *Nicotiana. CIBA Found. Symp., 133*, 120–135.

Erickson, F. L., Dinesh-Kumar, S. P., Holzberg, S., Ustach, C. V., Dutton, M., Handley, V., Corr, C., & Baker, B., (1999a). Interactions between tobacco mosaic virus and the tobacco *N* gene. *Phil. Trans. R. Soc. Lond. B, 354*, 653–658.

Erickson, F. L., Holzberg, S., Calderon-Urrea, A., Handley, V., Axtell, M., Corr, C., & Baker, B., (1999b). The helicase domain of the TMV replicase proteins induces the *N*-mediated defense response in tobacco. *Plant J., 18*, 67–75.

Fedorkin, O. N., Denisenko, O. N., Sitkov, A. S., Zelenina, D. A., Lukashova, L. I., Morozov, S. Y., & Atabekov, J. G., (1995). The tomato mosaic virus small gene product forms a stable complex with translation elongation factor EF-1alpha. *Doklady Akademii Nauk, 343*, 703, 704.

Fraenkel-Conrat, H., & Williams, R. C., (1955). Reconstitution of active tobacco mosaic virus from its inactive protein and nucleic acid components. Proc. Natl. Acad. Sci. U.S.A., 41, 690–698.

Fraser, R. S. S., (1985). Genetics of host resistance to viruses and of virulence. In: *Mechanisms of Resistance to Plant Disease* (pp. 62–79). Martinus Nijhoff/W. Junk, Dordrecht.

Gilliland, A., Singh, D. P., Hayward, J. M., Moore, C. A., Murphy, A. M., York, C. J., Slator, J., & Carr, J. P., (2003). Genetic modification of alternative respiration has differential effects on antimycin A *versus* salicylic acid-induced resistance to tobacco mosaic virus. *Plant Physiol., 132*, 1518–1528.

Grdzelishvili, V. Z., Chapman, S. N., Dawson, W. O., & Lewandowski, D. J., (2000). Mapping of the tobacco mosaic virus movement protein and coat protein subgenomic RNA promoters *in vivo. Virology, 275*, 177–192.

Hagiwara, Y., Komoda, K., Yamanaka, T., Tamai, A., Meshi, T., Funada, R., Tsuchiya, T., et al., (2003). Subcellular localization of host and viral proteins associated with tobamovirus RNA replication. *EMBO J., 22*, 344–353.

Harris, K. R., & Bradley, R. H., (1973). Importance of leaf hairs in the transmission of tobacco mosaic virus by aphids. Virology, 52, 295–300.

Heinlein, M., Padgett, H. S., Gens, J. S., Pickard, B. G., Casper, S. J., Epel, B. L., & Beachy, R. N., (1998). Changing patterns of localization of the tobacco mosaic virus movement protein and replicase to the endoplasmic reticulum and microtubules during infection. *Plant Cell, 10*, 1107–1120.

Hull, R., (2002). *Matthews' Plant Virology* (4th edn.). Academic Press NY.

Keith, J., & Fraenkel-Conrat, H., (1975). Tobacco mosaic virus RNA carries 5′-terminal triphosphorylated guanosine blocked by 5′-linked 7-methylguanosine. *FEBS Letters, 57*(1), 31–33.

Koh, H. K., Song, E. K., Lee, S. Y., Park, Y. I., & Park, W. M., (1992). Nucleotide sequence of cDNA of the tobacco mosaic virus RNA isolated in Korea. Nucleic Acids Research, 20(20), 5474.

Koonin, E. V., & Dolja, V. V., (1993). Evolution and taxonomy of positive-strand RNA viruses: Implications of comparative analysis of amino acid sequences. *Crit. Rev. Biochem. Mol. Biol., 28*, 375–430.

Kozak, M., (2002). Pushing the limits of the scanning mechanism for initiation of translation. *Gene., 299*, 1–34.

Laliberté, J. F., & Sanfaçon, H., (2010). Cellular remodeling during plant virus infection. *Annu. Rev. Phytopathol., 48*, 69–91.

Liu, Y. L., Schiff, M., Marathe, R., & Dinesh-Kumar, S. P., (2002). Tobacco *Rar1*, *EDS1* and *NPR1/NIM1*like genes are required for *N*-mediated resistance to tobacco mosaic virus. *Plant J., 30*, 415–429.

Malamy, J., Carr, J. P., Klessig, D. F., & Raskin, I., (1990). Salicylic acid: A likely endogenous signal in the resistance response of tobacco to viral infection. *Science, 250*, 1002–1004.

Mandahar, C. L., (2006). *Multiplication of RNA Plant Viruses* (1st edn., p. 1). Springer, Printed in the Netherlands, Chapter 1.

Marathe, R., Anandalakshmi, R., Liu, Y., & Dinesh-Kumar, S. P., (2002). The tobacco mosaic virus resistance gene, N. *Mol. Plant Pathol., 3*, 167–172.

Más, P., & Beachy, R. N., (1999). Replication of tobacco mosaic virus on endoplasmic reticulum and role of the cytoskeleton and virus movement protein in intracellular distribution of viral RNA. *J. Cell Biol., 147*, 945–958.

Matthieu, L., Bartoli, J., Shmakova, L., Sandra, J., Labadie, K., Adrait, A., Magali, L., et al., (2014). Thirty-thousand-year-old distant relative of giant icosahedral DNA viruses with a *Pandoravirus* morphology. *Proc. Natl. Acad. Sci. USA., 111*(11), 4274–4279.

Métraux, J. P., Signer, H., Ryals, J., Ward, E., Wyssbenz, M., Gaudin, J., Raschdorf, K., et al., (1990). Increase in salicylic acid at the onset of systemic acquired resistance in cucumber. *Science, 250,* 1004–1006.

Morozov, S. Y., Denisenko, O. N., Zelenina, D. A., Fedorkin, O. N., Solovyev, A. G., Maiss, E., Casper, R., & Atabekov, J. G., (1993). A novel open reading frame in tobacco mosaic virus genome coding for a putative small, positively charged protein. *Biochimie., 75*, 659–665.

Murphy, A. M., & Carr, J. P., (2002). Salicylic acid has cell-specific effects on tobacco mosaic virus replication and cell-to-cell movement. *Plant Physiol., 128*, 552–563.

Murphy, A. M., Gilliland, A., Wong, C. E., West, J., Singh, D. P., & Carr, J. P., (2001). Induced resistance to viruses. *Eur. J. Plant Pathol., 107*, 121–128.

Nadège, P., Matthieu, L., Gabriel, D., Yohann, C., Olivier, P., Magali, L., Arslan, D., et al., (2013). *Pandoraviruses*: Amoeba viruses with genomes Up to 2.5 Mb reaching that of parasitic eukaryotes. *Sci., 341,* 281–286.

Nozu, J., Okada, J., & Ohno, T., (1969). Demonstration of the universality of the genetic code *in vivo* by comparison of the coat proteins synthesized in different plants by tobacco mosaic virus RNA. *Genetics, 63*, 1189–1195.

Okada, Y., (1999). Historical overview of research on the tobacco mosaic virus genome: Genome organization, infectivity and gene manipulation. *Phil. Trans. R. Soc. Lond. B., 354*, 569–582.

Peart, J. R., Cook, G., Feys, B. J., Parker, J. E., & Baulcombe, D. C., (2002). An *EDS1* orthologue is required for *N*-mediated resistance against tobacco mosaic virus. *Plant J., 29*, 569–579.

Ross, A. F., (1961a). Localized acquired resistance to plant virus infection in hypersensitive hosts. *Virology, 14*, 329–339.

Ross, A. F., (1961b). Systemic acquired resistance induced by localized virus infections in plants. *Virology, 14*, 340–358.

Shaw, J. G., (1999). Tobacco mosaic virus and the study of early events in virus infections. *Proc. Roy. Soc. Lond. B., 354*, 603–611.

Siegel, A., & Wildman, S. G., (1954). Some natural relationships among strains of tobacco mosaic virus. *Phytopathology, 44, 277*–282.

Skuzeski, J. M., Nichols, L. M., Gesteland, R. F., & Atkins, J. F., (1991). The signal for a leaky UAG stop codon in several plant-viruses includes the two downstream codons. *J. Mol. Biol., 218*, 365–373.

Sulzinski, M. A., Gabard, K. A., Palukaitis, P., & Zaitlin, M., (1985). Replication of tobacco mosaic virus. VIII. Characterization of a third subgenomic TMV RNA. *Virology, 145*, 132–140.

Taraporewala, Z. F., & Culver, J. N., (1996). Identification of an elicitor active site within the three-dimensional structure of the tobacco mosaic tobamovirus coat protein. *Plant Cell, 8*, 169–178.

Taraporewala, Z. F., & Culver, J. N., (1997). Structural and functional conservation of the tobamovirus coat protein elicitor active site. *Mol. Plant-Microbe Interact., 10*, 597–604.

Weber, H., & Pfitzner, A. J. P., (1998). Tm-22 resistance in tomato requires recognition of the carboxy terminus of the movement protein of tomato mosaic virus. *Mol. Plant-Microbe Interact., 11*, 498–503.

Weststeijn, E. A., (1981). Lesion growth and virus localization in leaves of *Nicotiana tabacum* cv. Xanthi nc. After inoculation with tobacco mosaic virus and incubation alternately at 22°C and 32°C. *Physiol. Plant Pathol., 18*, 357–368.

Whitham, S., Dinesh-Kumar, S. P., Choi, D., Hehl, R., Corr, C., & Baker, B., (1994). The product of the tobacco mosaic virus resistance gene *N*: Similarity to toll and the interleukin-1 receptor. *Cell, 78*, 1101–1115.

Whitham, S., McCormick, S., & Baker, B., (1996). The *N* gene of tobacco confers resistance to tobacco mosaic virus in transgenic tomato. *Proc. Natl. Acad. Sci. USA, 93*, 8776–8781.

Wilson, T. M. A., (1984). Co-translational disassembly of tobacco mosaic virus *in vitro*. *Virology, 137*, 255–265.

Wright, K. M., Duncan, G. H., Pradel, K. S., Carr, F., Wood, S., Oparka, K. J., & Santa, C. S., (2000). Analysis of the *N* gene hypersensitive response induced by a fluorescently tagged tobacco mosaic virus. *Plant Physiol., 123*, 1375–1385.

Wu, X., Xu, Z., & Shaw, J. G., (1994). Uncoating of tobacco mosaic virus RNA in protoplasts. *Virology, 200,* 256–262.

Zimmern, D., (1975). The 5′ end group of tobacco mosaic virus is m7G5′ppp5′Gp. *Nucl. Acids Res., 2*, 1189–1201.

CHAPTER 10

MOLECULAR APPROACHES TO CONTROL THE ROOT-KNOT NEMATODE IN TOMATO

RIMA KUMARI,[1] PANKAJ KUMAR,[1] DAN SINGH JAKHAR,[2] and ARUN KUMAR[3]

[1]*Department of Agricultural Biotechnology and Molecular Biology, Dr. Rajendra Prasad Central Agricultural University, Pusa (Samastipur) – 848125, Bihar, India, E-mail: rimakumari1989@gmail.com (R. Kumari)*

[2]*Department of Genetics and Plant Breeding, Institute of Agricultural Sciences, Banaras Hindu University, Varanasi, Uttar Pradesh, India*

[3]*Department of Agronomy, Bihar Agricultural University, Sabour, Bihar, India*

ABSTRACT

Tomato (*Solanum lycopersicum* L.) is one of the most common horticultural crops worldwide. It is a good source of vitamins A and C widely grown around the world. However, root-knot nematodes in most cultivated tomato plants in the subtropical, tropical, and temperate zones have been found to be one of the most damaging nematode groups in the world. To agricultural scientists and the farmer population, the treatment of root-knot nematodes in tomatoes is a major challenge. For agricultural survival and food protection, management of nematodes below the threshold point is therefore quite critical. Thus, a number of management strategies have been applied for this purpose, including cultural, chemical, biological, molecular, and transgenic techniques methods.

10.1 INTRODUCTION

Tomato belongs to the Solanaceae family. This is the second main crop in food after potato production worldwide. It's a storehouse for micronutrients. This provides important nutrients, proteins, and enzymes for human diets. It also contains high levels of lycopene, an antioxidant that reduces many cancers and neurological problems. Growing plant diseases affect the efficiency and quantity of tomato production. One of them (El-Sappah et al., 2019) is a root-knot nematode (*Meloidogyne* spp.). These nematodes are obligatory endoparasites that invade the roots of tomatoes in particular and inflict substantial loss in crop yield. The presence of galls or root-knots varies from contaminated tomato plants. The signs indicate the poor yield of berries, stunted development, as well as susceptibility to wilting and other pathogens (Williamson, 1994). The management of nematode root-knot in tomatoes is considered a major challenge for agricultural scientists and the farmer population. For food protection and sustainable agriculture, control of root-core nematodes below the threshold is crucial. For this reason, many conceptual approaches were used, such as economic, mechanical, biochemical, molecular, and transgenic techniques. Taking all the above into consideration, the present book chapter is undertaken to study the molecular approaches for controlling the root-knot nematodes (RKNs) in tomato crops.

10.2 MANAGEMENT OF ROOT-KNOT NEMATODES (RKNS) IN TOMATOES

Tomatoes (*Solanum lycopersicum* L.) are the only vegetables in the world that have their special place. A broad variety of biotic pressures are faced by the vegetative and reproductive processes. The nematodes are susceptible to causing the most heavy and common stress of biotic stress (Khan et al., 2012). Such nematodes not only affect crop production, they also cause tomatoes more prone to be targeted by bacteria and fungi (Zhou et al., 2016). Prevention of RKNs in tomato crops (RKNs) requires numerous methods, including seed rotation, immune varieties and chemical-nematicidal treatment (Khan et al., 2014). Plant resistance to root-knot nematodes is unstable and leads often to lower annual yields. Their yield losses are considerable for all vegetables, particularly tomatoes, which can range between 10 and 35% according to their infection severity (Radwan et al., 2012). Several management strategies may be used for RKNs safety, including traditional methods such as chemical regulation of infested regions. Nevertheless, the application of substances such as these may have detrimental environmental consequences. Furthermore, their longer use made them impractical usage or was forbidden, which implied an urgent need for more effective and secure alternatives (Zukerman and Esnard, 1994). Their use was restricted or fully barred. The most promising strategy is biological control. A root-knot nematode (*Meloidogyne incognita*) which infects tomatoes is regulated by the nematicidal action of the biocontrol products, namely, *Bacillus megaterium*, *Trichoderma harzianum* and *Ascophyllum Nodosum*. Biological control is therefore, considerably higher than other controls. In the root zone of tomato plants, root knots, and root structures, the handling of soils by actinomycetes often has a noticeable impact on cultural species as well as notable improvements in the community of nematodes. By treatment with *Streptomyces*, the tomato plant's disease index is significantly reduced. The index of disease in tomatoes has been reported to decrease by 37% in many scientists when treated with *Streptomyces* sp. (Ma et al., 2017). In addition, when nematodes feed on bacteria present in the root zone of tomato crops, 14% of nematodes population declines. The fresh shoot and root weight thus increased dramatically. The cultivable root knots microflora is also significantly altered. In addition, the growing number of populations of nematicidal bacteria as well as PGPR is significantly increased, while the number of plant pathogenic bacteria is decreasing rapidly. Due to the activation of systemic resistance and defensive mechanism in tomato plants against infection with nematodes, the root-knot disease significantly decreases (Ma et al., 2017). The use of rhizobacteria for the growth of plants belongs to *Pseudomonas* spp. and *Bacillus* spp. The effective management of root-knot nematode infestation of tomato crops is also very helpful. By inducing systemic resistance against pathogens, PGPR (plant growth-promoting rhizobacteria) protects tomato plants (Molinari and Baser, 2010). A very different range of microbes with different physiological requirements is more favorable than the use of a single biocontrol agent, for example, *Piriformospora indicia*. It is an endophytic fungus that effectively suppresses the root-knot of nematodic infections in conjunction with two plants that stimulate the development of rhizobacteria (*Bacillus pumilus* and *Pseudomonas fluorescens*). Cyanobacteria are a simple microorganism group of around 40 toxic organisms. It is used for the function of biocontrol. The *Microcoleus* and *Oscillatoria* endospores were discovered because of their potential to destroy the nematodes. They inhibited and eventually destroyed the hatching of young people from the 2nd level. It can thus be utilized as the most powerful biocontrol agent (Khan et al., 2007).

In addition, several other management approaches, such as cultural, molecular, and transgenic, were used to control nematode root-knot in tomatoes.

10.3 CULTURAL CONTROL

In the event of cultural influence, tolerant crops, cover crop (Ali, 2017), solar (Katan, 1981) surface, flooding (Brodie and Murphy, 1975), seed rotation, and a mixture of these are becoming attractive management strategies for farmers (Westphal, 2011; Dababat et al., 2015). Resistant diversity is one of the most powerful management mechanisms for root node nematodes (Ferraz and Mendes, 1992). The insensitive type prevents or inhibits nematode development, according to Jenkins (1960). For example, infested vegetable plants. A non-host crop can be necessary for *M. hapla* to be produced. However, this lowers farmers' short-term profits. In contrast, rotation of crops is very successful if farmers use a high economic return for a non-host alternative crop.

In comparison, cover crops can often be planted beyond the usual crop area. When a covering crop is not a host to the Nematodes, the existence of cover crops, nematodes cannot relocate to another area as these nematodes will switch from each other quite shortly (Gill and Mcsorley, 2011). Cowpea (*Vigna unguiculata*), sorghum Sudan grass, sunshine hemp (*Crotalaria juncea*), are several sources for cover plants. These include cows. Moreover, a number of legumes are also one of the main cover crops, as nitrogen can be delivered to subsequent crops (Hartwig and Ammon, 2002). There are also benefits of growing cover crops. It leads to the rise in soil productivity, surface composition and degradation of the land. Insects, nematodes, and other plant pests are also removed (Ralmi et al., 2016).

In fact, another approach to reduce root-knot nematode harm by the soil solarization cycle (Tisserat, 2006). This process typically takes place mid-summer to optimize the benefits of oil heating. First, for 3 weeks, the soils have been coated with plastic tape, which then destroys the nematodes' eggs and gradually decreases the root nude nematodes population. Noling (2009) has suggested that soil solarization is very effective in loamy (heavier) soils instead of sandy soils. The heat transfer to deeper soil horizons raises soils with a strong water-keeping ability. The root-knot-nematodes are also influenced by ground depth (Ralmi et al., 2016).

10.4 CHEMICAL CONTROL

Chemicals are used in this method to treat the RKNs in tomato crops. There are three major categories of compounds, namely, non-fumigant nematicides, fumigant nematicides, and nematicides with multipurpose fumigants (Noling, 2012). It is necessary to uniformly distribute non-fumigant nematicides to the soil in a way that they can easily get in contact with nematodes (Castillo et al., 2017). Setting nematicides within the top 2–5 inches of soil would establish the protection zone for seed germination, transplantation, and new roots (Saikia et al., 2013). Non-fumigant nematicides, however, acquire integrated control of other chemical pests for the management of other weeds, pests, and insects. Noling (2009) conducted an experiment and observed that application of soil fumigants to control the RKNs in tomato crops is more effective than non-fumigants. Additionally, the most effective fumigations occur when the soils are well-drained in seedbed condition and above 55°F at temperatures. These fumigants must be applied 3 weeks before the planting of crops because fumigants are phytotoxic to tomato plants (Noling, 2009). Nowadays, most nematicides are banned because of their high cost and health and environmental issues (Sorribas et al., 2005).

10.5 BIOLOGICAL CONTROL

Some researchers have sought biological control approaches utilizing RKN antagonistic cells. The fungi and bacteria (Bhuiyan et al., 2018) are the most important biological control substances. Nematophageal fungi occur in several various forms. Some fungi use mycelial traps or sticky spores, for example, *Monacrosporium* spp. to attract nematodes and *Arthrobotrys* spp. Some other fungi parasite female eggs and nematode chains, e.g., to *Paecilomyces lilacinus* and *Chlamydosporia Pochonia. Pasteuria penetrans* and certain strains of *Bacillus* are the major bacterial antagonists. *P. penetrans* endospores bind to the juvenile cuticle and then send complexes of penetration that penetrate the nematodes. In both greenhouse and field studies, several nematode antagonists have been identified. Many commercial products are available to control the nematode root-core and some other nematodes based on biocontrol agents. Nevertheless, the production of successful biological control agents remains a severe challenge. The concern is that the vast amounts of biological material required to be utilized in broad regions are not economically produced. In fact, several studies have been performed to eradicate the symptoms of RKNs from biologic antagonistic microorganisms, including fungi and bacteria (Dababat et al., 2015). Sayre (1986) stated that the most effective nematodes for controlling nematodes are several nematodes such as *Nematophthora gynophila*, *Dactylella oviparasitica* and *Paecilomyces lilacinus*, together with a *Pasteuria penetrans* bacterium. Likewise, a *Fusarium oxysporum* strain was found by Martinuz et al. (2012), Fo162 is promising to control *M. incognita* in numerous studies.

10.6 MOLECULAR CONTROL

A linking sequence of molecular markers in related distributions in tomato plant phenotype is used as a marker-assisted search. The creation of the embryo indicates the presence of a genetic marker directly related to the gene of interest. There are several benefits of cultivating modern resistant tomato plants. Two very important benefits occur. The first benefit of molecular genetics is less biologically damaging than chemicals. The second advantage is that it's cheaper. It has been observed that tomato plants are one of the largest plants used in agricultural breeding through molecular markers (Foolad, 2007). In addition, molecular markers linked to the *Mi*-1 gene have demonstrated the rapid resistance screening alleles without nematodes being needed. Consequently, molecular markers are a positive path to progression in the cultivation of tomatoes in conventional breeding techniques.

Host resistance is the most effective and practical method of controlling the RKN disease, however, as other applications, such as the use of nematicides, environmental, and human health risks and high-cost limits for the use of nematicides, are considered to be host resistance to the illness (Ali and Sharma, 2002). Currently, it was confirmed that only the *Mi* gene provides resistance to RKNs, and from its wild relative *L. Peruvicanum*, it was introgressed with *Lycopersicon esculentum* by embryo rescue technique in the early 1940's. This hybrid plant is the source of resistance to both nematode nodes in fresh and refined tomato cultivars currently active. Early breeding was focused on direct experiments to combine *the Mi* gene with nematode resistance. More recently, the related markers Aps-1 and Rex-1 have aided breeders in introducing *Mi* in various conventional tomato cultivars, first with an acid phosphatase isozyme marker (Williamson, 1994). Such and other markers were also beneficial to reliably project *Mi* through the chromosome 6 short arm. *Meloidogyne javanica*, *Meloidogyne incognita* and *Meloidogyne Arenaria* are three forms of root-knots resistance. The Mi-1 gene provides resistance. Nonetheless, *Mi*-1 controlled root node resistance can lose its efficacy when soil temperatures reach 28°C (Wang et al., 2009). Resistance degradation communities may also be affected (Devran and Elekcioglu, 2004).

There are several approaches to track certain proteins, including molecular markers, after discovering proteins in the genomic system linked with disease tolerance (Secgin et al., 2018). The identification of disease resistance genes and the identification of markers associated with these genes can enhance molecular strategies.

10.7 GENETIC ENGINEERING OF TOMATOES FOR ROOT-KNOT NEMATODE (RKN) RESISTANCE

Recent advancement in biotechnology has enabled indigenous and heterologous proteins to be incorporated and transmitted from organism to organism (Goggin et al., 2006). With the advent of the Green Revolution (1960), this cycle is a new era of crop growth. In addition to the production of tolerance to multiple pressures, the genetic manipulation of tomato has shown increased quality and quantity of crop products (Li et al., 2007). During genetic engineering phases, many pests and disease resistant genes were applied to tomato plants (Tai et al., 1999). Throughout this respect, Bt cotton immune to bollworms, weed control herbicide resistance, and other biotic and abiotic tolerances are the most prominent manifestations of this in many societies. There you can find a variety of approaches for genetic modification to inhibit the impact of nematode species in soils below the threshold point on resistance genes in tomato crops. Many transgenic approaches were used in previous years to develop nematode resistance in tomato plants. Natural resources nematode resistance genes have been cloned from different crop species and may be passed to other crop species, e.g., *Mi* tomato gene for resistance to *M. Incognita*, sugar beet *Hs1pro1* (*Beta vulgaris*) vs *H. Schachtii*, *Gpa-2* potato versus *Globodera pallida*, and tomato versus *Hero A*, *G. Rostochiensis* (Fuller et al., 2008; Ali et al., 2017). Overexpression in nematode-resistant tomato plants (Lilley et al., 2012) of specific protease inhibitors (PIs), such as CpTI, cystatin (CpTI) and serine proteases. The selective elimination of essential nematode-effectors from tomatoes via RNA interference (RNAi) method is a further effective approach. Some scientists have recently proposed that the expression of certain genes may enhance tolerance to root-node nematode in tomatoes.

10.8 CONCLUSION

Tomato has family of Solanaceae. This is the second-largest crop in food processing worldwide after potatoes. Tomato is a wealthy micronutrient source. This provides the necessary vitamins, minerals, and antioxidants for the diets of men. It also includes strong lycopene (an antioxidant), which reduces the risk of different cancers and neurological disorders. The tomatoes are host to different forms of root knot nematodes; however, nematode infestation can contribute to significant loss of yield for the plant. Tomatoes (*Solanum lycopersicum* L.), however, are host for specific species of RKNs, and infestation with nematodes may result in significant yield loss for this crop. Root-knot nematode from the genus *Meloidogyne* which includes over 90 species and some of them have separate races. The four species, namely, *M. incognita*, *M. hapla*, *M. javanica,* and *M. arenaria* are considered to be the most economically devastating of the entire world. They are the larvae of biotrophy. More than 2000 plant organisms are contaminated. Nematodes of the root-knot will cause severe damage in tomatoes. The RKN infections worldwide have been faced by most countries with the same problem, especially in tropical, subtropical, and temperate regions. Nematode invasion starts with penetration into the cultivation roots and then migrates to the vascular ring, causing a series of damage to the root and finally galls formation. The transfer of nutrients and water in tomato plants is impaired. However, this form of root system harm done by this pathogen contributes to a balance in nutrient intake and water use, with low yields of tomatoes. Keeping these facts in mind, the present book chapter is undertaken to clarify many management practices like cultural, chemical, biological, molecular, and transgenic technique for controlling the root-knot nematodes in tomatoes.

KEYWORDS

- **Mi gene**
- **plant growth-promoting rhizobacteria**
- **protease inhibitor**
- **RNA interference**
- **root-knot nematode**
- **transgenic technique**

REFERENCES

Ali, M. A., Azeem, F., Abbas, A., Joyia, F. A., Li, H., & Dababat, A. A., (2017). Transgenic strategies for enhancement of nematode resistance in plants. *Front. Plant Sci., 8*, 750. doi: 10.3389/fpls.2017.00750.

Ali, S. S., & Sharma, S. B., (2002). Distribution and importance of plant-parasitic nematodes associated with chickpea in Rajasthan state. *Indian J. Pulses Res., 15*, 57–65.

Bhuiyan, S. A., Garlick, K., Anderson, J. M., et al., (2018). Biological control of root-knot nematode on sugarcane in soil naturally or artificially infested with *Pasteuria penetrans*. *Australas Plant Pathol.,* 45–52. https://doi.org/10.1007/s13313-017-0530-z.

Brodie, B. B., & Murphy, W. S., (1975). Population dynamics of plant nematodes as affected by combinations of fallow and cropping sequence. *J. Nematol., 7*, 91–92.

Castillo, G. X., Ozores-Hampton, M., & Navia, G. P. A., (2017). Effects of fluensulfone combined with soil fumigation on root-knot nematodes and fruit yield of drip-irrigated fresh-market tomatoes. *Crop Prot., 98*, 166–171. https://doi.org/10.1016/j.cropro.2017.03.029.

Dababat, A., Imren, M., Erginbas-Orakci, G., Ashrafi, S., Yavuzaslanoglu, E., & Toktay, H., (2015). The importance and management strategies of cereal cyst nematodes, *Heterodera* spp., in Turkey. *Euphytica., 202*, 173–188.

Devran, Z., & Elekcioglu, I. H., (2004). The screening of F2 plants for the root-knot nematode resistance gene, *Mi* by PCR in tomato. *Turk J. Agric For., 28*(4), 253–257.

El-Sappah, A. H., Islam, M. M., El-awady, H. H., Yan, S., Qi, S., Liu, J., Cheng, G., & Liang, Y., (2019). Tomato natural resistance genes in controlling the root-knot nematode. *Genes., 10*, 925. doi: 10.3390/genes10110925.

Ferraz, S., & Mendes, M. L., (1992). The root knot nematode. *Informacion Agropecuaria., 16*, 43–45.

Foolad, M. R., (2007). Genome mapping and molecular breeding of tomato. *Int. J. Plant Genom., 10*, 1–52.

Fuller, V. L., Lilley, C. J., & Urwin, P. E., (2008). Nematode resistance. *New Phytologist., 180*, 27–44.

Gill, H. K., & Mcsorley, R., (2011). *Cover Crops for Managing Root-Knot Nematodes* (pp. 1–6). University of Florida, IFAS Extension, ENY-063.

Goggin, F. L., Jia, L., Shah, G., Hebert, S., Williamson, V. M., & Ullman, D. E., (2006). Heterologous expression of the *Mi-1.2* gene from tomato confers resistance against nematodes but not aphids in eggplant. *Mol. Plant-Microbe Interact., 19*, 383–388.

Hartwig, N. L., & Ammon, H. U., (2002). Cover crops and living mulches. *Weed Sci., 50*, 688–699.

Katan, J., (1981). Solar heating (solarization) of soil for control of soilborne pests. *Annu. Rev. Phytopathol., 19*, 211–236.

Khan, M. R., Haque, Z., & Kausar, N., (2014). Management of the root-knot nematode *Meloidogyne graminicola* infesting rice in the nursery and crop field by integrating seed priming and soil application treatments of pesticides. *Crop Prot., 63*, 15–25. https://doi.org/10.1016/j.cropro.2014.04.024.

Khan, Z., Kim, Y. H., Kim, S. G., & Kim, H. W., (2007). Observations on the suppression of root-knot nematode (*Meloidogyne Arenaria*) on tomato by incorporation of cyanobacterial powder (*Oscillatoria chlorina*) into potting field soil. *Bioresour. Technol., 98*, 69–73. https://doi.org/10. 1016/j.biortech.2005.11.029.

Khan, Z., Tiyagi, S. A., Mahmood, I., & Rizvi, R., (2012). Effect of N fertilization, organic matter, and biofertilizers on growth and yield of chili in relation to management of plant-parasitic nematodes. *Turk J. Bot., 36*, 73–81.

Li, X. Q., Wei, J. Z., Tan, A., & Aroian, R. V., (2007). Resistance to root-knot nematode in tomato roots expressing a nematicidal *Bacillus thuringiensis* crystal protein. *Plant Biotechnol. J., 5*, 455–464.

Lilley, C. J., Davies, L. J., & Urwin, P. E., (2012). RNA interference in plant-parasitic nematodes: A summary of the current status. *Parasitol., 139*, 630–640.

Ma, Y. Y., Li, Y. L., Lai, H. X., et al., (2017). Effects of two strains of *Streptomyces* on root-zone microbes and nematodes for biocontrol of root-knot nematode disease in tomato. *Appl. Soil Ecol., 112*, 34–41. https://doi.org/10.1016/j.apsoil.2017.01.004.

Martinuz, A., Schouten, A., & Sikora, R. A., (2012). Post-infection development of *Meloidogyne incognita* on tomato treated with the endophytes *Fusarium oxysporum* strain Fo162 and *Rhizobium etli* strain G12. *Biocontrol., 58*, 95–104.

Molinari, S., & Baser, N., (2010). Induction of resistance to root-knot nematodes by SAR elicitors in tomato. *Crop Prot., 29*(11), 1354–1362.

Noling, J. W., (2009). *Nematode Management in Tomatoes, Peppers and Eggplant* (pp. 1–15). University of Florida, IFAS Extension, ENY-032.

Noling, J. W., (2012). *Nematode Management in Carrots* (pp. 1–13). University of Florida, IFAS Extension, ENY-021.

Radwan, M., Farrag, S., Abu-Elamayem, M., & Ahmed, N., (2012). Biological control of the root-knot nematode, *Meloidogyne incognita* on tomato using bioproducts of microbial origin. *Appl. Soil Ecol., 56*, 58–62. https://doi.org/10.1016/j.apsoil.2012.02.008.

Ralmi, N. H. A., Khandaker, M. M., & Mat, N., (2016). Occurrence and control of root-knot nematode in crops: A review. *Aust. J. Crop Sci., 10*(12), 1649–1654.

Saikia, S. K., Tiwari, S., & Pandey, R., (2013). Rhizospheric biological weapons for growth enhancement and *Meloidogyne incognita* management in *Withania somnifera* cv. poshita. *Biol. Control., 65*, 225–234. https://doi.org/10.1016/j.biocontrol.2013.01.014.

Secgin, Z., Arvas, Y. E., Sendawula, S. P., & Kaya, Y., (2018). Selection of root-knot nematode resistance in inbred tomato lines using CAPS molecular markers. *Int J. Life Sci. Biotechnol., 1*(1), 10–16.

Sorribas, F., Ornat, C., Verdejo-Lucas, S., Galeano, M., & Valero, J., (2005). Effectiveness and profitability of the *Mi*-resistant tomatoes to control root-knot nematodes. *Eur. J. Plant Pathol., 111*, 29–38.

Tai, T. H., Dahlbeck, D., Clark, E. T., Gajiwala, P., Pasion, R., Whalen, M. C., Stall, R. E., & Staskawicz, B. J., (1999). Expression of the Bs2 pepper gene confers resistance to bacterial spot disease in tomato. *Proc. Natl. Acad. Sci. USA, 96*, 14153–14158.

Tisserat, N., (2006). *Root-Knot Nematode of Tomato.* Fact sheets tomato-extension plant pathology. Kansas State University, Manhattan. Retrieved from: https://www.plantpath.k-state.edu/extension/ (accessed on 1 March 2021).

Varkey, S., Anith, K. N., Narayana, R., & Aswini, S., (2018). A consortium of rhizobacteria and fungal endophyte suppress the root-knot nematode parasite in tomato. *Rhizosphere, 5*, 38–42. https://doi.org/10.1016/j.rhisph.2017.11.005.

Wang, Y., Bao, Z., Zhu, Y., & Hua, J., (2009). Analysis of temperature modulation of plant defense against biotrophic microbes. *Mol. Plant-Microbe Interact., 22*, 498–506.

Westphal, A., (2011). Sustainable approaches to the management of plant-parasitic nematodes and disease complexes. *J. Nematol., 43*, 122–125.

Williamson, V. M., Ho, J. Y., Wu, F. F., Miller, N., & Kaloshian, I., (1994). A PCR-based marker tightly linked to the nematode resistance gene, *Mi* in tomato. *Theor. Appl. Genet., 87*(7), 757–763.

Zhou, L., Yuen, G., Wang, Y., et al., (2016). Evaluation of bacterial biological control agents for control of root-knot nematode disease on tomato. *Crop Prot., 84*, 8–13. https://doi.org/10.1016/j.cropro.2015.12.009.

Zukerman, B. M., & Esnard, J., (1994). Biological control of plant nematodes current status and hypothesis. *Jpn. J. Nematol., 24*, 1–13.

CHAPTER 11

MOLECULAR APPROACHES FOR THE CONTROL OF FRUIT BORER IN TOMATO

SHIRIN AKHTAR[1] and ABHISHEK NAIK[2]

[1]*Department of Horticulture (Vegetable and Floriculture), Bihar Agricultural University, Sabour, Bhagalpur, Bihar – 813210, India, E-mail: shirin.0410@gmail.com (Shirin Akhtar)*

[2]*Product Development Manager, Nath Bio-Genes (I)Pvt. Ltd., Kalyani, Nadia - 741235, West Bengal, India*

ABSTRACT

Tomato (*Solanum lycopersicum* L.) is one of the most popular and high valued vegetable crops grown throughout the world for fresh market purposes and processing. Tomato fruit borer, *Helicoverpa armigera* (Hübner), is a notorious pest hampering the production and productivity of tomatoes to a massive extent, besides having a large host range including vegetables, agronomic crops, even trees, and many weeds. The approximate loss of crop due to this pest has been reported to as high as 55% by various workers. Chemical control is most common due to convenience of obtaining them and usage, but their excessive and non-judicious use has led to resurgence of minor insect pests, development of insecticidal resistance in insects, devastation of population of natural enemies and non-target species, besides high levels of pesticide residue in harvested produce resulting in different health hazards as well as environmental pollution. Naturally occurring insecticidal proteins and molecules may be a safer alternative. Spray formulations of *Bacillus thuringiensis* (*Bt*) that producing an array of insecticidal crystal proteins (ICP) toxic to lepidopteran insects have been used for killing *Helicoverpa*. The ICP genes of *Bt (viz., Cry1Ac, Cry1Ab, Cry2A, Cry2Ax1)* have been successfully engineered into tomato for resistance against tomato fruit borer. However, large scale cultivation of *Bt* crop brings with it the terror of resistant strain of the insect, hence, other strategies should also be in hand. Proteinase inhibitors in many plant species also offer a good source of resistance to insect pests, and a combination of such proteinase inhibitors has been suggested by researchers for control of tomato fruit borer. RNA interference (RNAi) technology may be used for silencing the β-actin gene of the *Helicoverpa armigera* leading to hampered development and death in larvae. Besides, discovery of novel insecticidal proteins differing in mechanism of action of toxicity has enabled creation of strategy to control the pest without development of resistance in insect. The insect control through transgenic may be durable through incorporation of integrated pest management strategies during cultivation.

11.1 INTRODUCTION

Tomato fruit borer *Helicoverpa armigera* (Hübner) (synonym: *Heliothis armigera* Hubner) is a widespread lepidopteran pest infesting various crops worldwide. This serious pest is also called African bollworm, corn earworm, cotton bollworm, Old World bollworm, gram pod borer, tobacco budworm, etc., and is widely distributed between the 45° North and South latitudes in most parts of Africa, Asia, Oceania, South America and some regions of North America and Europe, while in India all the insect is found to affect crops in every state (EPPO, 2020). The insect is a highly polyphagous species and causes significant damage in crops (Queiroz-Santos et al., 2018). The host range comprises of about 200 plant species, mostly annuals, but also perennial horticultural crops as well as wild plant species. EPPO has listed *Helicoverpa armigera* as an A2 quarantine pest (EPPO, 1981).

11.2 TAXONOMIC CLASSIFICATION, MORPHOLOGY, AND LIFE CYCLE OF INSECT

Tomato fruit borer is a lepidopteran insect of the Noctuidae family. The taxonomic hierarchy of the insect is as follows (CAB International, 1993):

- **Domain:** Eukaryota;
- **Kingdom:** Metazoa;

- **Phylum:** Arthropoda;
- **Subphylum:** Uniramia;
- **Class:** Insecta;
- **Order:** Lepidoptera;
- **Family:** Noctuidae;
- **Genus:** *Helicoverpa*;
- **Species:** *armigera*.

The insect has been described by several authors *viz*. Cayrol (1972); Hardwick (1965); Dominguez Garcia-Tejero (1957) and several others. The eggs are glossy yellowish-white initially, and before hatching, they slowly become dark brown in color, 0.4–0.6 mm in diameter, and generally the adults lay them on flowering plants. On hatching, the color of the first and second instar larvae are yellowish-white to reddish-brown, possessing dark brown to blackish head and prothoracic legs. The fully grown larvae are about 30–40 mm long with brown and mottled head, pale brown legs with black claws and spiracles. The color may be brown to straw yellow to shades of green to pinkish or even reddish-brown or black depending on the host plant with a narrow, dark, median dorsal band and pale underside. The pupae are mahogany brown in color, about 14–18 mm in length, having a smooth surface and round at both anterior and posterior ends along with two tapering parallel spines at the posterior tip. The adults are stout moth having typical noctuid appearance, 14–18 mm in length with a wingspan of 35–40 mm. Color of the moth varies, but generally, the males are greenish-gray, and females are orange-brown in color. There are 7–8 black spots in a line on the margin of the forewings, along with a wide, asymmetrical, transverse brown band. The color of the hindwings is pale straw yellow, and they have a dark brown border with a pale patch and yellowish margins besides strongly marked veins. There is also a dark comma-shaped marking in the middle. Fine hairs cover the antennae.

Like other members of the Noctuidae family, the adult moths are active after sunset, at nightfall. The female moths mate several times, thus laying 300–3000 eggs in their lifetime of about 3 weeks, on stems, leaves, flowers, and young fruits. On hatching, the larvae appear which undergo six stages of development and damage the crop massively. The young caterpillars feed only on leaves; later, when they reach the third instar stage, they feed on flower buds and fruits, boring holes in them in the process. In tomato, the larvae bore into immature, ripening, and ripe fruits and thus cause secondary infection of bacteria and fungi and cause rotting. In summers, the insect develops in 25–40 days, while in cool climate 6–7 months is required, and often 4–5 generations are completed in a year (Figure 11.1) (Broza, 1986; Ragab et al., 2014).

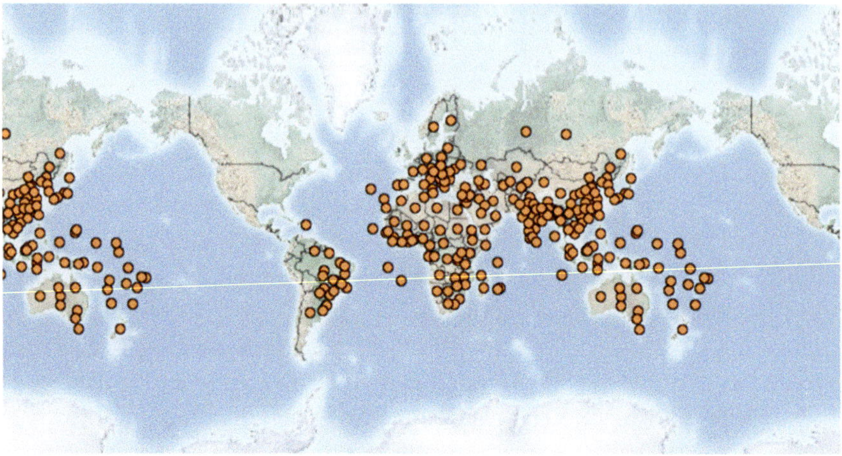

FIGURE 11.1 Distribution of *Helicoverpa armigera*.
Source: Reprinted from with permission from CABI, 2020. Helicoverpa armigera (cotton bollworm), https://www.cabi.org/isc/datasheet/26757. In: Invasive Species Compendium. Wallingford, UK: CAB International. https://www.cabi.org/isc/datasheet/26757

11.3 EXTENT OF HOST AND DAMAGE

Tomato fruit borer is a major pest damaging not only tomatoes, but also an array of diverse other important crops, including food, fodder, oilseeds, and fiber. Its polyphagous nature, rapid reproductive rate, high mobility and adaptability to different ecosystems makes it a very dangerous pest and the damages caused by the crop is very high every year. Apart from tomatoes, the major crops affected are cotton, chickpea, pigeon pea, cowpea, sorghum, okra, peas, and different beans, groundnuts, soybean, tobacco, potato, maize, etc. Besides, various fruit crops, particularly, *Prunus*, and *Citrus* and a wide range of forest trees are also affected.

It is the most serious insect pest of tomato and reports of 40–50% damage in South India and 30% in North India are there (Talekar et al., 2005), while on an average up to 46% loss has been noted (Reddy and Zehr, 2004). Almost all above-ground parts of the plant are affected by the pest from early vegetative phase to fruit maturation and ripe stage (Tripathy et al., 1999; Lal and Lal, 1996). In the case of massive infestation necrotic areas in chlorophyllous leaf tissue, flower drop and bored fruits that become rotten and unfit for consumption have been observed (Jallow et al., 2001).

11.4 STRATEGIES TO CONTROL THE INSECT

The major control measures of this devastating pest are based on broad spectrum insecticides used to control lepidopteran pests. Besides, some cultural practices like crop rotation, use of trap crop like African marigold (*Tagetes erecta* L.) (Srinivasan et al., 1994) and tropical soda apple (*Solanum viarum* Dunal) (AVRDC, 2000, 2001) have been used to some extent. Biological control using egg parasitoids (*T. pretiosum* Tiley and *T. chilonis*) and larval parasitoids (*Campoletis chloridae* Uchida) have been used effectively (Ballal and Singh, 2003; Gupta et al., 2004), but availability of these biocontrol agents on a mass scale is a problem. Sex pheromone traps can be used to trap male adults of the tomato fruit borer, but its efficacy is less due to high polyphagous nature of the insect (Srinivasan, 2013). There are commercial formulations of insecticidal crystal protein (ICP) of *Bacillus thuringiensis* used for spray in the field to control tomato fruit borer. Besides *Helicoverpa armigera* nucleopolyhedrovirus (HaNPV) and neem-based formulations have also been utilized in a limited way to control this notorious pest of tomato. But chemical pesticides have been the most common way to check tomato fruit borer. This pest has developed resistance to several groups of pesticides (Kranthi et al., 2002), which has necessitated the increase in dosage and frequency of their application (Sharma, 2012). The non-judicious use of pesticides has become a threat to the soil, the beneficial fauna, the environment as well as health of the human community as a whole. Resistant varieties seem to be the best option for the management of the insect. The sources of resistance to tomato fruit borer occur only in wild tomato species, especially, *Solanum habrochaites* and *Solanum penellii*, which have small fruits, but introgression of resistance genes from these wild species through conventional breeding results in small fruits which is not desirable trait (Talekdar et al., 2006). Biotechnological interventions have been recently been very effective in control of this notorious insect.

11.5 MOLECULAR APPROACHES FOR CONTROL OF TOMATO FRUIT BORER

Genetic modification through genetic engineering by developing transgenics have a higher scope of expression of plant defense molecules, particularly through introgression of novel genes having insecticidal properties. The ingested transgenic product expressing the transgene specific protein mostly target insect midgut and the pleiotropic membrane thereby disrupting digestion or nutrition (Czapla and Lang, 1990; Murdock et al., 1990; Eisemann et al., 1994; Harper et al., 1998; Hopkins and Harper, 2001). Lesser assimilation of the required nutrients resulted in hampered growth (Williams, 1999; Lopes et al., 2004; Zavala and Baldwin, 2004; Silva et al., 2006).

The first transgenic plant expressing insecticidal gene was produced way back in 1987 in tobacco expressing cowpea trypsin inhibitor (CpTI) and in the same year *Bt* endotoxins were transformed in tomato (Vaeck et al., 1987; Fischhoff et al., 1987; Barton et al., 1987). Ever since, *Bt* toxins have been the focal point for development of insecticidal genes on commercial scale and also prime interest of researchers (Pigott and Ellar, 2007; Bravo et al., 2007). However, reliance on a single insect resistance trait has led to the problem of development of *Bt*-resistance in insects (Heckel et al., 2007). Therefore, current research aims at identification of new insecticidal proteins (Haq et al., 2004; Lynch et al., 2003) including α-amylase inhibitors (Franco et al., 2002; Carlini and Grossi-de-Sa, 2002), vegetative insecticidal proteins (VIP) (Bhalla et al., 2005; Fang et al., 2007), protease inhibitors (PIs) (Maheswaran et al., 2007; Ferry et al., 2005), chitinases (Kabir et al., 2006) and many other proteins targeting insect gut.

11.5.1 ENGINEERING PLANTS WITH INSECTICIDAL CRYSTAL PROTEIN (ICP) GENES FROM BACILLUS THURINGIENSIS

Engineering genes for *Bt* toxins into plants is a method of delivering these toxins to the pests that feed on the plants. From the viewpoint of integrated pest management, it is basically a resistance breeding strategy (Thomas and Wage, 1996), through biotechnological intervention.

Bacillus thuringiensis (*Bt*) is a soil borne gram positive spore forming bacterium. This bacterium produces proteinaceous parasporal crystalline inclusion during sporulation (Schnepf et al., 1998), known as ICP that are toxic to the lepidopteran pests (Kumar et al., 1996). Genetic engineering has successfully been utilized to transform many crop plants introgressed with the ICP genes to cater resistance against the lepidopteran pests (Kumar and Sharma, 1994).

There are two types of *Bt* toxins, *viz*., *Cry*, and *Cyt*. These are further classified into classes *Cry1* to *Cry55* and *Cyt1* to *Cyt2* (van Frankenhuyzen, 2009; Crickmore et al., 1998; Höfte and Whiteley, 1989). The *Cry* toxins are distributed into three larger phylogenetically unrelated families. A three-domain family is the largest *Cry* family, and most of the commercialized *Bt* crops have genes from this family (Tabashnik et al., 2009). The *Cry* toxins of *Bt* crops mostly target lepidopteran insects like *Bombyx mori* (Nishiitsutsuji-Uwo, 1980), *Helicoverpa armigera* (Estela et al., 2004), *Heliothis virescens* (Ryerse et al., 1990; MacIntosh et al., 1981), *Manduca sexta* (Lane et al., 1989; Knight et al., 1994), *Ostrinia nubilalis* (Hua et al., 2001; Li et al., 2004; Siqueira et al., 2004; Tang et al., 1996), *Plutella xylostella* (Wright et al., 1997; Gonzalez-Cabrera et al., 2006), *Sesamia nonagrioides* (Gonzalez-Cabrera et al., 2006), *Spodoptera exigua* (Moar et al., 1995), *Spodoptera frugiperda* (Adamczyk et al., 1998) and *Spodoptera littoralis* (Avisar et al., 2004), while some effect on coleopteran insects have also been observed (Tabashnik et al., 2009).

11.5.1.1 MODE OF ACTION OF BT PROTEIN

When the insects feed on the Bt crops, the crystalline protein is solubilized in the insect midgut (Gill et al., 1992), which is alkaline since a pH of 9.5 is essential for solubilizing of *Bt* protoxins and effectiveness of the *Bt* toxins. These protoxins having a molecular weight of 130 kDa become active by insect gut proteases that cleave from both C- and N-terminus resulting in 43–65 kDa protease-resistant active core (Nagamatsu et al., 1984; Tojo and Aizawa, 1983; Diaz-Mendoza et al., 2007; Rukmini et al., 2000). There are two models explaining the mechanism of *Bt* toxin action. The first is the pore formation model which states that there is binding of the activated toxins with primary receptors in the brush border membrane of the midgut epithelium (Bravo et al., 2007), which is mainly cadherin-like proteins (Vadlamudi et al., 1993, 1995; Jurat-Fuentes and Adang, 2004; Jurat-Fuentes et al., 2004; Nagamatsu et al., 1998), thereafter facilitating proteolytic cleavage of toxin resulting in oligomer formation (Bravo et al., 2004; Nagamatsu et al., 1998). This is followed by interaction of these toxins with secondary receptors in the insect midgut larval membrane which are GPI anchored proteins, aminopeptidases or alkaline phosphatase in nature (Gill et al., 1992; Bravo et al., 2004; Jurat-Fuentes, and Adang, 2004; Knight et al., 1994), and the toxin inserts into membrane and pores are formed (Bravo et al., 2004). The pores disrupt membrane integrity resulting in electrolyte imbalance leading to the death of the insect by septicemia or starvation (Knowles, 1994; Jiménez-Juárez et al., 2007). In insects lacking cadherin-like proteins receptor, there are more receptors involved in the process and the insects are still killed by the Bt toxins (Soberón et al., 2007; Tabashnik et al., 2011). The second model is that of signal transduction pathway. The initial step is similar to the first model but here instead of oligomerization of the toxin and following steps, here on binding to the cadherin receptor a Mg^{++} dependent signal cascade pathway is initiated that involves a guanine nucleotide-binding protein, adenyl cyclase and protein kinase A leading to cell death (Stevens et al., 2012).

11.5.1.2 EXPRESSION OF NATIVE AND SYNTHETIC BT GENES IN TRANSFORMED PLANTS

Transformation of various plant species with native or synthetic *Bt* genes have been attempted with the help of constitutive promoters such as CaMV 35S and have exhibited diverse levels of foreign protein expression. The level of *Cry3A* in potato has been reported to be 0.1% of total soluble protein (Adang et al., 1993), *Cry1Ab* in tobacco has been found to be 0.03% of total soluble protein (Perlak et al., 1991), *Cry1Ab* in rice has been observed to be 0.05% of total soluble protein (Fujimoto et al., 1993), *Cry1Ab* in maize has been reported to be 0.4% of total soluble protein (Koziel et al., 1993), *Cry1Ac* in canola has been noted to be 0.4% of total soluble protein (Stewart et al., 1996) and *Cry1Ac* in tomato has been observed to be up to 0.41% of total soluble protein (Mandaokar et al., 2000). Besides, a very high level of expression of *Cry1Ac* synthetic gene in transgenic rice (3% of the total soluble protein) has been observed (Cheng et al., 1998). McBride et al. (1995) introgressed a native *Cry1Ac* into tobacco chloroplasts, and the transformed plants showed 3–5% of *Cry1Ac* protein, while high levels of expression of *Cry2Aa* protein (up to 2–3% of total soluble protein) has also been observed (Kota et al., 1999).

Cry1Ac has been found to be highly efficient against tomato fruit borer due to its binding ability to different receptors (Bravo et al., 2004; Purcell et al., 2004), though the rate of pore formation is lesser as compared to *Cry1Ab* (Kato et al., 2006). It has broader specificity being class 1 and 3 aminopeptidase N (APN) binding proteins, while *Cry1Aa* and *Cry1Ab* are class 1 APN binding proteins (Pigott and Ellar, 2007). However, overexpression of *Cry1Ac* is toxic to *in vitro* regeneration and development of the transformed plants and hence selection of transgenics with high *Cry1Ac* expression is less documented (Diehn et al., 1996; De Rocher et al., 1998; Mehrotra et al., 2011; Rawat et al., 2011). *Cry1Ab* protein interacts with brush border membrane vesicles by various modes in monomeric or oligomeric forms for formation of the pores (Pardo Lopez et al., 2013; Vachon et al., 2012; Padilla et al., 2006; Kato et al., 2006; Zhang et al., 2005, 2006), It is quite homologous to *Cry1Ac* (Schnepf et al., 1998), besides being non-toxic at the time of in vitro transgenic plant regeneration and development even on expression of high levels of the toxin (Estela et al., 2004; Padilla et al., 2006; Pacheco et al., 2009). The toxicity towards the pest as expressed by the plant has not been up to the expected high level when the native genes in the original form have been used (Perlak et al., 1991). However, truncated version of the gene with modified coding sequence, like removal of potential RNA processing and polyadenylation signals, besides optimizing the codon usage to achieve higher plant expression of the genes in the plants have led to higher expression of the genes in the target plants and these modified genes conferred significant protection against insects in different crops like tomato, eggplant, potato, rice, maize, cotton, etc., (Mandaokar et al., 2000; Kumar et al., 1998; Jansens et al., 1995; Fujimoto et al., 1993; Koziel et al., 1993; Perlak et al., 1990). A number of modifications in the native *cry1Ac* and *cry1Ab* genes have been done for getting stable transcripts (Table 11.1) (De Rocher et al., 1998).

TABLE 11.1 Successful *Bt* Gene Transformation in Tomato for Control of Tomato Fruit Borer *Helicoverpa armigera* Hubner by Various Researchers

***Bt* Gene Used**	**Explant**	**Transformation Mode**	**Salient Results**	**References**
Truncated version of *cry1Ab* gene	Cotyledonary explant of tomato cultivar Pusa early dwarf	*Agrobacterium tumefaciens* carrying binary vector pBIN200	Stable transgenic tomato lines with high levels of expression were obtained in T_4 generation with high resistance to insects	Koul et al. (2014)
Synthetic *cry1Ac* gene	Cotyledonary leaf explant of tomato cultivar Pusa Ruby	*Agrobacterium tumefaciens* strain carrying pBinBt3	Limited field trial of T_1 plants showed high insect resistance	Mandaokar et al. (2000)
cry1Ab gene	Italian genotype LEPA	*Agrobacterium tumefaciens*	Controlled field test showed significantly lesser incidence of insect	Kumar and Kumar (2004)
cry2Ab gene	Cotyledonary explants of tomato cultivar Moneymaker	*Agrobacterium tumefaciens*	100% mortality of *Helicoverpa armigera* larvae feeding on transformed plants	Saker et al. (2011)
Cry2A gene	Hypocotyl explant of tomato cultivar Arka Vikas	*Agrobacterium* strain EHA 105	100% mortality in neonate larvae within 4 days of feeding was observed in the transformants	Hanur et al. (2015)
cry2AX1 gene	Leaf explant of 8-day old tomato seedlings of Pusa Ruby	Biolistic gun method	Two plants were regenerated expressing the gene confirmed through ELISA	Chaithra et al. (2015)

Much work has been done to develop modified tomato with *Cry1Ac* and Mandaokar et al. (2000) used a synthetic *cry1Ac* gene-modified for plant codon usage and possessing G+C content of 47.7% and introgressed into tomato variety Pusa Ruby by *Agrobacterium tumefaciens* strain carrying pBinBt3 to transform the cotyledonary explants. High level of insect resistance was obtained in the putative transformants.

Koul et al. (2014) used a modified truncated version of *cry1Ab* gene for the development of transgenic plants overexpressing *Bt Cry1Ab* toxin in the popular commercial tomato variety Pusa early dwarf (PED) through *Agrobacterium tumefaciens* carrying binary vector pBIN200 harboring modified and truncated 1845 bp *cry1Ab* gene followed by extensive screening and selection of progenies initiating at T_0 transgenic plants. Stable transgenic tomato lines with high levels of expression were obtained in T_4 generation with high resistance to insects and the T_6 there was no retardation in plant growth, development, and fruit yield in comparison to non-transgenic plants.

Hanur et al. (2014) transformed Arka Vikas, a popular tomato cultivar of India, using hypocotyl explant through *Agrobacterium* strain EHA105 mediated transformation. Bt Cry2A gene was introgressed through *Agrobacterium* along with 35S CaMV promoter, OCS terminator, and the selectable nptII marker. The laboratory assays revealed that moderate to high insect resistance in the transgenic *Bt* plants.

Chaithra et al. (2015) transformed tomato cultivar Pusa Ruby by biolistic gun method to introgress *Cry2Ax* gene that was carried by binary vector pCAMBIA2300 with the help of EnCAMV35S promoter assembled on to disarmed *Agrobacterium tumefaciens* strain LBA4404. Leaf explants from 8-day old seedling were used and callus from them were transformed with GUS reporter gene by biolistic gun method. The putative transformants has similar morphology as their non-transgenic counterparts, but showed slow growth, and finally, regeneration of two plants was possible.

Different tomato cultivars have been transformed to express *cry1Ab*, *cry1Ac*, *cry2A*, *cry2Ab* genes giving effective control of tomato fruit borer, but the final outcome of transformation frequency varied as reported as 6% in Pusa Ruby (Vidya et al., 2000), 40% in Micro-Tom (Sun et al., 2006), 18% in Arka Vikas (Hanur et al., 2015) through Agrobacterium-mediated transformation. The success of transformation is determined by the gene integration and its optimum expression in the plant and its successive generations. Biolistic gun method has been tried by some workers and up to 75% transformation has been reported when bombarded twice at a distance of 9 cm at 1100 psi (Chaithra et al., 2015), although low regeneration of transformed plants may be the result of injury following the technique (Sidhu et al., 2014) which has not made this technique much popular.

11.5.2 USE OF PROTEINASE INHIBITORS

One of plant's natural defense mechanism against pests and diseases are the proteinase inhibitors, where plants directly protect themselves by expression of proteinase inhibitors as a response to insect infestation and wounding (Ryan, 1990; Jongsma et al., 1996; Ryan, 1989), a mechanism of direct defense of plants.

Volatile compounds may be released that attract predators (Tamayo et al., 2000) that are a mechanism of indirect defense of the plant (Kessler and Baldwin, 2000). Besides, some volatile compounds like methyl jasmonate may be produced that induce proteinase inhibitors in the neighboring undamaged plants and thereby causing an alarm for the surrounding population for an upcoming insect attack (Farmer and Ryan, 1990). Local and systemic defense are activated via signaling pathways involving jasmonate, systemin, hydrogen peroxide and oligogalacturonic acid (Figure 11.2) (Ferry et al., 2004).

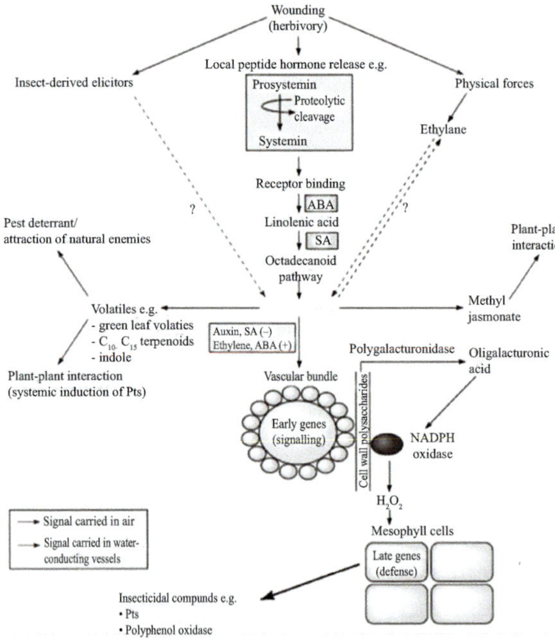

FIGURE 11.2 The general view of plant wounding response and how signaling molecules control it.

Source: Reprinted with permission from Ferry et al. (2004). © Elselvier.

Proteinase inhibitors bind to insect digestive proteases, restricting proteolysis that prevents digestion of protein (Johnston et al., 1991). The larvae are thus starved of protein and essential amino acids necessary for their normal growth and development processes (Broadway, 1995; Bown et al., 1997; Markwick et al., 1998; Zhu-Salzman et al., 2003). There is enhancement of production of proteases to overcome the ingested proteinase inhibitors (Broadway and Duffey, 1986; De Leo et al., 1998), leading to limited bioavailability essential amino acids for protein synthesis, hampering normal development and even lead to death (Broadway, 1995; Broadway and Duffey, 1986).

Despite different works on enhanced plant protection after transforming plants with genes encoding for PIs, there are no commercial application due to chiefly the limitations, viz., level of expression of proteinase inhibitor in the transgenic plant and the capacity of the insect to react to proteinase inhibitor ingestion (Stevens et al., 2012). The use of single transgene producing a proteinase inhibitor targeting one single proteinase or a class of proteinase in the insect midgut is the major limiting factor.

11.5.3 RNA INTERFERENCE (RNAI) TECHNOLOGY

The discovery of dsRNA that has the property of silencing genes has made RNA interference (RNAi) to be efficiently used for regulation of gene expression in plants and animals, and it has been used to inhibit virus replication and spread in the modified plants (Shimizu et al., 2009). But insect control through RNAi has been less explored. Cytochrome P450 monooxygenase has been found to make insects capable of tolerance to otherwise inhibitory concentrations of the cotton metabolite, gossypol. *H. armigera* fed on plants expressing cytochrome P450 dsRNA showed hampered growth, which was more evident in the presence of gossypol (Mao et al., 2007). dsRNA of sufficiently high concentration may be injected or administered orally through artificial diet to the insect, or through soaking. Chickpea-based semi-synthetic diet prepared with 0.1% DEPC with elevated levels of dsRNA has been used to silence β-actin gene of tomato fruit borer up to 93% and weight reduction in the larvae has been reported by Sharath Chandra et al. (2012). However, all these can be done only at lower scale only and application of RNAi technology in large scale or in field is difficult until efficient method of delivering dsRNA is standardized (Mao et al., 2007; Bettencourt et al., 2005).

11.5.4 NOVEL INSECTICIDAL PROTEINS

There is always a chance of development of resistance to *Bt* toxins by the insects when transgenics are produced utilizing ICPs from Bacillus thuringiensis only. This requires exploring other sources of toxins, preferably by foreign genes from plants like enzyme inhibitors and lectins (Ceci et al., 2003; Rahbe´ et al., 2003; Tinjuangjun et al., 2000; Loc et al., 2002), besides other animal sources including insects, viz., biotin-binding proteins (Burgess et al., 2002), neurohormones (Fitches et al., 2002) and enzyme inhibitors (Christeller et al., 2002). Biotin binding proteins, avidin, which is a water-soluble tetrameric glycoprotein from chicken egg binding strongly to biotin, and streptavidin, a homologous protein extracted from the culture supernatant of *Streptomyces avidinii*, have been used to control tomato fruit borer *Helicoverpa armigera*. Transgenic plants having leaves expressing avidin in the vacuole stopped growth and lead to the death of *H. armigera* larvae (Burgess et al., 2002).

Toxins from other insect pathogens are also being discovered that may be used and recently a 283 kDa protein from *Photorhabdus luminescens* expressed in *Arabidopsis* exhibited resistance to insect (Liu et al., 2003). One recent strategy has been the development of hybrid *Cry* proteins (Naimov et al., 2001) as well as other fused protein to enhance the spectrum and durability of resistance (Fitches et al., 2002, 2004; Bohorova et al., 2001).

11.6 ROLE OF BIOTECHNOLOGY IN BIOPESTICIDE DEVELOPMENT

As biological control of tomato fruit borer, there are several commercial products, of which *Bt* formulations are very much in demand. The *Bt* formulations are very less harmful to the predators and beneficial insects than the chemical pesticides and hence a market for the product is upscale. Its advantages are that it may be applied easily as a chemical pesticide, sans the ill effects on the predators and beneficial insects and the environment as a whole. They can kill the pests, which is functional response, and reproduce at the expense of pest which is numeric response and hence also controls the future generations of the pest population (Talukdar, 2013). The focus is basically on enhancing the insecticidal toxin. Besides *Bt* formulations, commercial formulations of *Helicoverpa armigera* NPV (nucleopolyhedrovirus), a group of Baculoviruses is also very popular (Ranga Rao et al., 2015).

11.7 CONCLUSION

The durability of resistance to tomato fruit borer in transgenic tomatoes depends on the utilization of various mechanisms to evade or hinder the incidence of the insect infestation (Kumar et al., 1996). The strategy for control of the devastating insect should include high level of ICP expression (Schuler et al., 1998). This, on the other hand, leads to high selection pressure that may eventually result in development of resistance in towards the transgenic line. To check this, a proper strategy has to be built that includes the use of refuge crops, pyramiding of genes encoding insecticidal proteins differing in their mode of action (Gould, 1998) as well as regulation of pathways for signaling various compounds responsible for indirect defense mechanism of the plants. Diverse genes encoding for inhibitors of insect proteinases and VIP and other novel insecticidal proteins can be alternative choice for development of transgenic tomato coupled with *Bt* genes that may be a long-lasting and resilient solution for control of the obnoxious pest.

KEYWORDS

- ***Helicoverpa armigera***
- **insecticidal crystal protein**
- **nucleopolyhedrovirus**
- **proteinase inhibitor**
- **RNAi**
- **vegetative insecticidal proteins**

REFERENCES

Adamczyk, J. J., Holloway, J. W., Church, G. E., Leonard, B. R., & Graves, J. B., (1998). Larval survival and development of the fall armyworm (*Lepidoptera*: Noctuidae) on normal and transgenic cotton expressing the *Bacillus thuringiensis* CryIA (δ)-endotoxin. *J. Econ. Entomol., 91*, 539–545.

Avisar, D., Keller, M., Gazit, E., Prudovsky, E., Sneh, B., & Zilberstein, A., (2004). The role of *Bacillus thuringiensis* Cry1C and Cry1E separate structural domains in the interaction with *Spodoptera littoralis* gut epithelial cells. *J. Biol. Chem., 279*, 15779–15786.

AVRDC, (2000). *AVRDC-The World Vegetable Center, 2000* (p. 152). AVRDC Progress Report-1999. Asian Vegetable Research and Development Center, Shanhua, Taiwan.

AVRDC, (2001). *AVRDC-The World Vegetable Center, 2001* (p. vii, 152). AVRDC Report 2000. Asian Vegetable Research and Development Center, Shanhua, Tainan, Taiwan.

Ballal, C. R., & Singh, S. P., (2003). The effectiveness of *Trichogramma chilonis*, *Trichogramma pretiosum*, and *Trichogramma brasiliense* (Hymenoptera: Trichogrammatidae) as parasitoids of *Helicoverpa armigera* (*Lepidoptera*: Noctuidae) on Sunflower (*Helianthus annuus*) and redgram (*Cajanus cajan*). *Biocontrol Sci. Tech., 13*, 231–240.

Barton, K. A., Whiteley, H. R., & Yang, N. S., (1987). *Bacillus thuringiensis* §-endotoxin expressed in transgenic *Nicotiana tabacum* provides resistance to lepidopteran insects. *Plant Physiol., 85*, 1103–1109.

Bettencourt, R., Terenius, O., & Faye, I., (2002). Hemolin gene silencing by ds-RNA injected into *Cecropia pupae* is lethal to next generation embryos. *Insect Mol. Biol., 11*, 267–271.

Bhalla, R., Dalal, M., Panguluri, S. K., Jagadish, B., Mandaokar, A. D., Singh, A. K., & Kumar, P. A., (2005). Isolation, characterization and expression of a novel vegetative insecticidal protein gene of *Bacillus thuringiensis*. *FEMS Microbiol. Lett., 243*, 467–472.

Bohorova, N., Frutos, R., Royer, M., Estanol, P., Pacheco, M., Rascon, Q., McLean, S., & Hoisington, D., (2001). Novel synthetic *Bacillus thuringiensis cry1B* gene and the *cry1B-cry1Ab* translational fusion confer resistance to southwestern corn borer, sugarcane borer and fall armyworm in transgenic tropical maize. *Theor. Appl. Genet., 103*, 817–826.

Bown, D. P., Wilkinson, H. S., & Gatehouse, J. A., (1997). Differentially regulated inhibitor-sensitive and insensitive protease genes from the *Phytophagous* insect pest, *Helicoverpa armigera*, are members of complex multigene families. *Insect Biochem. Mol. Biol., 27*, 625–638.

Bravo, A., Gill, S. S., & Soberón, M., (2007). Mode of action of *Bacillus thuringiensis* Cry and Cyt toxins and their potential for insect control. *Toxicon, 49*, 423–435.

Bravo, A., Gómez, I., Conde, J., Muñoz-Garay, C., Sánchez, J., Miranda, R., Zhuang, M., et al., (2004). Oligomerization triggers binding of a *Bacillus thuringiensis* Cry1Ab pore-forming toxin to aminopeptidase N receptor leading to insertion into membrane microdomains. *Biochem. Biophys. Acta, 1667*, 38–46.

Broadway, R. M., & Duffey, S. S., (1986). Plant proteinase inhibitors: Mechanism of action and effect on the growth and digestive physiology of larval *Heliothis zea* and *Spodoptera exiqua*. *J. Insect Physiol., 32*, 827–833.

Broadway, R. M., (1995). Are insects resistant to plant proteinase inhibitors? *J. Insect. Physiol., 41*, 107–116.

Broza, M., (1986). Seasonal changes in population of *Heliothis armigera* (Hb.) (*Lepidoptera*; Noctuidae) in cotton fields in Israel and its control with a *Bacillus thuringiensis* preparation. *J. Appl. Ento., 102*, 363–370.

Burgess, E. P. J., Malone, L. A., Christeller, J. T., Lester, M. T., Murray, C., Philip, B. A., Phung, M. M., & Tregidga, E. L., (2002). Avidin expressed in transgenic tobacco leaves confers resistance to two noctuid pests, *Helicoverpa armigera* and *Spodoptera litura*. *Transgenic Res., 11*, 185–198.

CABI, (2020). *Helicoverpa armigera (Cotton Bollworm)*. In: Invasive Species Compendium. Wallingford, UK: CAB International. https://www.cabi.org/isc/datasheet/26757 (accessed on 1 March 2021).

Cao, J., Shelton, A. M., & Earle, E. D., (2008). Sequential transformation to pyramid two Bt genes in vegetable Indian mustard (*Brassica juncea* L.) and its potential for control of diamondback moth larvae. *Plant Cell Rep., 27*, 479–487.

Carlini, C. R., & Grossi-de-Sa, M. F., (2002). Plant toxic proteins with insecticidal properties. A review on their potentialities as bioinsecticides. *Toxicon, 40*, 1515–1539.

Cayrol, R. A., (1972). Famille des Noctuidae. Sous-famille des Melicleptriinae. Helicoverpa armigera Hb. In: Balachowsky, A. S., (ed.), *Entomologie Appliquée à L'agriculture* [Family Noctuidae, Sub-family Melicleptriinae, *Helicoverpa armigera* Hb. In: Balachowsky, A. S., ed. Entomology Applied to Agriculture (Vol. 2, pp. 1431–1444). Paris, France: Masson et Cie.

Ceci, L. R., Volpicella, M., Rahbe, Y., Gallerani, R., Beekwilder, J., & Jongsma, M. A., (2003). Selection by phage display of a variant mustard trypsin inhibitor toxic against aphids. *Plant J., 33*, 557–566.

Chaithra, N., Gowda, P. H. R., & Guleria, N., (2015). Transformation of tomato with *Cry2ax1* by biolistic gun method for fruit borer resistance. *Int. J. Agri. Env. Biotech., 8*, 795–803. doi: 10.5958/2230-732X.2015.00088.1.

Christeller, J. T., Burgess, E. P. J., Mett, V., Gatehouse, H. S., Markwick, N. P., Murray, C., Malone, L. A., Wright, M. A., Philip, B. A., Watt, D., et al., (2002). The expression of a mammalian proteinase inhibitor, bovine spleen trypsin inhibitor in tobacco and its effects on *Helicoverpa armigera* larvae. *Transgenic Res., 11*, 161–173.

Crickmore, N., Zeigler, D. R., Feitelson, J., Schnepf, E., Van, R. J., Lereclus, D., Baum, J., & Dean, D. H., (1998). Revision of the nomenclature for the *Bacillus thuringiensis* pesticidal crystal proteins. *Microbiol. Mol. Biol. Rev., 62*, 807–813.

Czapla, T. H., & Lang, B. A., (1990). Effects of plant lectins on the larval development of European corn borer (Lepidoptera: Pyralidae) and Southern corn rootworm (Coleoptera: Chrysomelidae). *J. Econ. Entomol., 83*, 2480–2485.

De Leo, F., Bonade-Bottino, M. A., Ceci, L. R., Gallerani, R., & Jouanin, L., (1998). Opposite effects on *Spodoptera littoralis* larvae of high expression level of a trypsin proteinase inhibitor in transgenic plants. *Plant Physiol., 118*, 997–1004.

De Rocher, E. J., Vargo-Gogola, T. C., Diehn, S. H., & Green, P. J., (1998). Direct evidence for rapid degradation of *Bacillus thuringiensis* toxin mRNA as a cause of poor expression in plants. *Plant Physiol., 117*, 1445–1461.

Diaz-Mendoza, M., Farinós, G. P., Castañera, P., Hernández-Crespo, P., & Ortego, F., (2007). Proteolytic processing of native Cry1Ab toxin by midgut extracts and purified trypsins from the Mediterranean corn borer *Sesamia nonagrioides*. *J. Insect Physiol., 53*, 428–435.

Diehn, S. H., De Rocher, E. J., & Green, P. J., (1996). Problems that can limit the expression of foreign genes in plants: Lessons to be learned from Bt.-toxin genes. In: Setlow, J. K., (ed), *Genetic Engineering: Principles and Methods* (Vol. 18, pp. 83–99). Plenum Press, New York.

Eisemann, C. H., Donaldson, R. A., Pearson, R. D., Cadogan, L. C., Vuocolo, T., & Tellam, R. L., (1994). Larvicidal activity of lectins on *Lucilia cuprina*-mechanism of action. *Entomol. Exp. Appl., 72*, 1–10.

Endo, Y., & Nishiitsutsuji, U., (1980). Mode of action of *Bacillus thuringiensis* δ-endotoxin: Histopathological changes in the silkworm midgut. *J. Invertebr. Pathol., 36*, 90–103.

EPPO, (2020). EPPO Global Database. In: *EPPO Global database*, Paris, France, EPPO.

Estela, A., Escriche, B., & Ferré, J., (2004). Interaction of *Bacillus thuringiensis* toxins with larval midgut binding sites of *Helicoverpa armigera* (Lepidoptera: Noctuidae). *Appl. Environ. Microbiol., 70*, 1378–1384.

Fang, J., Xu, X. L., Wang, P., Zhao, J. Z., Shelton, A. M., Cheng, J., Feng, M. G., & Shen, Z. C., (2007). Characterization of chimeric *Bacillus thuringiensis* Vip3 toxins. *Appl. Environ. Microbiol., 73*, 956–961.

Farmer, E. E., & Ryan, C. A., (1990). Interplant communication: Airborne methyl jasmonate induces synthesis of proteinase inhibitors in plant leaves. *Proc. Natl. Acad. Sci. USA, 87*, 7713–7716.

Ferry, N., Edwards, M. G., Gatehouse, J. A., & Gatehouse, A. M. R., (2004). Plant-insect interactions: Molecular approaches to insect resistance. *Curr. Opin. Biotech., 15*, 155–161.

Ferry, N., Jouanin, L., Ceci, L. R., Mulligan, E. A., Emami, K., Gatehouse, J. A., & Gatehouse, A. M. R., (2005). Impact of oilseed rape expressing the insecticidal serine protease inhibitor, mustard trypsin inhibitor-2 on the beneficial predator *Pterostichus madidus*. *Mol. Ecol., 14*, 337–349.

Fischhoff, D. A., Bowdish, K. S., Perlak, F. J., Marrone, P. G., Mccormick, S. M., Niedermeyer, J. G., Dean, D. A., et al., (1987). Insect tolerant transgenic tomato plants. *Bio-Technology, 5*, 807–813.

Fitches, E., Audsley, N., Gatehouse, J. A., & Edwards, J. P., (2002). Fusion proteins containing neuropeptides as novel insect control agents: Snowdrop lectin delivers fused allatostatin to insect hemolymph following oral ingestion. *Insect Biochem. Mol. Biol., 32*, 1653–1661.

Fitches, E., Edwards, M. G., Mee, C., Grishin, E., Gatehouse, A. M. R., Edwards, J. P., & Gatehouse, J. A., (2004). Fusion proteins containing insect-specific toxins as pest control agents: Snowdrop lectin delivers fused insecticidal spider venom toxic to insect hemolymph following oral digestion. *J. Insect Physiol., 50*, 61–71.

Franco, O. L., Rigden, D. J., Melo, F. R., & Grossi-de-Sa, M. F., (2002). Plant alpha-amylase inhibitors and their interaction with insect alpha-amylases. Structure, function and potential for crop protection. *Eur. J. Biochem., 269*, 397–412.

Fujimoto, H., Itoh, K., Yamamoto, M., Kyozuka, J., & Shimamoto, K., (1993). Insect resistant rice generated by introduction of a modified d-endotoxin gene of *Bacillus thuringiensis* Cry1Ab insecticidal crystal protein. *Biotechnology, 11*, 1151–1155.

Garcia-Tejero, F. D., (1957). Bollworm of tomato, Heliothis armigera Hb. (= absoleta F). In: Dossat, S. A., (ed.), *Plagas y Enfermedades de las Plantas Cultivadas* [Pests and Diseases of Cultivated Plants (in Spanish)] (pp. 403–407). Madrid, Spain.

Gill, S. S., Cowles, E. A., & Pietrantonio, P. V., (1992). The mode of action of *Bacillus thuringiensis* endotoxins. *Annu. Rev. Entomol., 37*, 615–636.

Gonzalez-Cabrera, J., Farinos, G. P., Caccia, S., Diaz-Mendoza, M., Castanera, P., Leonardi, M. G., Giordana, B., & Ferre, J., (2006). Toxicity and mode of action of *Bacillus thuringiensis* cry proteins in the Mediterranean corn borer, *Sesamia nonagrioides* (Lefebvre). *App. Environ. Microbiol., 72*, 2594–2600.

Gould, F., (1998). Sustainability of transgenic insecticidal cultivars: integrating pest genetics and ecology. *Annu. Rev. Entomol., 43*, 701–726.

Gupta, R. K., Raj, D., & Devi, N., (2004). Biological and impact assessment studies on *Campoletis chlorideae* Uchida: A promising solitary larval endoparasitoid of *Helicoverpa armigera* (Hübner). *J. Asia Pacific Entomol., 7*, 239–247.

Hanur, V. S., Reddy, B., Arya, V. V., & Rami, R. P. V., (2015). Genetic transformation of tomato using Bt *Cry2A* gene and characterization in Indian cultivar Arka Vikas. *J. Agr. Sci. Tech., 17*, 1805–1814.

Haq, S. K., Atif, S. M., & Khan, R. H., (2004). Protein proteinase inhibitor genes in combat against insects, pests, and pathogens: Natural and engineered photoprotection. *Arch. Biochem. Biophys., 431*, 145–159.

Hardwick, D. F., (1965). The corn earworm complex. *Mem. Entomol. Soc. Canada, 40*, 1–247.

Hardwick, D. F., (1970). A generic revision of the North American Heliothidinae (*Lepidoptera*: Noctuidae). *Mem. Entomol. Soc. Canada, 73*, 1–59.

Harper, M. S., Hopkins, T. L., & Czapla, T. H., (1998). Effect of wheat germ agglutinin on formation and structure of the peritrophic membrane in European corn borer (*Ostrinia nubilalis*) larvae. *Tissue and Cell, 30*, 166–176.

Heckel, D. G., Gahan, L. J., Baxter, S. W., Zhao, J. Z., Shelton, A. M., Gould, F., & Tabashnik, B. E., (2007). The diversity of Bt resistance genes in species of *Lepidoptera*. *J. Invertebr. Pathol., 95*, 192–197.

Höfte, H., & Whiteley, H. R., (1989). Insecticidal crystal proteins of *Bacillus thuringiensis*. *Micro. Rev., 53*, 242–255.

Hopkins, T. L., & Harper, M. S., (2001). Lepidopteran peritrophic membranes and effects of dietary wheat germ agglutinin on their formation and structure. *Arch. Insect Biochem. Physiol., 47*, 100–109.

Hua, G., Masson, L., Jurat-Fuentes, J. L., Schwab, G., & Adang, M. J., (2001). Binding analyses of *Bacillus thuringiensis* cry δ-endotoxins using brush border membrane vesicles of *Ostrinia nubilalis*. *Appl. Environ. Microbiol., 67*, 872–879.

Jallow, M. F. A., Matsumura, M., & Suzuki, Y., (2001). Oviposition preference and reproductive performance of Japanese *Helicoverpa armigera* (Hübner) (Lepidoptera: Noctuidae). *Appl. Entomol. Zool., 36*, 419–426.

Jansens, S., Cornellissen, M., Deelerg, R., Reynaerts, A., & Peferoen, M., (1995). *Phthorimea operculella* resistance in potato by expression of the *Bacillus thuringiensis* Cry1ab insecticidal crystal protein. *J. Econ. Entomol., 88*, 1469–1476.

Jiménez-Juárez, N., Muñoz-Garay, C., Gómez, I., Saab-Rincon, G., Damian-Almazo, J. Y., Gill, S. S., Soberón, M., & Bravo, A., (2007). *Bacillus thuringiensis* Cry1Ab mutants affecting oligomer formation are non-toxic to *Manduca sexta* larvae. *J. Biol. Chem., 282*, 21222–21229.

Johnston, K. S., Lee, M. J., Gatehouse, J. A., & Anstee, J. H., (1991). The partial purification and characterization of serine protease activity in midgut of larval *Helicoverpa armigera. Insect Biochem., 21*, 389–397.

Jongsma, M. A., Stiekema, W. J., & Bosch, D., (1996). Combatting inhibitor-insensitive proteases of insect pests. *Trends Biotechnol., 14*, 331–333.

Jurat-Fuentes, J. L., & Adang, M. J., (2004). Characterization of a Cry1ac-receptor alkaline phosphatase in susceptible and resistant *Heliothis virescens* larvae. *Eur. J. Biochem., 271*, 3127–3135.

Jurat-Fuentes, J. L., Gahan, L. J., Gould, F. L., Heckel, D. G., & Adang, M. J., (2004). The HevCaLP protein mediates binding specificity of the Cry1A class of *Bacillus thuringiensis* toxins in *Heliothis virescens. Biochemistry, 43*, 14299–14305.

Kabir, K. E., Sugimoto, H., Tado, H., Endo, K., Yamanaka, A., Tanaka, S., & Koga, D., (2006). Effect of *Bombyx mori* chitinase against Japanese pine sawyer (*Monochamus alternatus*) adults as a biopesticide. *Biosci. Biotech. Biochem., 70*, 219–229.

Kato, T., Higuchi, M., Endo, R., Maruyama, T., Haginoya, K., Shitomi, Y., Hayakawa, T., Mitsui, T., Sato, R., & Hori, H., (2006). *Bacillus thuringiensis* Cry1Ab, but not Cry1Aa or Cry1Ac, disrupts liposomes. *Pesticide Biochem. Physiol., 84*, 1–9.

Kessler, A., & Baldwin, I. T., (2002). Plant responses to insect herbivory: The emerging molecular analysis. *Annu. Rev. Plant Biol., 53*, 299–328.

Knight, P. J., Crickmore, N., & Ellar, D. J., (1994). The receptor for *Bacillus thuringiensis* CrylA(c) delta-endotoxin in the brush border membrane of the lepidopteran *Manduca sexta* is aminopeptidase N. *Mol. Microbiol., 11*, 429–436.

Knowles, B. H., (1994). Mechanism of action of *Bacillus thuringiensis* insecticidal endotoxins. *Adv. Insect Physiol., 24*, 275–308.

Kota, M., Daniell, H., Varma, S., Garczynski, S. F., Gould, F., & Moar, W. J., (1999). Overexpression of the *Bacillus thuringiensis* (*Bt*) Cry2Aa2 protein in chloroplasts confers resistance to plants against susceptible and *Bt*-resistant insects. *Proc. Natl. Acad. Sci. USA, 96*, 1840–1845.

Koul, B., Srivastava, S., Sanyal, S., Tripathi, B., Sharma, V., & Amla, D. V., (2014). *Transgenic Tomato Line Expressing Modified Bacillus Thuringiensis cry1Ab Gene Showing Complete Resistance to Two Lepidopteran Pests* (Vol. 3, p. 84). Springer Plus.

Koziel, M. G., Beland, G. L., Bowman, C., Carozzi, N. B., Crenshaw, R., Crossland, R., Dawson, L., et al., (1993). Field performance of elite transgenic maize plants expressing an insecticidal protein gene derived from *Bacillus thuringiensis. Biotechnology, 11*, 1151–1155.

Kranthi, K. R., Jadhav, D. R., Kranthi, S., Wanjari, H. R., Al, S. S., & Hussel, D. A., (2002). Insecticide resistance in five major insect pests of cotton in India. *Crop Protection. 21*, 449–460.

Kumar, H., & Kumar, V., (2004). Tomato expressing Cry1A(b) insecticidal protein from *Bacillus thuringiensis* protected against tomato fruit borer, *Helicoverpa armigera* (H.ubner) (*Lepidoptera*: Noctuidae) damage in the laboratory, greenhouse and field. *Crop Protection, 23*, 135–139.

Kumar, P. A., Mandaokar, A., Sreenivasu, K., Chakrabarti, S. K., Bisaria, S., Sharma, S. R., Kaur, S., & Sharma, R. P., (1998). Insect-resistant transgenic brinjal plants. *Mol. Breed., 4*, 33–37.

Kumar, P. A., Sharma, R. P., & Malik, V. S., (1996). Insecticidal crystal proteins of *Bacillus thuringiensis*. *Adv. Appl. Microbiol., 42*, 1–43.

Lal, O. P., & Lal, S. K., (1996). Failure of control measures against *Heliothis armigera* (Hubner) infesting tomato in heavy pesticidal application areas in Delhi and satellite towns in western Uttar Pradesh and Haryana (India). *J. Entomol. Res., 20*, 355–364.

Lane, N. J., Harrison, J. B., & Lee, W. M., (1989). Changes in microvilli and Golgi-associated membranes of lepidopteran cells induced by an insecticidally active bacterial δ-endotoxin. *J. Cell Sci., 93*, 337–347.

Li, H. R., Oppert, B., Higgins, R. A., Huang, F. N., Zhu, K. Y., & Buschman, L. L., (2004). Comparative analysis of proteinase activities of *Bacillus thuringiensis*-resistant and susceptible *Ostrinia nubilalis* (*Lepidoptera* : Crambidae). *Insect Biochem. Mol. Biol., 34*, 753–762.

Liu, D., Burton, S., Glancy, T., Li, Z., Hampton, R., Meade, T., & Merlo, D., (2003). Insect resistance conferred by 283-kDa *Photorhabdus luminescens* protein TcdA in *Arabidopsis thaliana*. *Nat. Biotechnol., 21*, 1222–1228.

Loc, N. T., Tinjuangjun, P., Gatehouse, A. M. R., Christou, P., & Gatehouse, J. A., (2002). Linear transgene constructs lacking vector backbone sequences generate transgenic rice plants which accumulate higher levels of proteins conferring insect resistance. *Mol. Breed., 9*, 231–244.

Lopes, A. R., Juliano, M. A., Juliano, L., & Terra, W. R., (2004). Coevolution of insect trypsins and inhibitors. *Arch. Insect Biochem. Physiol., 55*, 140–152.

Lynch, R. E., Guo, B., Timper, P., & Wilson, J. P., (2003). United states department of agriculture-agricultural research service: Research on improving host-plant resistance to pests. *Pest Manage. Sci., 59*, 718–727.

MacIntosh, S. C., Stone, T. B., Jokerst, R. S., & Fuchs, R. L., (1991). Binding of *Bacillus thuringiensis* proteins to a laboratory-selected line of *Heliothis virescens*. *Proc. Natl. Acad. Sci. USA, 88*, 8930–8933.

Maheswaran, G., Pridmore, L., Franz, P., & Anderson, M. A., (2007). A proteinase inhibitor from *Nicotiana alata* inhibits the normal development of light-brown apple moth, *Epiphyas postvittana* in transgenic apple plants. *Plant Cell Rep., 26*, 773–782.

Mandaokar, A. D., Goyal, R. K., Shukla, A., Bisaria, S., Bhalla, R., Reddy, V. S., Chaurasia, A., et al., (2000). Transgenic tomato plants resistant to fruit borer (*Helicoverpa armigera* Hubner). *Crop Protection, 19*, 307–312.

Mandaokar, A. D., Kumar, P. A., Sharma, R. P., & Malik, V. S., (1999). BT transgenic crop plants-progress and prospectus. In: Chopra, V. L., Malik, V. S., & Bhat, S. R., (eds.), *Applied Plant Biotechnology* (pp. 285–300). Oxford IBH Publishing Co, New Delhi.

Mao, Y. B., Cai, W. J., Wang, J. W., Hong, G. J., Tao, X. Y., Wang, L. J., Huang, Y. P., & Chen, X. Y., (2007). Silencing a cotton bollworm P450 monooxygenase gene by plant-mediated RNAi impairs larval tolerance of gossypol. *Nat. Biotechnol., 25*, 1307–1313.

Markwick, N. P., Laing, W. A., Christeller, J. T., McHenry, J. Z., & Newton, M. R., (1998). Overproduction of digestive enzymes compensates for inhibitory effects of protease and -amylase inhibitors fed to three species of leafrollers (*Lepidoptera*: Tortricidae). *J. Econ. Entomol., 91*, 1265–1276.

McBride, K. E., Svab, Z., Schaaf, D. J., Hogan, P. S., Stalker, D. M., & Maliga, P., (1995). Amplification of a chimeric *Bacillus* gene in chloroplasts leads to an extraordinary level of an insecticidal protein in tobacco. *Biotechnology, 13*, 362–365.

Mehrotra, M., Singh, A. K., Sanyal, I., Altosaar, I., & Amla, D. V., (2011). Pyramiding of modified cry1Ab and cry1Ac genes of *Bacillus thuringiensis* in transgenic chickpea (*Cicer arietinum* L.) for improved resistance to pod borer insect *Helicoverpa armigera. Euphytica, 82*, 87–102.

Moar, W. J., Pusztaicarey, M., Vanfaassen, H., Bosch, D., Frutos, R., Rang, C., Luo, K., & Adang, M. J., (1995). Development of *Bacillus thuringiensis* CryIC resistance by *Spodoptera exigua* (Hubner) (Lepidoptera, Noctuidae). *Appl. Environ. Microbiol., 61*, 2086–2092.

Murdock, L. L., Huesing, J. E., Nielsen, S. S., Pratt, R. C., & Shade, R. E., (1990). Biological effects of plant lectins on the cowpea weevil. *Phytochemistry, 29*, 85–89.

Nagamatsu, Y., Itai, Y., Hatanaka, C., Funatsu, G., & Hayashi, K., (1984). A toxic fragment from the entomocidal crystal protein of *Bacillus thuringiensis. Agric. Biol. Chem., 48*, 611–619.

Nagamatsu, Y., Toda, S., Yamaguchi, F., Ogo, M., Kogure, M., Nakamura, M., Shibata, Y., & Katsumoto, T., (1998). Identification of *Bombyx mori* midgut receptor for *Bacillus thuringiensis* insecticidal CryIA(a) toxin. *Biosci. Biotechnol. Biochem., 62*, 718–726.

Naimov, S., Weemen-Hendriks, M., Dukiandjiev, S., & De Maagd, R. A., (2001). *Bacillus thuringiensis* d-endotoxin Cry1 hybrid proteins with increased activity against the Colorado potato beetle. *Appl. Environ. Microbiol., 67*, 5328–5330.

OEPP/EPPO, (1981). Datasheets on quarantine organisms No. 110, *Helicoverpa armigera. Bulletin OEPP/EPPO, 11*.

Pacheco, S., Gómez, I., Arenas, I., Saab-Rincon, G., Rodríguez-Almazán, C., Gill, S. S., Bravo, A., & Soberón, M., (2009). mechanism with *Manduca sexta* aminopeptidase-N and cadherin receptors. *J. Biol. Chem., 284*, 32750–32757.

Padilla, C., Pardo-López, L., De La Riva, G., Gómez, I., Sánchez, J., Hernandez, G., Nuñez, M. E., et al., (2006). Role of tryptophan residues in toxicity of Cry1Ab toxin from *Bacillus thuringiensis*. *Appl. Environ. Microbiol., 72*, 901–907.

Pardo, L. L., Soberon, M., & Bravo, A., (2013). *Bacillus thuringiensis* insecticidal three-domain cry toxins: Mode of action, insect resistance and consequences for crop protection. *FEMS Microb. Rev., 37*, 3–22.

Perlak, F. J., Deaton, R. W., Armstrong, T. A., Fuchs, R. L., Sims, S. R., Greenplate, J. T., & Fischhoff, D. A., (1990). Insect-resistant cotton plants. *BioTechnology, 8*, 939–943.

Perlak, F. J., Fuchs, R. L., Dean, D. A., McPherson, S. L., & Fischhoff, D. A., (1991). Modification of the coding sequence enhances plant expression of insect control protein gene. *Proc. Natl. Acad. Sci. USA., 88*, 3324–3328.

Perlak, F. J., Oppenhuizen, M., Gustafson, K., Voth, R., Sivasupramaniam, S., Herring, D., Carey, B., et al., (2001). Development and commercial use of bollgard® cotton in the USA-early promises versus today's reality. *Plant J., 27*, 489–501.

Pigott, C. R., & Ellar, D. J., (2007). Role of receptors in *Bacillus thuringiensis* crystal toxin activity. *Microbiol. Mol. Biol. Rev., 71*, 255–281.

Purcell, J. P., Oppenhuizen, M., Wofford, T., Reed, A. J., & Perlak, F. J., (2004). The story of Bollgard cotton. In: Christou, P., & Klee, H., (ed), *Handbook of Plant Biotechnology* (pp. 1147–1163). John Wiley & Sons.

Queiroz-Santos, L., Casagrande, M. M., & Specht, A., (2018). Morphological characterization of *Helicoverpa armigera* (Hübner) (Lepidoptera: Noctuidae: Heliothinae). *Neotrop. Entomol., 47*, 517–542. https://doi.org/10.1007/s13744-017-0581-4.

Ragab, M. G., El-Sayed, A. A., & Nada, M. A., (2014). The effect of some biotic and abiotic factors on seasonal fluctuations of *Helicoverpa armigera* (Hub.). *Egyptian J. Agril. Res., 92*, 101–119.

Rahbe, Y., Deraison, C., Bonade, B. M., Girard, C., Nardon, C., & Jouanin, L., (2003). Effects of cysteine protease inhibitor *Oryza cystatin* (OC-1) on different aphids and reduced performance of *Myzus persicae* on OC-1 expressing transgenic oilseed rape. *Plant Sci., 164*, 441–450.

Ranga, R. G. V., Kumar, C. S., Sireesha, K., & Kumar, P. L., (2015). Role of nucleopolyhedroviruses (NPVs) in the management of lepidopteran pests in Asia. In: Sree, K. S., & Varma, A., (eds.), *Biocontrol of Lepidopteran Pests: Use of Soil Microbes and Their Metabolites* (pp. 16, 17). Springer.

Rawat, P., Singh, A. K., Ray, K., Chaudhary, B., Kumar, S., Gautam, T., Kanoria, S., et al., (2011). Detrimental effect of expression of Bt endotoxin Cry1Ac on *in vitro* regeneration, *in vivo* growth and development of tobacco and cotton transgenics. *J. Biosci., 36*, 363–376.

Reddy, K. V. S., & Zehr, U. B., (2004). Novel strategies for overcoming pests and diseases in India 2004 "new directions for a diverse planet". *Proceedings of the 4th International Crop Science Congress*. Brisbane, Australia. Published on CDROM, Web site: www.cropscience.org.au (accessed on 1 March 2021).

Rukmini, V., Reddy, C. Y., & Venkateswerlu, G., (2000). *Bacillus thuringiensis* crystal δ- endotoxin: Role of proteases in the conversion of protoxin to toxin. *Biochimie, 82*, 109–116.

Ryan, C. A., (1989). Proteinase inhibitor gene families: Strategies for transformation to improve plant defenses against herbivores. *Bioessays, 10*, 20–22.

Ryan, C. A., (1990). Protease inhibitors in plants: genes for improving defenses against insects and pathogens. *Annu. Rev. Phytopathol., 28*, 425.

Ryerse, J. S., Beck, J. R. Jr., & Lavrik, P. B., (1990). Light microscope immunolocation of *Bacillus thuringiensis kurstaki* delta-endotoxin in the midgut and Malpighian tubules of the tobacco budworm, *Heliothis virescens*. *J. Invertebr. Pathol., 56*, 86–90.

Saker, M. M., Salama, H. S., Salama, M., El-Banna, A., & Abdel, G. A. M., (2011). Production of transgenic tomato plants expressing Cry 2Ab gene for the control of some lepidopterous insect's endemic in Egypt. *J. Genet. Engg. Biotech., 9*, 149–155.

Schnepf, E., Crickmore, N., Van, R. J., Lereclus, D., Baum, J., Feitelson, J., Zeigler, D. R., & Dean, D. H., (1998). *Bacillus thuringiensis* and its pesticidal crystal proteins. *Microbiol. Mol. Biol. Rev., 62*, 775–806.

Schuler, T. H., Poppy, G. M., Kerry, B. R., & Denholm, I., (1998). Insect resistant transgenic plants. *Trends Biotechnol., 16*, 168–175.

Sharath, C. G., Asokan, R., Manamohan, M., & Sitaramamma, T., (2012). Control of lepidopteran insect pest, *Helicoverpa armigera* (Hubner) through ingested RNA interference *Agrotechnol., 1*, 2 http://dx.doi.org/10.4172/2168-9881.S1.002.

Sharma, H. C., (2012). Conventional and biotechnological approaches for pest management: Potential and limitations. In: *Environmental Safety of Biotech and Conventional IPM Technologies* (pp. 3–25). Studium Press LLC, Texas, USA.

Shimizu, T., Yoshii, M., Wei, T., Hirochika, H., & Omura, T., (2009). Silencing by RNAi of the gene for Pns12, a viroplasm matrix protein of rice dwarf virus, results in strong resistance of transgenic rice plants to the virus. *Plant Biotechnol. J., 7*, 24–32.

Sidhu, M. K., Dhatt, A. S., Sandhu, J. S., & Gosal, S. S., (2014). Biolistic transformation of *cry1Ac* gene in eggplant (*Solanum melongena* L.). *Int. J. Agri. Env. Biotech., 7*, 679–687.

Silva, F. C., Alcazar, A., Macedo, L. L., Oliveira, A. S., Macedo, F. P., Abreu, L. R., Santos, E. A., & Sales, M. P., (2006). Digestive enzymes during development of *Ceratitis capitate* (Diptera: Tephritidae) and effects of SBTI on its digestive serine proteinase targets. *Insect Biochem. Mol. Biol., 36*, 561–569.

Siqueira, H. A. A., Nickerson, K. W., Moellenbeck, D., & Siegfried, B. D., (2004). Activity of gut proteinases from Cry1Ab-selected colonies of the European corn borer, *Ostrinia nubilalis* (Lepidoptera: Crambidae). *Pest Manage. Sci., 60*, 1189–1196.

Soberón, M., Pardo-López, L., López, I., Gómez, I., Tabashnik, B. E., & Bravo, A., (2007). Engineering modified Bt toxins to counter insect resistance. *Science, 318*, 1640–1642.

Srinivasan, K., Krishna, M. P. N., & Raviprasad, T. N., (1994). African marigold as a trap crop for the management of the fruit borer, *Helicoverpa armigera* on tomato. *International J. Pest Manage., 40*, 56–63.

Srinivasan, R., (2013). Safe management of borers in vegetables. In: Peter, K. V., (ed.), *Biotechnology in Horticulture: Methods and Application* (pp. 333–357). New India Publishing Agency, New Delhi.

Stevens, J., Dunse, K., Fox, J., & Anderson, M., (2012). Biotechnological approaches for the control of insect pests in crop plants. In: *Pesticides-Advances in Chemical and Botanical Pesticides* (pp. 269–308).

Stewart, C. N., Adang, M. J., All, J. N., Raymer, P. L., Ramachandran, S., & Parrott, W. A., (1996). Insect control and dosage effects in transgenic canola containing a synthetic *Bacillus thuringiensis* crylAc gene. *Plant Physiol., 112*, 115–120.

Sun, H. J., Uchii, Watanabe, S., & Ezura, H., (2006). A highly efficient transformation protocol for micro-tom: A model cultivar for tomato functional genomics. *Plant Cell Physiol., 47*, 426–431.

Tabashnik, B. E., Huang, F., Ghimire, M. N., Leonard, B. R., Siegfried, B. D., Rangasamy, M., Yang, Y., et al., (2011). Efficacy of genetically modified Bt toxins against insects with different genetic mechanisms of resistance. *Nat. Biotechnol., 29*, 1128–1131.

Tabashnik, B. E., Van, R. J. B. J., & Carrière, Y., (2009). Field-evolved insect resistance to Bt crops: Definition, theory, and data. *J. Econ. Entomol., 102*, 2011–2025.

Talekar, N. S., Opena, R. T., & Hanson, P., (2005). *Helicoverpa armigera* management: A review of AVRDC's research on host plant resistance in tomato. *Crop Protection, 25*, 461–467.

Talukdar, D., (2013). Modern biotechnological approaches in insect research, *Int. Res. J. Sci. Engg., 1*, 71–78.

Tamayo, M. C., Rufat, M., Bravo, J. M., & San, S. B., (2000). Accumulation of a maize proteinase inhibitor in response to wounding and insect feeding, and characterization of its activity toward digestive proteinases of *Spodoptera littoralis* larvae. *Planta, 211*, 62–71.

Tang, J. D., Shelton, A. M., Vanrie, J., Deroeck, S., Moar, W. J., Roush, R. T., & Peferoen, M., (1996). Toxicity of *Bacillus Thuringiensis* spore and crystal protein to resistant diamondback moth (*Plutella xylostella*). *Appl. Environ. Microbiol., 62*, 564–569.

Thomas, M. B., & Waage, J. K., (1996). *Integrating Biological Control and Host Plant Resistance Breeding: A Scientific and Literature Review* (p. 99). Wageningen: Technical Center for Agricultural and Rural Cooperation of the European Union (CTA).

Tinjuangjun, P., Loc, N. T., Gatehouse, A. M. R., Gatehouse, J. A., & Christou, P., (2000). Enhanced insect resistance in Thai rice varieties generated by particle bombardment. *Mol. Breed., 6*, 391–399.

Tojo, A., & Aizawa, K., (1983). Dissolution and degradation of *Bacillus thuringiensis* δ-endotoxin by gut juice protease of the silkworm *Bombyx mori*. *Appl. Environ. Microbiol., 45*, 576–580.

Tripathy, M. K., Kumar, R., & Singh, H. N., (1999). Host range and population dynamics of *Helicoverpa armigera* Hübn. In eastern Uttar Pradesh. *J. Appl. Zool. Res., 10*, 22–24.

UK, CAB International, (1993). *Helicoverpa armigera*. [Distribution map]. In: *Distribution Maps of Plant Pests*. Wallingford, UK: CAB International. Map 15 (2nd Revision).

Vachon, V., Laprade, R., & Schwartz, J. L., (2012). Current models of the mode of action of *Bacillus thuringiensis* insecticidal crystal proteins: A critical review. *J. Invertebrate Pathol., 111*, 1–12.

Vadlamudi, R. K., Ji, T. H., & Bulla, L. A. Jr., (1993). A specific binding protein from *Manduca sexta* for the insecticidal toxin of *Bacillus thuringiensis* subsp. *Berliner*. *J. Biol. Chem., 268*, 12334–12340.

Vadlamudi, R. K., Weber, E., Ji, I., Ji, T. H., & Bulla, L. A. Jr., (1995). Cloning and expression of a receptor for an insecticidal toxin of *Bacillus thuringiensis*. *J. Biol. Chem., 270*, 5490–5494.

Van, F. K., (2009). Insecticidal activity of *Bacillus thuringiensis* crystal proteins. *J. Invertebr. Pathol., 101*, 1–16.

Vidya, C. S. S., Manoharan, M., Kumar, C. T. R., Savathri, H. S., & Sita, G. L., (2000). *Agrobacterium*-mediated transformation of tomato (*Lycopersicum esculentum* var. Pusa Ruby) with *Coat Protein* Gene of *Physalis mottle* tymovirus. *J. Plant Physiol., 156*, 106–110.

Williams, I. S., (1999). Slow-growth, high-mortality: A general hypothesis, or is it? *Ecol. Entomol., 24*, 490–495.

Wright, D. J., Iqbal, M., Granero, F., & Ferré, J., (1997). A change in a single midgut receptor in the diamondback moth (*Plutella xylostella*) is only in part responsible for field resistance to *Bacillus thuringiensis* subsp. *kurstaki* and *B. thuringiensis* subsp. *aizawai*. *Appl. Environ. Microbiol., 63*, 1814–1819.

Zavala, J. A., & Baldwin, I. T., (2004). Fitness benefits of trypsin proteinase inhibitor expression in *Nicotiana attenuata* are greater than their costs when plants are attacked. *BMC Ecol., 4*, 11.

Zhang, X., Candas, M., Griko, N. B., Rose-Young, L., & Bulla, L. A. Jr., (2005). Cytotoxicity of *Bacillus thuringiensis* Cry1Ab toxin depends on specific binding of the toxin to the cadherin receptor BT-R1 expressed in insect cells. *Cell Death Differ., 2*, 1407–1416.

Zhang, X., Candas, M., Griko, N. B., Taussig, R., & Bulla, L. A. Jr. (2006). A mechanism of cell death involving an adenylyl cyclase/PKA signaling pathway is induced by the Cry1Ab toxin of *Bacillus thuringiensis*. *Proc. Natl. Acad. Sci. USA, 103*, 9897–9902.

Zhu-Salzman, K., Koiwa, H., Salzman, R. A., Shade, R. E., & Ahn, J. E., (2003). Cowpea bruchid *Callosobruchus maculatus* uses a three-component strategy to overcome a plant defensive cysteine protease inhibitor. *Insect Mol. Biol., 12*, 135–145.

CHAPTER 12

MOLECULAR APPROACHES FOR CONTROL OF SUCKING PEST IN TOMATO

PANKAJ KUMAR,[1] ANUPAM ADARSH,[2] RIMA KUMARI,[1] DAN SINGH JAKHAR,[3] ARUN KUMAR,[4] and ANUPMA KUMARI[2]

[1]*Department of Agricultural Biotechnology and Molecular Biology, Dr. Rajendra Prasad Central Agricultural University, Pusa (Samastipur) – 848125, Bihar, India, E-mail: pankajcocbiotech@gmail.com (P. Kumar)*

[2]*Krishi Vigyan Kendra, Saraiya, Dr. Rajendra Prasad Central Agricultural University, Pusa (Samstipur) – 848125, Bihar, India*

[3]*Department of Genetics and Plant Breeding, Institute of Agricultural Sciences, Banaras Hindu University, Varanasi – 221005, Uttar Pradesh, India*

[4]*Department of Agronomy, Bihar Agricultural University, Sabour – 813210, Bihar, India*

ABSTRACT

Tomato (*Solanum lycopersicum* L.) is one of the crucial vegetable crops, which has been attacking by huge numbers of insect pests and disease. Tomato yield distorted with biotic elements, particularly insect-pests at the reproductive stage. The negative impact of agrochemicals on the environment, animals, related plant species, and on human fitness have raised the demand for new tools of insect-pest. Thus, there is a need to use new tools for the development of resistance in tomatoes. Among these, several molecular methods like RNA interference (RNAi), *Agrobacterium*-mediated transformation, transcriptomic, CRISPR/CAS9 gene can be a valid tool to manage techniques towards this pest. Several literatures are available on biotechnical processes for manipulate of sucking pest. Therefore, in this chapter, we will speak about molecular approaches for manipulate of sucking pest in tomato.

12.1 INTRODUCTION

Tomato is one of the prime vegetable crops belonging to the family of nightshades, having a strong attraction for farmers because of its more yield and relatively short growing time (Naika et al., 2005; Alam et al., 2007). Because of its taste, nutritional values, and excellent processing quality, the plant often called business crop. Tomato is one of a few vegetables' having large vegetation due to its diverse consumption, such as an element in many dishes, sauces, salads, and beverages (Sarkar et al., 2018). At times, it can end up with very vital emergency measures; the entire yield may need to be reduced. Tomatoes has threat to attack by large sponges at times, it can end up with very vital and emergency measures, and the entire yield may need to be reduced. Tomatoes are challenge to attack via a massive spectrum of insect pests. Tomato have been suffering from insects like Colorado potato beetle (CPB), tomato fruit worm, beet armyworm, yellow striped armyworm, tomato pinworm, two noticed spider mites, flea beetles, thrips, aphids, whiteflies, potato leaf hooper and fruit and shoot borer (Brust, 2013; Bessin, 2013).

Some resistances are powerful against a broad variety of pest species, i.e., due to Herbivore-specific reactions. Exploitation of herbal resistances, repeatedly determined by wild relatives to fight against pest species that consume a selected plant organ or tissue (i.e., aphids, whiteflies, and other phloem-feeding insects). R-gene-based resistance is governed on the basis of "gene-for-gene" interaction, where a compound secreted by way of insect which exclusively allowing the plant to provoke a protection response. R-gene-mediated resistance has not been mounted for the plant tissue-chewing insects (i.e., Lepidoptera and Coleopteran). Few dominant R-genes offer resistance towards phloem-feeders, i.e., has been cloned (e.g., Mi-1.2, VAT, and BPH16) and drastically utilized in agricultural settings with marker-assisted breeding (Doorn and Vos, 2013).

Sucking pests present a serious challenge, because transgenic technologies that are available are not strong. One way to control these pests is to supply fusion protein toxins. The negative results of fusion proteins in the *Phenacoccus solenopsis* (mealybug) phloem-flowering insect-pest were assessed translationally using two protein pollutants, Hvt (the pur-atracotoxin of *Hadronyche versuta*) and onion leaf lectin. Hvt was cloned in the fusion protein form, both N-terminally (HL) and C-terminally (LH); the use of a potato virus X (PVX) vector in Nicotiana tobacco was expressed transiently (Quillis et al., 2014). It is shown that the HL fusion protein is stronger than Search Results *Phenacoccus solenopses* have a death rate of 83% as compared to LH, which has caused a death rate of 65%. Hvt and lectin alone triggered 42 and 45% of the protein, respectively, under the same conditions. The HL protein has been shown to be more stable than the LH protein by computational studies with both fusion proteins. The translational fusion of two insecticide proteins has advanced the insecticide activity in relation to every protein and is possible for effective sucking pest handling in transgenic plants (Javaid et al., 2018). Secondary plant metabolites can be of remarkable value in order to provide robust plant life.

The phrases of the currently used insecticides such as pyrethroid, neonicotinoid, and butenolides contain small lipophilic molecules (SLMs) and similar phrases that offer extra powerful control than patterns of resistance used by breeding techniques. Crop flora does not have the SLMs that resist pesticides its wild ancestors. Biosynthesis-based resistance trends in SLMs promise new opportunities for crop resistance. The genetic production of secondary metabolite pathways creates insecticidal compounds; however, they give special processes that are extra troubling as previous genetic manipulation techniques involved in plant colonization and improvements such as insect pheromones. The use of resistance patterns not articulated in the context of a sustainable seed approach is the basis of the development of perennial arable vegetation blanked in sentinel technology (Birkett and Pickett, 2009).

12.2 COLORADO POTATO BEETLE (CPB)

One of the plagues that significantly lower production is CPB. The CPB 's fast, flexible, and different lifestyles round has particularly adverse feeding habits and excessive adaptability to various environmental stresses (Radcliffe and Lagnaoui, 2007). Adults, who have wintered in their soil, migrate, and feed regularly on the subject margins into the tomato fields. The human being and the larvae feed on the leaves, the final boom and gentle stems and the younger faces of the plant. It can defoliate tomato vegetation under severe infestations and cause significant cultivation loss (Hare and Moore, 1988).

Traces of tomato introgression evolved through the collection of *Solanum pennelli* genome introgressions of the *Solanum* genome cv. *lycopersicum* M82. These traces contain introduction of the segments described in the *S. pennellii* genome. *S. Pennellii's* introgression population Lycopersicum has been developed for the examination of genetic components of tomato production and fruit quality, and was widely used for mapping various critical, biological, and abiotic pressure reactions (QTL) traits (Alseekh et al., 2013). The use of the S was found for QTLs. The fitness, salt tolerance, and antioxidant capacity strains of *pennellii*. Tomato defensives which have been mapped to trace introgression to investigate the range of trichome-symptomatically synthesized compounds, production of monoterpenes, acyl sugar biosynthetic enzymes and the biosynthetic pathway of glycoalkaloids (Schilmiller et al., 2009, 2015).

Populations of tomato genetic mapping produce natural CPB resistance type. CPB bioassays with *Solanum pennellii* introgressions in *S. lycopersicum*. S. *pennellii* introgressions recognized by M82. *S. pennellii* conferred CPB susceptibilities on chromosomes 1 and 6 while introgressions conferred greater chromosome resistance on chromosomes 1, 8, and 10. Mapping of CPB Resistance to the use of a cross between S on 113 recombinant inbred lines (RILs). The massive quantitative loci characteristics for chromosomes 6 and 8 were identified by *Lycopersicum* cv. UC-204B and *Solanum galapagense*. The allele of *S. galapagense* was linked to decreased leaf harm and larval rise. Both the techniques of genetic mapping converged in chromosome 6 at that place, which could also have significant functions in the protection of tomatoes towards CPB herbivory.

Although quantitative feature-related experiments with acyl sugar near-isogenic lines (NILs) and transgenic glycoalkaloid poor, GAME9 glycoalkaloid overproduction traces have been identified as a source of genetic mapping, they show no great effect on CPB performance from insect-protective metabolites, any other time (Vargas-Ortiz et al., 2018).

However, pesticide strategy is ineffective because the pesticide resistance to all pesticides is improved. Max Planck Institutes of Molecular Plant Physiology in Potsdam-Golm and Chemical Ecology in Jena have proven that potato plant life can be blanketed from herbivory using RNA interference (RNAi). RNAi is a sort of gene law that naturally happens in eukaryotes. In plant life, fungi, and bugs, it is also used for safety in opposition to sure viruses. During infection, many viral pathogens transfer their genetic statistics into the host cells as double-stranded RNA (dsRNA). Replication of viral RNA leads to high amounts of dsRNA that is diagnosed with the aid of the host's RNAi machine and chopped up into smaller RNA fragments, known as siRNAs (small interfering RNAs). However, the RNAi mechanism also can be exploited to knock down any desired gene, by means of tailoring dsRNA to goal the gene's messenger RNA (mRNA). When the centered mRNA destroyed, synthesis of the encoded protein may be diminished or blocked completely. They genetically modified plants to allow their chloroplasts to build up double-stranded RNAs (dsRNAs) targeted against vital beetle genes (Zhang et al., 2015). For the reorganization of the genetic foundation of herbivory and insecticide resistance, structural, and useful genomic changes related to diverse arthropod species has been investigated by using genome sequencing, transcriptomics, and community annotation. There are two aspects that would help rapid evolutionary change consist of transposable elements, which comprise at the least 17% of the genome and are hastily evolving in contrast to other Coleoptera. High levels of nucleotide diversity in unexpectedly developing pest populations have also been reported. Adaptations to plant feeding are glaring in gene expansions and differential expression of digestive enzymes in intestine tissues, in addition to expansions of gustatory receptors for bitter tasting. Surprisingly, the suite of genes involved in insecticide resistance is similar to other beetles. Finally, duplications inside the RNAi pathway might provide an explanation for why *Leptinotarsa decemlineata* has excessive sensitivity to dsRNA. The *L. decemlineata* genome offers possibilities to examine a broad range of phenotypes and to increase sustainable modes to control this important pest of tomato (Schoville et al., 2018).

CPB is immune to all main groups of pesticides, consisting of organophosphates (OP) and carbamates. The target site of OP and carbamate insecticides is the same as they inhibit the activity of AChE. The feature of acetylcholinesterase (AChE) is the degradation of acetylcholine (ACh-neurotransmitter) in the insect cholinergic synapses. Mutations inside the

AChE-encoding locus were proven to confer target website insensitivity to OP and carbamate insecticides, leading to modification of AChE (MACE). A range of amino acid substitutions in AChE confer insecticide resistance, and those mutations generally reside close to or within the lively website of the enzyme. Such AChE mutations, associated with insecticide resistance, mostly referred to as Ace in Drosophila, which include *L. decemlineata*. Based on bioassays and literature, modified/insensitive AChE confers two essential patterns of resistance, i.e., OPs/carbamates. Resistance is characterized through notably higher resistance ratio (RR) (much extra reduction in the sensitivity of AChE at the biochemical level) to carbamates than to OP pesticides. Pattern II resistance is characterized by means of resistance ratios (and/or reductions in the sensitivity of AChE) which are approximately equal for each carbamates and Ops (Kostic et al., 2015).

12.3 TOMATO FRUIT WORM (*HELICOVERPA ZEA*)

It is an important polyphagous insect pest of tomato and other crops, depending at the host, which consist of tomato fruit worm, corn earworm. The coloration of the adults varies; however, forewings appear yellowish-brown with a darkish spot close to the center. On the hind wings, a darkish band flows towards the wing margins, but the wing margin and closing element seems creamy white (Gilligan and Passoa, 2014). Eggs are faded green, turning yellow near hatching, and appear as flattened sphere with numerous ridges bobbing up from the top center. High tiers of crop damage have observed late within the season (Flint, 1998). Weather regimes and wind styles are the most influential matters for migration of the adults (Sandstrom et al., 2007). The moths can travel numerous hundred to kilometers in a single night (Westbrook, 2008).

Heavy and frequent losses in fields because of the prevalence of harm in trees, ornamentals, and greenhouse crops. Larvae feed on many plant parts, preferring flowers and culmination. Late plantings may result in 100% fruit damage. A movement threshold regime with minimum fruit damage, reduced variety of insecticide sprays, and most yield returns became determined while the first insecticidal application turned into made when 10% of the sampled tomato flora were infested with *H. zea* eggs, and the second one spray turned into made whilst extra than three damaged culmination/one hundred market-sized unripe end result were determined (Kuhar et al., 2006).

The excessive fecundity rate, polyphagous feeding activity, excessive mobility, seasonal migration, and facultative pupal diapause of *H. zea* contribute to its unnecessary pest status (Capinera, 2001). A physiological edition to discuss choices for crops and pesticides. Specific date of the seedling, approximate soil damp, natural matter and nitrogen status is initialized. Specific date of the seedling, approximate soil damp, natural matter and nitrogen status is initialized. TOMSIM is initialized by variated postponed system to accommodate the developmental variability (Colligne et al., 2007) differs with the crop variation and two currently existing pest model (tomato fruit worm and beet armyworm). TOMSIM preliminary validation runs provide for the ability to use the decision-making process to optimize irrigation (and rate) selections, fertilization (nitrogen) selection and control of fruit worms and beets by armyworms (Wilson et al., 1987).

Lycopersicon hirsutum (PI 134417) vegetation grown below the long-day regime exhibited extra mortality from foliage larvae grown under a short-day regime. *Lycopersicon sexta* (L.) larvae restricted from foliage The *Lycopersicon hirsutum*. In the foliage of plants under the long rather than long-term regimes, the toxin 2-Tridecanone, essential within the insect resistance PI 134417, has evolved into a considerably more abundant product. No two-tridecanone ranges or resistance were driven by the intensity of light (Kennedy and Yamamoto, 1979). The presence or density of glandular trichomes is a part of the resistance of insects. The main elements of glandular trichome exudates, i.e., the acyl sugar and 2-tridecanones secreted by *Lycopersicon pennelli* type IV trichomes and L type VI trichomes. *Hirsutum* mediates the resistance of the insect. The genetics of these tomato characters appear complex. This means that the QTLs associated with the trichome density, the accretion of the acyl sugar and 2-tridecanone mediated insect resistance are mapped and marker-assisted. Tomato cultivars can develop sugar ester synthesis and accelerate the content of two tridecanons significantly increase the level of insect resistance in cultivated tomatoes (Sajjanar and Balikai, 2009). A pr1 gene is provided in tomato cultivar and this gene is used for the use of tomato mosaic virus (ToMV) resistance. The result was found that the transgenic lines with gene NPR1 were all common in morphology in four generations. Tropovirus inflects a limited component on the tomato plant and is conducted according to an experiment to resist tospovirus. Tomatoes have shown proven resistance to leaf curl disease with the help of the antisense method. The flora produced by the transgenic technology is very advantageous for tomatoes (*Solanum esculentum*) from insect attack. Cry1A, an especially powerful protein against *Helicoverpa Nigeria* (Iqbal et al., 2019).

12.4 BEET ARMYWORM

A well-sized pest occurs every year in tomato fields in Beet Armyworms. The most important caterpillar attacking tomatoes is the Beet Armyworm. Beet armyworm attacks both the fruit and the leaf, creating round or irregular holes, either single or closely grouped. Fruit food is often shallow and superficial for the processing of tomatoes, as ultimately the maximum wounds are dry. Little loss will result from damage caused by feeding while the percentage of processing is used for paste or juice. But loss is larger, as decaying organisms enter wounds and redden the fruit at once, or as faces and caterpillars are kept in the fruit. Eggs put in clusters, i.e., on the leaves. Covered by the lady moth (extra 100 eggs per cluster) with skin-like scales. Newly hatched larvae feed on the leaves near the egg cluster and systematically disperse with their development. Older larvae feed on fruit and leaves. Larvae are normally inexperienced stupid with many fine wave stripes of light color on both sides of the lower back and a wider strip; often have a dark spot on the thorax's face on the second right leg. The color of some populations is different, and the spot is missing. In a melancholy manner made on or just under the soil, the pupa of a tomato fruit computer virus pupates. The adult moth has an approximately 1-inch range of mottled gray and brown. In warm weather, the life cycle lasts about a month and 3 or 5 generations a year. Beet armyworm ate far fewer OP-flowers than SP-plant life foliages (both young and old leaves). OP feeding larvae generally showed reduced weight gains compared to those feeding on SP leaves (Bhonwong et al., 2009). Beet armyworms, the use in the same genetic background of genetically modified tomato vegetation with modified PPO expression. Beet armyworm infestation suggests that the effect of PPO on younger larvae is stronger than on older larvae and that PPO results on older tomato leaves are stronger than in younger leaves. Beet armyworm feeding is subjected to numerous secondary reactions, forming ROS in plant tissue under acidic conditions (Guyot et al., 1996). This could be a significant reaction in tomatoes, where extensive PPO and phenolic compounds have been stored in trichomes, the rupture of which could rapidly generate large quinones as soon as possible or in low oxygen midgut (Barbehenn et al., 2007).

S. exigua transcriptome consists of the use of *Illumina solexa* platform by eggs, larvae, pupae, adults, and girls. About 31,414 contigs culminated in the transcriptome of Trinity. Of these, 18,592 have been annotated by BLAST searches as protein-coding genes against the NCBI database. The knockdown by dsRNAs or siRNAs of critical insect genes is a viable mechanism for the management of insect pests. A successful RNAi-mediated method of pest control is to find suitable target genes. We have selected nine candidate genes to show effective target genes in the beet armyworm. The use of the RACE approach to chemical synthesized siRNAs has been amplified in the sequences of these genes. Injection into 4th instars larvae 2 μl siRNA (2 μg/μl) to disable their respective target genes. The abundance of mRNA in target genes decreased in the injection of siRNAs to specific levels (CH 20–94.3%). A significant excessive degree of death was observed compared to bad control ($P<.05$), due to the knockdown of 8 genes including chitinase7, PGCP, chitinase 1, ATPase, tubulin1, arf2 andarf1.

Approximately 80% of the existing five genes (PGCP, chitinose1, tubulin1, tubulin2, and helicase) insects in the siRNA-treated cluster were delayed. Around 12.5% of survivors showed "half-ecdysis" in the chitinase1-siRNA and chitinase7-siRNA teams given. The color of the body was black until injected in arf1-siRNA and arf2-siRNA teams. In a nutshell, the genetic resource for the genes in *S. exigua* is transcriptome. In summary, the genetic resource for the S genetic identification can be a useful transcriptome (Hang et al., 2013). Insecticidal toxins produced during the vegetative process by the successful selection against certain insect pests by *Bacillus thuringensis* (*Bt*). The current knowledge, a *Bt* protein cluster is a motivationally different from Bt *Cry* toxins that do not share the same mode of action. Evaluation of organic phenomenon changes due to the sublethal dose therapy of Vip3Aa (causing an increase of 99%) at 8–24 hours once feeding. A good transcriptional reaction was determined to be aggravated by the poison with 19% of the unigenous array. There was a terribly similar amount of up- and down-regulated unigenes. The number of genes with a regulated expression at eight number, i.e., same as the number of genes whose expression was regulated once in twenty-four-hour. The up-regulated sequences were enriched for genes involved in innate response, in infective agent response adore antimicrobial peptides (AMPs), and repeat genes. The down-regulated sequences were in the major unigenes with similarities to genes involved in metabolism. Gens were found to be slightly exhausted concerning the mode of action of *Bt Cry* proteins. The genome-wide analysis of Vip3Aa insect lepidopterous insects can allow for the planning of simple gadfly management methods (Bel et al., 2013).

It's been found in S. male-specific antenna OR6 had been broadly triggered by OR13 and OR16 with multiple secretional elements, but OR6 did not respond to any secretional compound (Liu et al., 2015). OBP (OBP1-OBP11) expression profiles in *Solanum*. In general, the OBP1-4 and OBP10 are restricted to female antennas, while OBP7 is directed to male-antenna expression (Zhu et al., 2013). In addition, PBP/GOBPs have been demonstrated to have primary signals, however, with entirely different patterns of sex-biased communication, in both male and female antennae (Liu et al., 2013). For insect combinations and oviposition, section, and plant odorants are required. In the *Spodoptera exigua*, molecular mechanisms of the control and odorant signaling using qRT-PCRs for ABP, OBP, PBP, and organic processes and their reactions before exposure to sex secretion or volatilities. Biological time, as is their behavioral rhythm, within the profiles of some chemo-sensorial antennas. OR3, OR6, OR11, OR13, OR16, OR18, Orco, ABP2, OBP1, OBS7 and PBP1 and EAG expression profiles were also significantly affected when exposed to artificial sexual pheromones and plant variability.

Their organic phenomenon is the sensitivity to sexual secretions in chemo-sensory genes and modality of meaning through insect preexposure to plant and pheromone compounds. It understands the molecular mechanisms of secretion and the application by pairing interruptions or mass attacks of sex pheromones and seed volatiles (OECD, 2017). Large percentage of N together increased the density of *Spodoptera exigua* with two inhibitor enzymes, enzyme (CAT) and enzyme (SOD). The next proportion of P plays an additional role in enhancing CAT and SOD activities with multiple saprozoic intensities. Higher P undoubtedly increased the activity of neurotranetesterases (AChE) at lower defoliation levels, but the next N proportion increased the activity of AChE at higher defoliation levels. Higher N-rate led to decreased short-run defoliation carboxylesterase (CarE) activity. The CarE activity distinction between fertilizer groups was very low when defoliation intensity increased. The relation between the damage to *S. exigua*, the accumulation of biomass of its *Brassica rapa* host plant and also the influence on this interaction of the quantitative N/P ratio in the plant fertilizer. Systematic analysis was provided on the biomass of *B. rapa* and also the activity of metabolic enzymes of *S. exigua* beneath totally different treatments (Wang et al., 2018).

Jasmonates deal with plant safety and work with various host specializations in prompt induction of defensive responses against insect herbivores of different food guilds. Transgenic plants in a synthesis of jasmonic acid (JA) that are required as a defense feature in the tomato, poignant oviposition and feeding behavior of the expert genus Manduca 6ta for the formation of organ trichome and trichome-borne metabolites. JA was not necessary for the indigenous induction of organic defense phenomenon once wounded. JA precursor oxophytodianoic acid (OPDA) substituted organic defense phenomenon regulator in JA-deficient plants that maintain strong resistance to M. 6 larvae. 6ta larvae. Several researchers found that JA and OPDA demonstrate their *Spodoptera demandua* defense mechanisms (Bosch et al., 2014). Under laboratory, growth, and greenhouse conditions, the transgenic crops showed traditional growths, development, and replicas. The PPO expression of antibiotics has significantly accumulated while the PPO over-expression of tomato plants increases resistance to the Pseudomonas syringae type. Similarly, the resistance of transgenic plants to various insects (*Spodoptera litura*, F.) and to beet armyworm (*Spodoptera exigua*, Hübner) increased with PPO-overexpressants. Larvae feeding on plants with repressed PPO are more rapidly exposed in larvae with more feeding. Similar results also observed in the increasing of body mass, foliage consumption, and survival were conjointly discovered with CPBs (*Leptinotarsa decemlineata* (Say) feeding on antisense PPO transgenic tomatoes. There are already mentioned the PPO's defensive mechanisms and its interaction with alternative defense proteins. In comparison to transformed controls, transgenic plants with suppressed PPO had a lot of favorable water connections, as well as mitigated photography inhibition, suggesting that PPO may have an ongoing role in developing plant water stress and the potential to inhibit and cause photooxidative damage, which will be unrelated to any impact on the Mahler response.

These findings substantiate the defensive function of PPO in the modulation of PDO activity in specific tissues, however, the effect of PPO on postharvest efficiency, as well as water stress physiology, should be considered, may produce broad-spectrum resistance to any unwealthy and bug pests at once. Relevance of antisense technology to decode the role of PPO in alternative plant species as well as industrial uses to the investigate the functional study of tomato PPO will open a new insight (Thipyapong et al., 2017).

12.5 YELLOW ARMYWORM

The yellow stripy armyworms are common in the North American country, the genus *Spodoptera ornithogalli* (Guenée). Production takes place between egg and adult in 30 to 90 days due to favorable environmental and temperature conditions. Females lay eggs with individual females able to birth > 2000 eggs in groups of two hundred and five hundred over the period. Egg larvae and six larval instars (stadiums between molting), Capinera EY216. EY216. There are forty to fifty linear units of old caterpillars (1.5 to 2.0 inches) in length and brown, gray to black with two distinct yellow-orange lines extending across all the abdomen sides. The prime of each abdominal segment is the black triangle form spots. The top is black and has light-colored reversed larvae for about 3 weeks and a gift of black's salt on the plant leaves. The top is brown with black markings. The yellow stripy armyworm overwinters with the animal's cell lined with silk, as an insect in the soil. Every year there are 3 to 4 generations. Adults in the Gregorian calendar emerge from pupae months to month. The front wings of adults are gray-brown with light and dark markings. The center of the wings is near the tan to the brown diagonal band. The hind wings are slim and white (Currey, 2018).

The most important tomato pests considered to be tomato pinworm (*Tuta absoluta*). These include, RNAi, a legitimate means of enshrining new management approaches against polymer interference. A pair of 5 μg of dsRNA were available to reduce the organic pheromone by dsRNAs target of AChE, nicotinic neurotransmitter Alpha vi (nAChR) injection, Ryanodine (RyR) receptors and root delivery during injection procedures. The root absorption dsRNA @5 showed the ability of this delivery mechanism to affect the 47–69% decrease of the sequence and thus mortality (from 67.1 to 80.5%) for the treated specimens. Each *Tuta absoluta* was affected by dsRNAs given (both injected and root administered). Pupae and utter prepupae. *Tuta absoluta* correct destinations may be the chosen AChE, nAChRs, and RyRs genes. Under absolute management, dsRNAs (Majediani et al., 2013) in the plant delivery. Mechanisms of pesticide resistance include the most frequent: (a) extended detoxification of organic compound substitutes or deletions that are comparable to P450s (CYP450), and of reduced activity response; and reduced penetration (Li et al., 2007; Feyereisen et al., 2015); (b) extended metabolic detoxification through the use of organic compound substitute or deletion that leads to lower sensitivity. The mechanisms most commonly increased the level of detoxification enzymes and changed target sites. Absolute to the proliferation of chemical pesticide categories. Pyrethroid resistance mechanisms during this species by showing the presence of the target-site mutations in the voltage-gated atomic channel number 11 to those of the other persecutor species equally (Haddi et al., 2012; Rinkevich et al., 2013). Resistance to three mutations commonly linked to resistance to pyrethroids across insect species (M917 T, T929I and L1014) has been recorded in para-type 11 channel IIS4-IIS6 bioassay for resistant strains region and genotyping of different *Tuta absoluta* populations. The presence of an L1014F mutation in the most frequent populations has been shown by from countries in South America and Europe (Haddi et al., 2012; Timber et al., 2015). In addition, two other significantly enhancing pyrethroid resistance (called super-cdr) were gift in connection with this mutation, which led to certain pyrethroid failures. Detoxifying enzymes, including esterases and CYP450s, may metabolize pyrethroids, but these mechanisms seem to be limited in *Tuta absoluta* completely. However, over-expressed CYP450 and target-site mutations will have a strong pyrethroid resistance effect, as demonstrated in the arthropod gambiae genus (Vontas et al., 2018). Esterase and glutathione S-transferases determined theatrical role. Instead of its activation, CYP450s probably mediate car-tapes demethylation and sulfoxidation as detoxification mechanism (Lee et al., 2004), which can justify removing piperonyl butoxide in *T. cartapes* resistance completely utter. Another early engine-assisted major involvement in abamectin resistance of CYP450 was earlier (Siqueira et al., 2001).

12.6 FRUIT AND SHOOT BORER

The borer of tomato fruits is one of the most dangerous tomato pests. The adult sets eggs in the high cover on top and bottom of the primary four leaves. Until early or late, the larvae scratch the tomato leaves. The borer bores the fruit and makes it unsuitable to be promoted. More than 80% of the fruits are split in extreme cases of infestation. In the event of larval injuries host fruit as seeds are lost and flesh.1 brute per fruit is used to produce the fruit that is not available on the market as the larvae develop inside the fruit. Injury of the larvae, in particular in the holes of fruit larvae that emerge from the fruit which can contribute to the entry of microorganisms that cause disease in the plant.

Conogethes genus is a large cosmopolitan moth taxon, which is taxonomically complex. To date, 24 species were identified and deposited in the BOLD (barcode of life data systems) (http:/www.boldsystems.org) barcodes by DNA barcodes. *Conogethes punctiferalis'* type locality in India, so many closely linked species can be included, but taxonomical overhaul of them has long been neglected. The integration of several approaches, such as conventional taxonomy, the DNA barcode, adult, and larvae electrophysiologic and behavioral responses, host-plant relations, hybridization experiments, biochemical analyzes, and pheromone reactions analyzes, confirmed that the population belongs to two different species, Conogethese pheromone composites, rivet, and cardamom. Conogethes have been well studied in bioecology and management. There were over 31 different host plants reported for the borer. Thus, adaptation in cultivated ecosystems is improved (Shashank et al., 2015). *Bacillus thuringiensis* (Bt) was passed to the tomato through co-cultivating cotyledonal explants using *Agrobacterium tumefaciens* for the synthetic cry1Ac gene coding for insecticidal crystal protein (ICP). Kanamycin-resistant transforming plants have been regenerated. Experiments with hybridization have shown the incorporation of genes and the transgenic plants gene number. Doppel-antibody ELISA test reveals that the leaves of transgenic plants contain high levels of Bt ICP content. The expression resulted in a high level of protection of transgenic plant leaves and fruits against the larvae of tomato fruit borer (Table 12.1) (Mandaokar et al., 2000).

TABLE 12.1 List of Transgenic Approaches for the Fruit Borer Control

SL. No.	Insect	Approaches	References
1.	Fruit borer	Cry1Ac gene code from *Bacillus thuringiensis* (Bt) has been transferred from cotyledonary explant cottage to tomato with *Agrobacterium tumefaciens* for insecticidal crystal protein (ICP). Hybridization in transgenic plants includes gene integration and gene copy number. The ELISA study reveals high Bt IPT expression rates on transgenic plant leaves. Double-antibody sandwich.	Mandaokar et al. (2000)
		Development in transgenic tomato fruit of human b-secretase (BACE1) acts as an immune response vaccine antigen. The Yersinia f1-v fusion gene-encoded transgenic tomato plants, an antigen fusion protein. F1-V. F1-V's immunogenicity in a subcutaneous Y challenge. Mice vaccinated orally in freeze-dried fruit had been confirmed.	Kim et al. (2012)
		GM tomatoes promise to enhance the quality of human life in field trials their possibilities have rarely been tested. Only if biotechnology plants contribute to the success of their development is the safety of GM crops and their benefit to breeders and consumers achieved.	Gerszberg et al. (2015)
		Synthetic cry genes, protease inhibitors, trypsin inhibitors, and cystatin genes have been used to incorporate insect and nematode resistance.	Parmar et al. (2017)
		Cry1Ab gene incorporation and ancestry in T0. PCR, RT-PCR, and Southern blot hybridization study determined transgenic plants and their progenies. Double enzyme-connected immuno-sorbent testing (DAS-ELISA) has controlled toxin expression. Analyzes of Southern blot. The cry1Ab copy of the cry1Ab gene was evident at the Western blot analysis in transgenical plants generated by T4, T5 and T6, with high-expression Ab25 E transgenic line and expression of Cry1Ab molecular toxin ~65 kDa. Binding test with the Cry1Ab protein of Ab25 E transgenic tomato line conducted with a partially purified receptor, confirmed effective protein/protein interaction with Cry1Ab.	Koul et al. (2014)

The truncated *Bacillus thuringiensis* factor Bt-cry1Ab was used for the events and for the selection of transgenic events expressed by an Agrobacterium-mediated sheet disk transformation process in a very commercially vital type of tomato (*Solanum* genus *lycopersicum* L). PCR RT-PCR has been determined to combine and inherit the cry1Ab element in transgenic T0 plants and their progenies, followed by Southern blot conjugation analysis. Double-protein enzyme enzyme-linked immunosorbent research (DAS-ELISA) has controlled the expression of poison. The transgenic line Ab25 E, showing a total soluble macromolecule (TSP) poison of 0.47 ± 0.01%, was finally chosen in the generation T4, which was a segregated population with mortality of 100 pc for the second larvae of arthropod *H. armigration* and *S. armigera*. Southern blot analysis Knowledge of a single copy of cry1Ab in high-pressure Ab25 E transgenic line and expression of Cry1Ab molecular mass poison ~65 kDa was evident in Western blot analysis of transgenic T4, T5, and T6 generation.

Binding of the receptor assay with partially pure Ab25 E transgenic Cry1Ab macromolecule, confirmed economical protein-protein interaction between Cry1Ab poison and each insect receptor(s). In the case of conventional transgenic regeneration, plant growth, and fruit production during this transgenic line, no impact was observed at the higher level of Cry1Ab poison (alternative to zero.4 7% ± 0.01%). This high-expression homozygous transgenic line of Cry1Ab is a helpful candidate for the introgression of essential scientific characteristics (Koul et al., 2014) in tomato breeding. Maximum genomic polymer from samples obtained and incubated from field moths and larvae. The QIAmp ® DNA Kit (Qiagen GmbH, Hilden, Germany) has been used for polymer extraction, in accordance with the protocol of the producer. Using the NanoDrop ® ND-1000 photometer (NanoDrop Technologies, Inc.) extracted polymer concentrations have been measured. In subsequent PCR amplification, one ng of polymer was used. The agreement sequence has been used to scan matching sequences using a BLASTN (Basic Native Alignment Search Tool). The biological process tree and COI series matched the *Tuta absoluta* Absolute (100% identity) KJ657881, KJ657680, KC852871 (first report from India, Asokan et al., 2015 (unpublished data)), KP814057 and JQ749676 sequences (First report from India, Asokan et al., 2015 (unpublished data)). The genetic differences between zero.00 and 0.59 between *Tuta absoluta* were demonstrated by biological process analytics. The GeneBank has an absolute specimen and different samples, ranging from 12.67 to 12.84 from T. Samples utter and subsequently the *Ephysteris promptella* outgroup. Sequencing by polymer barcodes of the mitochondrial cytochrome enzyme fractional currency unit I (COI) coupled Botswana's lure with T catches. Absolute samples from *Tuta absoluta* (accession variety JQ749676) and India (adhesion number KP793741), therefore, support a Morphological Description which was exhausted Botswana and Africa by the characteristic host harm symptoms and sex-specific secretion households. The lepidopterous insect identity has been identified as Tunisia. Absolutely, it suggests the very invasive alien cuss and its geographical distribution is increasing.

RNAi is currently commonly used in purposeful genetic studies as a cistron silencing mechanism involving the supply of polymer molecules that suit a specified target gene sequence. The potential use of RNAi-mediated management of farm insect pests has quickly emerged. A method for gene-building larvae that prey on these plants, leading to larval phenotypes, varying from loss of craving to death, could be used to assemble transgenic plants that express dsRNA molecules focused on essential insect cistrons. This study appears to show that the RNAi is aimed at a major threat to the production of commercial tomatoes (Tuta absoluta). Tomato leaf laborer (*Tuta absoluta*) The reportable RNAi reaction in alternative gadfly species has been assisted by the selection of 2 target genes (Vacuolar ATPase-A and Critical Amino acid kinase). The synthetic diet shortage evident for *Tuta absoluta*. Absolutely, two approaches are used for the delivery of dsRNA in tomato flyers. A primary approach to dsRNA use was supported by leaflets, and the second was a well-established methodology for silencing plant genes, 'plant-induced cistron silencing' (PITGS), which was used in platform transcribed dsRNA primarily for insect genes. Tuta absoluta larvae showing a reduction of 60% in target gene transcript accumulation, an increased larval mortality, and fewer leaf injuries in the associated nursing leaves with target cistrons dsRNA. The transgenic 'Micro-Tom' plants that express the pin sequences on each gene were then generated and the foliar T injury discount was established. The transgenic 'Micro-Tom' plants that express the pin sequences on each gene were then generated and the foliar T injury discount was established. Our results show the feasibility of RNAi to dominate the crucial tomato pest as an additional approach (Camargo et al., 2016).

The application of tomato plant models was important for the mechanisms of interactions between plant and microbe. In particular, models contributed to the development of insights into the plant system (Jones and Dangl, 2006). As the scientific community is prepared to dissect the dialog established between the host and its microorganism on a molecular basis and to understand which lies behind the susceptibility to disease and the resistance of plant disease because of the wide range of pathogens ready to infect these two plant models and their entirely different method of infection. However, the clearing of the PRR and R proteins has been activated, and the downstream signage has been activated to achieve resistance. Several key players focused on increasing resistance in crop plants. However, it has been rather restricted overall transfer of data from basic analysis to crop plants. There are few examples of how much economic development has been produced, and progress has been disappointingly slow in recent years despite how much we have learned about plant resistance. A consistent move is to characterize the restrictive protection sequence networks in model plants that can speed up the transfer of data to crops. Some improvements in plant breeders to engineering crops with inducible defenses against pathogens that lack the necessary fitness prices do not make it attractive to farmers through the modern technologies of artificial biology approaches. This growing information field will be used within the models to create resilient resistance in large crop plants in order to maintain yields in the face of favorable weather conditions, as a long-term challenge for plant pathology.

The genus of Conogethes could be a giant, taxonomical, and cosmopolitan advanced lepidopteron taxon known as DNA bar codes for the BOLD (http:/www.boldsystems.org) Barcode of Life Knowledge Systems. In the *Conogetes punctiferalis* neighborhood in India, a large number of closely related species could also be closed, although the categorization of the species was neglected for a prolonged period of time. Integration of numerous approaches such as typical taxonomy, DNA code, electrophysiological, and behavioral responses of adults and larvae, patterns of host-plan relationships, cross-breeding, organic chemical analysis, secretion element analysis and responses. *Agrobacterium* media transformation technique has reshaped the attack of these insects, changed BT genes (Cry1Ab), into highly productive industrial tomato selection. In reducing the attack of lepidopteran insects, this transformation has a polar role. Love apple arms are extremely sensitive and are ultimately susceptible to infectious attacks and cause hazardous diseases prevalence in tomatoes. CsiHPL1 is shown to be highly effective in strengthening the arms of tomato plants in the plastid localized tea sequence.

In addition, the processing of CsiHPL1 does not only protect tomato cultivation from all fungal attacks but jointly produce resistance to animals in tomatoes. In cuticular development, several transcription factors are involved, such as SlSHINE3, Associate of Arabidopsis *Arabidopsis* WIN/SHN3 Nursing Orthology. All the roles of the leaf cuticle are the contact between pathogens and the environment and serve as a barrier to physical stress and pathogens. In addition, it works jointly in developing and controlling the plant weapons as a chemical detergent. Several factors of transcription are listed previously to be concerned about the improvement of the combination of cuticle in the leaves that establish the resistance against all types of stresses. The most recent research isolated, cloned, and transferred into an industrial tomato selection micro-organism-resistant N-sequence, from transgenic tobacco plant ultimately produced resistance to mosaic virus in tomato. In the tomato crop losses are caused by ToMV. Return losses of 25–30% in tomato production are speculated as a result of this assault on microspecies. The generation of a 90% resistance to TMV and ToMV in tomatoes (CP genes) is found to be coat macromolecule genes. CP genes were introduced into Tomato genetics by the transformation of agrobacteria to avoid the microorganism. The effects of these CP genes on fruit yield are not negative. The effects of these CP genes on fruit yield are not negative. Tomato yellow leaf curl virus (TYLCV) causes conjoint loss of fruit in tomato crops in Brobdingnagian. The genetic transformation of the capsid protein (VI) genes in TYLCV was regulated by this microorganism disease. This strategy has reduced the risk of *Solanum spp.* curling leaves (Ahuja and Flaung, 2014; Xin et al., 2014; Buxdorf et al., 2014; Whitham et al., 1996; Kunik et al., 1994).

KEYWORDS

- **acetylcholinesterase**
- ***Agrobacterium* transformation**
- ***Bt* genes**
- **Colorado potato beetle**
- **double-stranded RNAs**
- **insect pests**
- **resistance**

REFERENCES

Ahuja, M., & Fladung, M., (2014). Integration and inheritance of transgenes in crop plants and trees. *Tree Genetics and Genomes,* 1–12.

Alam, T., Tanweer, G., & Goyal, G. K., (2007). Stewart postharvest review, packaging and storage of tomato puree and paste. *Res. Art.,* 3(5), 1–8.

Bel, Y., Jakubowska, A. K., Costa, J., Herrero, S., & Escriche, B., (2013). Comprehensive analysis of gene expression profiles of the beet armyworm *Spodoptera exigua* larvae challenged with *Bacillus thuringiensis* Vip3Aa toxin. *PLoS One, 8*(12), e81927. https://doi.org/10.1371/journal.pone.0081927.

Bessin, R., (2011). Extension specialist. *Tomato Insect IPM Guidelines.* Kentucky Agricultural College. University of Agriculture Food and Environment. Intact-313.

Bhonwong, A., Stout, M. J., Attajarusit, J., & Tantasawat, P., (2009). Defensive role of tomato polyphenol oxidases against cotton bollworm *Helicoverpa armigera* and beet armyworm *Spodoptera exigua. Journal of Chemical Ecology.* https://doi.org/10.1007/s10886-008-9571-7.

Birkett, M. A., & Pickett, J. A., (2014). Prospects of genetic engineering for robust insect resistance. *Current Opinion in Plant Biology, 19,* 59–67.

Bosch, M., Sonja, B., Andreas, S., & Annick, S., (2014). Jasmonate-dependent induction of polyphenol oxidase activity in tomato foliage is important for defense against *Spodoptera exigua* but not against *Manduca sexta. BMC Plant Biology.* https://doi.org/10.1186/s12870-014-0257-8.

Buxdorf, K., Rubinsky, G., Barda, O., Burdman, S., Aharoni, A., et al., (2014). The transcription factor SlSHINE3 modulates defense responses in tomato plants. *Plant Molecular Biology, 84*(1/2), 37–47.

Camargo, R. A., Barbosa, G. O., Possignolo, I. P., Peres, L. E. P., Lam, E., Lima, J. E., Figueira, A., & Marques-Souza, H., (2016). RNA interference as a gene silencing tool to control *Tuta absoluta* in tomato (*Solanum lycopersicum*). *Peer J., 4,* e2673. https://doi.org/10.7717/peerj.2673.

Capinera, J. L., (2001). *Handbook of Vegetable Pests* (p. 729). Academic Press, San Diego.

Collinge, D. B., Lund, O. S., & Thordal, H., (2007), What are the prospects for genetically engineered, disease-resistant plants. *European Journal of Plant Pathology, 121,* 217–231.

Currey, C. J., (2018). *Tomatoes 101, Part II: A Production Guide.* Departments-Hydroponic Production Primer.

Doorn, A. V., & Martin, D. V., (2013). Resistance to sap-sucking insects in modern-day agriculture. *Front. Plant Sci.* https://doi.org/10.3389/fpls.2013.00222.

Gerald, E. B., (2008). *IPM Vegetable Specialist.* Maryland Cooperative Extension. University of Maryland.

John, L. C., (2017). *Department of Entomology and Nematology.* UF/IFAS Extension, Gainesville, FL 32611.

Kostic, M., Stankovic, S., & Kuzevski, J., (2015). Role of AChE in Colorado potato beetle (*Leptinotarsa decemlineata* Say) resistance to carbamates and organophosphates. *Insecticides Resistance.* 10.5772/61460.

Koul, B., Srivastava, S., Sanyal, I., Tripathi, B., Sharma, V., & Amla, D. V., (2014). *Transgenic Tomato Line Expressing Modified Bacillus Thuringiensis Cry1Ab Gene Showing Complete Resistance to Two Lepidopteran Pests* (Vol. 3, p. 84). Springer Plus.

Kunik, T., Salomon, R., Zamir, D., Navot, N., Zeidan, M., et al., (1994). Transgenic tomato plants expressing the tomato yellow leaf curl virus capsid protein are resistant to the virus. *Nature Biotechnology, 12*(5), 500–504.

Li, H., Jiang, W., Zhang, Z., Xing, Y., & Li, F., (2013). Transcriptome analysis and screening for potential target genes for RNAi-mediated pest control of the beet armyworm, *Spodoptera exigua*. transcriptome analysis and screening for potential target genes for RNAi-mediated pest control of the beet armyworm, *Spodoptera exigua*. *PLoS One, 8*(6), e65931. https://doi.org/10.1371/journal.pone.0065931.

Liu, C., Liu, Y., Guo, M., Cao, D., Dong, S., & Wang, G., (2014). Narrow tuning of an odorant receptor to plant volatiles in *Spodoptera exigua* (Hubner). *Insect Mol. Biol., 23*, 487–496. doi: 10.1111/imb.12096.

Liu, N. Y., et al., (2015). Two subclasses of odorant-binding proteins in *Spodoptera exigua* display structural conservation and functional divergence. *Insect Mol. Biol., 24*, 167–182. doi: 10.1111/imb.12143.

Lu, W. H., Kennedy, G. G., & Gould, F., (1997). Genetic variation in larval survival and growth and response to selection by Colorado potato beetle (Coleoptera: Chrysomelidae) on tomato. *Environ Entomol., 26*, 67–75.

Majidiani, S., Pour, A. R. F., Laudani, F., Campolo, O., Zappalà, L., Rahmani, S., Aboalghasem, S. M., & Palmeri, V., (2019). RNAi in *Tuta absoluta* management: Effects of injection and root delivery of dsRNAs. *Journal of Pest Science*, *92*, 1409–1419.

Mandaokar, A. D., Goyal, R. K., Shukla, A., Bisaria, S., Bhalla, R., Reddy, V. S., Chaurasia, A., Sharma, R. P., Altosaa, I., & Kumar, P. A., (2000). Transgenic tomato plants resistant to fruit borer (*Helicoverpa armigera* Hubner). *Crop Protection, 19*, 307–312.

Mutamiswa, R., Machekano, H., & Casper, N., (2017). First report of tomato leaf miner, *Tuta absoluta* (Meyrick) (Lepidoptera: Gelechiidae), in Botswana. *Agriculture and Food Security, 6*(49), 1–10.

Natwick, E. T., Stoddard, C. S., Zalom, F. G., Trumble, J. T., Miyao, G., & Stapleton, J. J., (2013). UC IPM Pest Management Guidelines: Tomato. UC ANR Publication 3470.

OECD, (2017). Tomato (*Solanum lycopersicum*). In: *Safety Assessment of Transgenic Organisms in the Environment* (Vol. 7). OECD Consensus Documents, OECD Publishing, Paris. doi: https://doi.org/10.1787/9789264279728-6.

Quilis, J., López-García, B., Meynard, D., Guiderdoni, E., & San, S. B., (2014). Inducible expression of a fusion gene encoding two proteinase inhibitors leads to insect and pathogen resistance in transgenic rice. *Plant Biotechnology. J., 12*, 367–377.

Sajjanar, G. M., & Balikai, R. A., (2009). Molecular mapping and marker assisted selection for trichome mediated insect resistance in tomato. *International Journal of Agricultural Sciences*, *5*(1), 327–333.

Sarkar, P., Hembram, S., & Islam, S., (2018). Host plant preference of sucking pest to different tomato genotypes under West Bengal conditions. *Int. J. Curr. Microbiol. App. Sci., 7*(11), 3244–3252. doi: https://doi.org/10.20546/ijcmas.2018.711.374.

Schoville, S. D., Chen, Y. H., Richards, S., et al., (2018). A model species for agricultural pest genomics: The genome of the Colorado potato beetle, *Leptinotarsa decemlineata* (Coleoptera: Chrysomelidae). *Scientific Reports, 8*, 1931. doi: 10.1038/s41598-018-20154-1.

Shashank, P. R., Doddabasappa, B., Kammar, V., Chakravarthy, A. K., & Honda, H., (2015). Molecular characterization and management of shoot and fruit borer *Conogethes punctiferalis Guenee* (Crambidae: Lepidoptera) populations infesting cardamom, castor and other Hosts. *New Horizons in Insect Science: Towards Sustainable Pest Management*. doi: 10.1007/978-81-322-2089–3_20, © Springer India.

Skryabin, K., (2010). Do Russia and Eastern Europe need GM plants? *Nat. Biotechnol., 27*, 593–595.

Thipyapong, P., Stout, M. J., & Attajarusit, J., (2007). Functional analysis of polyphenol oxidases by antisense/sense technology. *Molecules, 12*(8), 1569–1595. https://doi.org/10.3390/ 12081569.

Vargas-Ortiz, E., Gonda, I., Smeda, J. R., Mutschler, M. A., Giovannoni, J. J., & Jander, G., (2018). Genetic mapping identifies loci that influence tomato resistance against Colorado potato beetles. *Scientific Reports 8*(1), 1–10.

Wang, S., Ding, T., Xu, M., & Zhang, B., (2018). Bidirectional interactions between beet armyworm and its host in response to different fertilization conditions. *PLoS One, 13*(1), e0190502. https://doi.org/10.1371/journal.pone.0190502.

Whitham, S., McCormick, S., & Baker, B., (1996). The N gene of tobacco confers resistance to tobacco mosaic virus in transgenic tomato. *Proceedings of the National Academy of Sciences, 93*(16), 8776–8781. 27.

Wilson, L. T., Tennyson, R., Gutierrez, A. P., Zalom, F. G., (1987). A physiological based model for processing tomatoes: Crop and pest management. *International Society for Horticultural Science*. https://doi.org/10.17660/ActaHortic.1987.200.11.

Xin, Z., Zhang, L., Zhang, Z., Chen, Z., & Sun, X., (2014). A tea hydroperoxide lyase gene, CsiHPL1, regulates tomato defense response against *Prodenia litura* (Fabricius) and *Alternaria Alternata* f. sp. *lycopersici* by modulating green leaf volatiles (GLVs) release and jasmonic acid (JA) gene expression. *Plant Molecular Biology Reporter, 32*(1), 62–69.

Zhang, J., Khan, S. A., Heckel, D. G., & Bock, R., (2015). Full crop protection from an insect pest by expression of long double-stranded RNAs in plastids. *Science*. doi: 10.1126/science.1261680.

Zhu, J. Y., Zhang, L. F., Ze, S. Z., Wang, D. W., & Yang, B., (2013). Identification and tissue distribution of odorant-binding protein genes in the beet armyworm, *Spodoptera exigua. J. Insect Physiol., 59*, 722–728. doi: 10.1016/j.jinsphys.2013.02.011.

CHAPTER 13

MOLECULAR APPROACHES FOR MULTIPLE GENES STACKING/ PYRAMIDING IN TOMATO FOR MAJOR BIOTIC STRESS MANAGEMENT

RAJA HUSAIN,[1] NITIN VIKRAM,[2] KUNVAR GYANENDRA,[3] VINEETA SINGH,[4] N. A. KHAN,[4] MD. SHAMIM,[5] and DEEPAK KUMAR[6]

[1]*Department of Agriculture, Himalayan University, Jullang, Itanagar, Arunachal Pradesh – 791111, India, E-mail: rajahusain02@gmail.com (Raja Husain)*

[2]*Department of Biochemistry, Zila Parishad Agriculture College, Banda – 210001, Uttar Pradesh, India*

[3]*Faculty of Agriculture Sciences &Technology, Madhyanchal Professional University Ratibad, Bhopal-462044, Madhya Pradesh, India*

[4]*Department of Plant Molecular Biology and Genetic Engineering, Acharya Narendra Deva University of Agriculture and Technology, Kumarganj, Ayodhya – 224229, Uttar Pradesh, India*

[5]*Department of Molecular Biology and Genetic Engineering, Dr. Kalam Agricultural College (BAU, Sabour), Kishanganj, Bihar – 855107, India*

[6]*Department of Manufacturing and Development, Nextnode Bioscience Pvt. Ltd., Kadi – 382725, Gujarat, India*

ABSTRACT

Tomato is one of the most important species of vegetables in the world. Tomatoes can be infected with about 200 diseases which reduce their production. Many resistant genes (R-genes) have been discovered in this plant, whose proteins recognize pathogens' avirulent proteins and activate the mechanisms for defense. Cf for *Cladosporium fulvum*, Ve for *Verticillium dahlia,* and Cmm for *Clavibacter michiganensis*, etc., are some of the known R-genes which has been identified and pyramided in tomato lines. Marker-assisted selection and gene pyramiding are very important breeding strategies to confer wide spectrum and long-lasting resistance to diseases that cause loss of tomato yields. Genetic engineering is one of the many methods for producing better, new crops. For the production of new resistant varieties, developments in biotechnological methods need to be reforms to the new genes anticipated from resistance tomato lineages.

13.1 INTRODUCTION

Tomato (*Solanum lycopersicum* L.) is the highly consumed vegetable crop in the world. It is the second most important vegetable crop in the world after potato. India is the second maximum tomato producing country after China. In 2014, the total production of tomato was 170 million MT in which 31% covered by China, i.e., followed by USA, India, and Turkey. Due to diploid and sequenced genome, In many earlier research, tomato was used as a model plant for genetic level study for stress tolerance genomics, yield attributes, fruit quality, and several other physiological parameters (Ranjan et al., 2012). In the present scenario, due to many biotic stresses, the production and quality are being affected (Ramyabharathi et al., 2012). Without an effective management system or antimicrobial agent, the microbial growth at the time of cropping in field, post-harvest storage and loading for marketing may lead to maximum economic loss. That's why this is a very urgent need of a reliable and stable approach for the management of plant pathogens.

Nowadays, the usefulness of many chemicals against many biotic stresses is decreasing because of cost or not affording by small-scale farmers (Brading et al., 2002). The maximum use of chemical spray may lead to a negative impact on human health as well as environment. It could also lead to the evolution of resistant pathogens against applied chemicals, weeds, and pest (Vincelli, 2016; Miedaner et al., 2013). In this context, many plant breeding and biotechnological approach may be the beneficial tools for transferring the resistance genes against many biotic stresses. The breeding approach proved the important method to transfer the resistant gene from resistant to susceptible cultivar which showed the resistance against stress (Kottapalli et al., 2010). So the modern breeding approach have been used to develop the cultivars containing several genes that could be resistant to multiple stresses, including high yielding, by assembling these different genes from various parents into single cultivar known as gene pyramiding/gene stacking (Suresh and Malathi, 2013). Malav et al. (2016) explained very well about gene pyramiding that it is a breeding approach that focused on assembling the multiple genes from different parents and transferring into a single genotype. This is a very useful techniques for developing the resistant varieties against many pathogens, pests, and other abiotic stresses (Joshi and Nayak, 2010). In this chapter, we will focus on several diseases of tomato and their management through the breeding approaches by incorporating the multiple resistant genes/QTLs to develop the durable resistant cultivar against various biotic stresses.

13.2 OBJECTIVES OF PLANT BREEDING/GENE PYRAMIDING

The plant breeding methods can provide new opportunities in the agriculture system that may be related to production and quality or nutritional parameters (high protein, Vitamin, for example); some secondary metabolites may be valuable in resistance to plant diseases. Some important parameters are:

- Increased the quality like protein, vitamins, and minerals, etc.;
- Increased the production of crop;
- Increased the tolerance against many abiotic stresses (drought, submergence, salinity, and heat);
- Increased tolerance against various types of insect pests;
- Increased tolerance of herbicides;
- Varieties suited to particular soils and climates;
- Varieties resistant to biotic (bacterial, fungal, and viral diseases);
- Novel or exotic varieties.

13.3 MARKER ASSISTED SELECTION (MAS) IN TOMATO

MAS are defined as that the process of indirect selection, where selection of a trait of interest is based on a marker (i.e., biochemical/morphological/DNA) linked to trait of interest (e.g.: biotic/abiotic stress resistance, yield, and quality) rather than the trait of itself (Ribaut and Hoisington, 1998). In this process, a marker can be directly used for the selection of genetic determinants of a trait including the biotic/abiotic stress tolerance. It can be used for phenotypic screening of developed varieties against many stresses. With the help of molecular markers, we can select the plants containing the desired gene of interest in their genomic region expressing the traits. After testing the parents and progenies with markers, we must have to know the exact mapping position of marker so that several genes contributing to many traits like resistance to many stresses, yield, and quality parameters could be easily pyramided into singe genotype by using the linked markers (Tuvesson et al., 2007).

The confirmation of heterosis for hybrid seed production could be possible through MAS (Reif et al., 2003). In tomato breeding, several advantages of MAS for purity testing of lots of tomato seeds, quick screening of germplasms against many disease and fruit quality and (Foolad and Panthee, 2012). It is simple and easy than phenotype screening which saves time, less labor and cost-intensive. With the help of MAS, breeders developed lot of crop varieties resistance to various biotic stresses and tolerance to drought, salinity, submergence, and cold, etc. (Table 13.2).

13.3.1 PROCEDURE FOR MAS

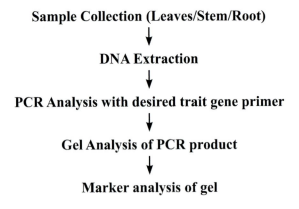

Sample Collection (Leaves/Stem/Root)

↓

DNA Extraction

↓

PCR Analysis with desired trait gene primer

↓

Gel Analysis of PCR product

↓

Marker analysis of gel

13.4 MARKERS USED IN TOMATO BREEDING

The markers played a significant role in MAS. The markers are categorized into three categories:

1. **Morphological Markers:** these markers can be detected by eye by simple visual inspection like color, size, shape, and height.
2. **Biochemical Markers:** These markers can be isolated and observed, for example, proteins and isozymes. Isozymes can be used for expression analysis of plant cell (Yang et al., 2015).
3. **Molecular Markers:** Defined as that the DNA fragment (i.e., associated the specific location of genome) used for the identification of particular sequence of DNA in unknown DNA pool. The DNA markers have been used for scanning of whole rice genome and also involved in assigning the landmarks in high density on every plant chromosome (Sorrells et al., 2003).

Among these markers, most of the molecular markers are PCR-based having great importance in MAS and other breeding programs that could be designed to detect the DNA polymorphism in the entire genome. In various earlier research, Many types of DNA based molecular markers have been used in tomato molecular breeding program against different diseases resistance (Table 13.2), which are following:

- RAPD (random amplified polymorphic DNA);
- CAPD (cleaved amplified polymorphic sequence);
- SSR (simple sequence repeat);
- RFLP (Restriction fragment length polymorphism);
- SNP (Single nucleotide polymorphism);
- SCAR (sequence characterized amplified region).

The characteristics of an ideal marker for molecular breeding must be the following:

- Marker should be polymorphic;
- Co-dominant;
- Easy to detect;
- Highly reproducible;
- Cheap in cost;
- Randomly distributed throughout the genome (Ragimekula et al., 2013).

13.4.1 APPLICATION OF MOLECULAR MARKER FOR BREEDING ANALYSIS

- Gene pyramiding (that is the main theme of chapter).
- Introgression of genes from wild varieties.
- DNA fingerprinting for genetic diversity analysis for best management of stresses.
- Selection of transgressive segregates in early generation for time saving and improve in the speed and efficiency to develop the new varieties.
- Identification of multiple different regions of chromosomes having genes containing alleles which governs the desirable traits.
- Use of molecular markers has reduced the time for improvement of tomato molecular breeding which leads to advance breeding techniques for developing the new varieties resistance to many biotic stresses (Tanksley, 1983).

13.5 GENE PYRAMIDING IN TOMATO BREEDING PROGRAMS

Gene pyramiding in tomato is the best technique which assumes the transfer of more than one gene (from multiple parent) resistance to specific disease(s) into single cultivar. Basically, it is defined as the incorporation of two or more genes from multiple parents into a single genotype to develop superior genotypes. In this process, molecular markers help in the selection of the best plant genotypes for preceding the research. The gene pyramiding has great importance for developing varieties having good quality traits, resistance to various biotic stresses and high yielding. Ji et al. (2014) described that gene pyramiding basically divided into two categories; the first one is pedigree which focuses on incorporation of target genes from multiple parents into single genotype called root genotype while another one is the fixation step that targets on the fixing the target genes in homozygous form for driving the ideal genotype.

Molecular markers played a significant role in gene pyramiding by saving the time after reducing the no of generations which can be evaluated by breeders to confirm the presence of desired gene. Pinta et al. (2013) explained that gene pyramiding is one of the best techniques for the improvement of germplasm. In this approach, more than one gene which is responsible for several characters like resistance to many stresses as well as yield parameters. With reference to MAS, several plant breeders used these techniques for developing the resistance varieties containing the genes resistance to many diseases like tomato spotted wilt virus (TSWV), tomato leaf curl virus (TLCV), Bacterial speck, Bacterial spot, late blight, fusarium wilt and many others (Table 13.1). This is a very urgent need to make some backcrossing with recurrent parent for developing the superior lines carrying the desired gene sets without unnecessary genes. In this regards, molecular markers played an important role for selection of tomato progenies containing the gene of interest for many backcrossing, while the progenies not carrying the desired genes could be easily discarded (Table 13.2).

TABLE 13.1 Tomato Plants with Mi-Genes for Root Nematodes Resistance

Genotype/Lines	Mi Genes	Notes	References
(*L. peruvianum*) PI128657	Mi-1	High-level resistance and the main source of resistance	Smith (1944)
VFNT of *L. esculentum*	Mi-1	Resistance	Yaghoobi et al. (1995); Seah et al. (2004)
(*L. esculentum*) Mobile	Mi-1	Resistance	Seah et al. (2004)
(*L. esculentum*) Ontario	Mi-1	Resistance	Seah et al. (2004)
Solanum lycopersicum cv. Amelia	Mi-1	Resistance	Desaeger and Csinos (2006)
(*L. esculentum*) CLN2026C	Mi-2	Resistance	Rani et al. (2009)
(*L. esculentum*) CLN2026E	Mi-2	Resistance	Rani et al. (2009)
(*L. esculentum*) CLN1464A	Mi-2	Resistance	Rani et al. (2009)
(*L. peruvianum*) 2R2-clone PI270435	Mi-2	Resistance to heat stress	Cap et al. (1993); Devran and Sogut (2010)
(*L. esculentum*) VWP2	Mi-3	Resistance to heat stress	Yaghoobi et al. (2005)
(*L. peruvianum*) 1MH-clone PI126443	Mi-3	Resistance to heat stress	Devran and Sogut (2010)

TABLE 13.1 *(Continued)*

Genotype/Lines	Mi Genes	Notes	References
(*L. peruvianum*) Maranon LA1708	Mi-4	Resistance to heat stress	Vermis and Roberts (1996)
1MH-clone PI126443of *L. peruvianum*)	Mi-5	Resistance to heat stress	Vermis and Roberts (1996)
3MH-clone PI270435 of *L. peruvianum*	Mi-6	Resistance to heat stress	Vermis and Roberts (1996)
3MH-clone PI270435 of *L. peruvianum*)	Mi-7	Resistance to RKN, including strains virulent on Mi	Vermis and Roberts (1996)
2R2-clone PI270435 of *L. peruvianum*	Mi-8	Resistance to RKN, including strains virulent on Mi	Vermis and Roberts (1996)
LA2157 of *S. arcanum*)	Mi-9	Heat stable resistance	Jablonska et al. (2007); Ammiraju et al. (2003)
ZN48	Mi-HT	Heat-stable resistance	Wang et al. (2013)
ZN17	Mi-HT	Heat-stable resistance	Wang et al. (2013)
LA0385	–	Heat-stable resistance	Wang et al. (2013)
CastlerockII	–	Resistance	Milligan et al. (1998)
Sun6082	–	Resistance	Milligan et al. (1998)
(*S. lycopersicum*) tomato Mongal T-11	–	High resistance	Jaiteh et al. (2012)
(*S. lycopersicum* L) Samrudhi F1	–	Resistance	Okorley et al. (2018)
(*L. esculentum*) LE812	–	Resistance	Rani et al. (2009)

Sources: Reprinted with permission from El-Sappah et al. (2019). (http://creativecommons.org/licenses/by/4.0/).

TABLE 13.2 List of Improved Tomato Genotypes by Using Marker-Assisted Selection

Pyramided Characters	Important Gene(s) and QTL(s)	Improved Genotypes	Donor Parents(s)	Linked Markers in the Study	References
Resistance against yellow leaf curl and spotted wilt virus	*Ty-1, Ty-3, Sw-5*	Cuban tomato (LD3)	LA2779, LA1969,	CAPS, RAPD (*Eco*R1), SCAR	Consuegra et al. (2015)
Resistance against yellow leaf curl virus and mottle virus	*Ty-1, Ty-3*	LA2779	LA1932, LA1938	CAPS, RAPD(*Eco*R1), SCAR	Ji et al. (2007)
Resistance against ate blight, yellow leaf curl virus, bacterial wilt, Fusarium wilt, gray leaf spot and tobacco mosaic virus	*Bwr-12, Ty-2, Ty-3, Tm22, Ph-2, Ph-3, Sm*	CLN3241	CLN2777G, G2-6-20-15B, LBR-11	CAPS, SCAR	Hanson et al. (2016)
Resistance against Fusarium wilt and late blight	*I1, Ph-3*	Accession 1008, 017878, 107868, 0101, and 1002	Not defined	SSR, SCAR	Akbar et al. (2016)
Resistance against bacterial spot and bacterial speck	*Pto, Rx3*	Crosses between parents	Ohio 9834, Ohio 8819, Ohio 981205	CAP (*Rsa1, BsrB1*)	Yang and Francis (2005)
Resistance against tomato yellow leaf curl viruses (monopartite and bipartite)	*Ty-2, Ty-3*	Crosses between: FLA478-6-1-11 and CLN2498C, CLN1621E	*Ty*-stock of: LA3473, CLN2585D, CA4, GC171, TY-172	CAPS, SCAR, SSR	Prasanna et al. (2014)
Resistance against tomato spotted wilt virus and late blight	*Sw-5, Ph-3*	Crosses between NC946 and NC592	NC946, NC592	CAPS, SCAR	Robbins et al. (2010)
Resistance against tomato leaf curl virus	*Ty-1, Ty-2*	Pbc, H-86 crosses with the parents	EC538408, EC520061, H-24	CAPS, SSR	Kumar et al. (2014)
Resistance against tobacco mosaic virus, Fusarium wilt, root-knot nematode disease and leaf mildew	*TM-2a, I-2, Mi-1, Cf-9*	L11, L19, L46, and L51	Longkeeper, Jia Powder	CAPS	Zhu et al. (2010)

TABLE 13.2 *(Continued)*

Pyramided Characters	Important Gene(s) and QTL(s)	Improved Genotypes	Donor Parents(s)	Linked Markers in the Study	References
Resistance against *Phytophthora infestans*	*Ph-2**	AD17, 137, P15250, SeI8	Heline, Momor, Motelle, Ontario, Pyrella, Stevens	CAPS	Barone A (not yet published. see reference section for link to this information)
Resistance against *Fusarium oxysporium* f. sp. radicis-lycopersici, Tobacco mosaic virus and *Verticillium dahlia*	*Frl, TM-2a, Ve**	AD17, 137, P15250, SeI8	Heline, Momor, Motelle, Ontario, Pyrella, Stevens	CAPS	
Resistance against *Fusarium oxysporium* f.sp. lycopersici, *Meloidogyne* spp. and *Verticillium dahlia*	*I2*, Mi, Ve**	AD17, 137, P15250, SeI8	Heline, Momor, Motelle, Ontario, Pyrella, Stevens	CAPS	
Pseudomonas syringae	*Pto*	AD17, 137, P15250, SeI8	Heline, Momor, Motelle, Ontario, Pyrella, Stevens	CAPS	
Pyrenochaeta lycopersici	*py-1*	AD17, 137, P15250, SeI8	Heline, Momor, Motelle, Ontario, Pyrella, Stevens	SCAR	
Tomato spotted wilt virus	*Sw-5*	AD17, 137, P15250, SeI8	Heline, Momor, Motelle, Ontario, Pyrella, Stevens	CAPS	

Sources: From Oladokun and Mugisa (2019).

13.5.1 STRATEGY FOR GENE PYRAMIDING

Identification of resistant genes/gene sources

↓

Transfer the gene into elite genotype which is deficient for that resistant gene through backcrossing

↓

Fixation of that genotype

13.6 MAJOR DISEASES IN TOMATO

In present scenario, approximately 200 pests and diseases was reported in tomato which induces the yield loss (Nowicki et al., 2013). The diseases caused by different microbes like fungi, bacteria, nematodes, and viruses caused the maximum severity in cereal and vegetable crops that not only affect their quality but also affect human health and overall economy. There are many diseases caused by fungi are late blight, Fusarium wilt, early blight (EB), sclerotinia rot and crown. Bacterial diseases are Bacterial speck, Bacterial spot, bacterial wilt and Bacterial canker, etc. Viral diseases are TLCV, TSWV and their management through the breeding approach.

13.7 MOLECULAR BREEDING FOR FUNGAL, BACTERIAL, NEMATODE, AND VIRAL DISEASES RESISTANCE

13.7.1 QTLS IDENTIFICATION FOR LATE BLIGHT

Late blight is one of the most destructive diseases of tomato caused by *Phytophthora infestans.* It causes the approximately 20–70% economic losses (Foolad et al., 2008; Nowicki et al., 2013). In earlier late blight research, Ph-1 is the late blight resistant gene which was reported in wild tomato (*Solanum pimpinellifolium*) and mapped at the distal end of chromosome 7 (Foolad et al., 2008). But due to new races of *Phytophthora infestans,* this gene was not found durable. Subsequently another gene such as Ph-2 was also reported in wild species of tomato (*Solanum pimpinellifolium*) that was mapped on chromosome 10 (Moreau et al., 1998).

Chunwongse et al. (2002) reported the third gene, i.e., Ph-3 in wild tomato (*Solanum pimpinellifolium*) at chromosome 9. The marker that is closely associated with Ph-3 is TG591A.

The QTLs unwillingness to late blight was reported on chromosome 12 of tomato by using the composite interval mapping. Several reciprocal backcross populations developed from *Solanum lycopersium* and *Solanum habrochaites*. RFLP marker was used for the construction of linkage maps for each backcross populations. Total 6 QTLs in BC-E (lb1a, lb2a, lb3, lb4, lb5b, and lb11b) and 2 QTLs in BC-H (lb5ab and lb6ab) were significantly identified in replicated experiments and beyond assay methods (Brouwer et al., 2004). Except lb2a allele, all alleles of *Solanum habrochaites* showed resistance. Near isogenic lines (NILs) is one of the important breeding method which helps in the development of disease resistance varieties. NILs was used for verification of these QTLs after mapping for lb5b, lb4, and lb11b through marker-assisted backcross breeding to *S. lycoperscum*. The evaluation of NILs populations against disease resistance were carried out at three different field locations. The disease resistance QLTs were observed in all the sets of NILs. Lb4 was mapped between TG182 and CT194 closest to TG609 on the chromosome 4 at break of 6.9 cM. Lb5b was mapped between TG69a and TG413 closest to TG23 on the chromosome 5 at the break of 8.8 cM while lb11b was mapped between TG194 and TG400 at the interval of 15.1 cM with the center peak between CT182 and TG147 (Brouwer et al., 2004). The excellent mapping of these QTLs developed the potential MAS for the resistance against late blight (Table 13.2).

13.7.2 QTLS IDENTIFICATION FOR EARLY BLIGHT (EB)

EB is the most common fungal disease in tomato that is caused by *Alternaria solai*. The major sources for EB resistance was reported in *Solanum habrochaites, Solanum pimpinellifolium,* and *S. peruvianum*. The mapping for resistance QTL for EB can be achieved by backcrossing population which has been developed by the crossing of NC84173 (susceptible cultivar) and *Solanum habrochaites* (accession PI126445 resistant cultivar). Unlike other QTL analysis studies, all the resistance QTLs was contributed by cultivar *S. harbrochaites*. There were total six QTLs mapped on chromosome 1, 2, 5, 6, 7, and 9 including three QTLs resistance to stem lesions in field which elucidated 35% phenotypic variation (Chaerani et al., 2007).

Many breeding lines against EB resistance have been developed, but if we are using traditional breeding, it is very laborious and challenging due to two reasons; like polygenic nature of resistance and connection of resistance with many characters like plant type, fruit loading, and physiological maturity. The resistant lines were developed by crossing between *Solanum lycopersium* and *Solanum habrochaites* whereas low yielding, late maturity and intermediate growth were found inherit from *Solanum habrochaites* accession (Chaerani and Voorrips, 2006). However, some of the RILs (Recombinant inbreed line) have been developed by the crossing between *Solanum pimpinellifolium* (accession LA2093) and *S. lycopersium.* Here tomato line NCEBR-1 was reported (Table 13.2) against EB resistance other characters like high yielding, early maturation and small plant size (Ashrafi and Foolad, 2015).

13.8 GENES/QTLS IDENTIFICATION FOR FUSARIUM WILT

Fusarium wilt is the most common disease, caused by *Fusarium oxysporium* f. sp. *lycopersici* (fol) which reduced the production and leads to yield loss about 10–80% in many tomatoes producing country (Kesavan and Chaudhary, 1977). There are three races of Fol namely race 1, race 2 and race 3 were reported for causing Fusarium wilt while the three loci I-1, I-2, and I-3 have also been reported which showed the resistance against *Fusarium oxysporium* in tomato crop where the I-2 was introgressed from wilt tomato species *Solanum pimpinellifolium* (Accession PI126915) to *S. lycopersium.* In this concern, TG105 is tightly associated with resistance gene for Fol on chromosome 11 (Sarfatti et al., 1989).

13.9 GENES/QTLS IDENTIFICATION FOR ROOT-KNOT NEMATODE (RKN)

RKN is the serious and destructive nematode disease in tomato that is caused by *Meloidogyne* sp. (Zhou et al., 2016). RKN is not only affect the crop yield but also makes the plants more susceptible to many bacterial and fungal diseases (Ashraf and Khan, 2010). RKN is most dangerous which affect the quality and production of tomato crop. Many methods have been used for controlling the RKN in tomatoes; R-genes are the important strategy which creates the resistance against nematodes and other diseases. There are many resistance genes which showed resistance against RKN, was reported as Mi-1, Mi-2, Mi-3, Mi-4, Mi-5, Mi-6, Mi-7, Mi-8, Mi-9, and Mi-HT in which only 5 genes have been mapped (Table 13.1). But one of the major problems is that the resistance mechanism breaks down at higher temperature while some genes could work at high temperatures. Some plant varieties which showed resistance at high temperature like PI126443, LA0385, and ZN48 (*L. peruvianum*) (Table 13.1).

13.10 VIRAL DISEASES OF TOMATO

Several viral disease have been identified in tomato which includes TSWV which can lead to plant death (Rossello et al., 1993). TLCV is also the most destructive disease worldwide, which reduces the yield loss up to 100%.

13.11 MARKER-ASSISTED GENE PYRAMIDING FOR TOMATO LEAF CURL VIRUS (TLCV) AND TOMATO MOTTLE VIRUS (TOMOV) DISEASE RESISTANCE IN TOMATO

No more research has been performed on marker-assisted gene pyramiding for TLCV, but a little bit part of this has been published by some scientists. Kumar et al. (2014) performed marker-assisted gene pyramiding for transferring TLCV resistance genes from resistant tomato varieties like EC-538408 (*Solanum chilnese*) and EC-520061 (*S. peruvianum*) into susceptible tomato varieties like Pbc and H-86. The Co-dominant SSR marker (SSR-218), i.e., associated with TLCV resistance gene was used to distinguish homozygotes from heterozygotes during seedling stage prior to pollination that discarded the non-target backcrosses and progenies of crosses Pbc × EC-538408 and H-86 × EC-520061, while SSRF306 was used for the cross pbc × EC-538408. Which discarded the non-target backcrosses and pyramiding progenies of the crosses PbcxEC-520061 and H-86xEC-520061, while SSR-306 was used for the cross PbcxEC-538408. *Ty-2* genes CAPS marker was used for the cross H-86 × H-24. The pyramiding lines containing both the pyramided R-gens were found resistant against TLCV. In another research, the admittance of *Ty-1, Ty-2*, and *Ty-3* genes developed the resistance mechanism against TLCV in an advance backcrossed population of tomato which leads to develop the multiple diseases resistant line. These lines may be helpful for breeders in the future in pre-breeding tomato varietal development projects (Kumar et al., 2019). In one more research, Gill et al. (2019) mapped the additional locus *Ty-6* on chromosome 10 of tomato. *Ty-6* is an effective gene that showed resistance against monopartite TLCV and bipartite tomato mottle virus (ToMoV) (Table 13.2).

13.12 DEVELOPMENT OF MARKERS FOR TOMATO YELLOW LEAF CURL VIRUS (TYLCV) RESISTANCE

In many research, several efforts have been made for the characterization of R-genes in tomato. *Ty-3* was reported in wild species of tomato by introgression breeding against TYLCV that may be used for further breeding of tomato. The resistance cultivar has been developed through the use of MAS. The newly developed markers ACY were screened against *Ty-3* linked markers P6-25 by the screening of resistant as well as susceptible hybrid tomato lines and genetic segregation using F2 population obtained from AG208 (a commercial hybrid contains resistant). ACY is a co-dominant indel-based marker that showed the strong polymorphic banding pattern for distinguishing the resistant plant varieties from susceptible ones. So *Ty-3* gene sequence donated for the development of ACY molecular marker aiding the phenotype selection. This marker is best for MAS to develop the TLCV resistant varieties and thus recommended to breeders for future (Nevame et al., 2018).

13.13 GENES/QTLS IDENTIFICATION FOR BACTERIAL SPECK

Bacterial speck is economically most destructive bacterial disease of tomato, infected by *Pseudomonas syringae* (Pst). The resistant source of bacterial speck was recognized in wild species of tomato. Four Bacterial speck resistance genes have been reported such as Pto (known as Pto-1) from *S. pimpinellifolium* accession PI370093, Pto-2 from *S. pimpinellifolium* accession PI126430 and Pto-3 from *S. habrochaites* accession PI34417 confer resistance to race 0, Pto-4 from accession PI34417 confer resistance to race 1 (Yang and Francis, 2007). Thapa et al. (2015) reported the QTLs in *S. habrochaites* accession LA1777 conferring the resistance against race 1. Martin et al. (1993) cloned the gene Pto which showed the resistance to race 0 (Pedly and Martin, 2003). Because of resistance, it is used in many breeding programs and developed the resistance cultivar in entire world (Yang and Francis, 2007). Sun et al. (2011) developed the molecular markers for the gene Pto and used in many breeding research of MAS. The molecular marker for observed as a good performer to develop more than 1000 elite breeding lines.

13.14 GENES/QTLS IDENTIFICATION FOR BACTERIAL SPOT

Bacterial spot is the most common disease of tomato that is mainly found in wet and humid environment worldwide (Jones et al., 2014). Four species of *Xanthomonas* which cause the bacterial spot are *Xanthomonas euvesicatoria, Xanthomonas vesicatoria, Xanthomonas perforans, Xanthomonas gandneri* (Jones et al., 2014). On the basis of virulence characteristics on different genotypes, the four races such as T_1 in *Xanthomonas euvesicatoria,* T_2 in *Xanthomonas vesicatoria,* T_3 and T_4 in *Xanthomonas perforans* was reported by Jones et al. (2005). Resistance source against all four species of *Xanthomonas* have reported (Table 13.2) and validated in wild and cultivated tomato genotype (Yang and Francis, 2007; Liabeuf et al., 2015). The resistance source was identified in Hawaii7998, Hawaii79981, PI114490, and PI128216 (*S. pimpinellifolium*) and incorporated into susceptible cultivar through breeding programs. The resistance against T_1 developed in Hawaii7998 containing the major QTL on chromosome 5 accounting for 41% phenotypic variation identified in advance backcross lines (Yang and Francis, 2005). PI128216 was identified as resistant cultivar against race T_3 that is regulated by the dominant gene Rx4 located on 45.1 kb region of chromosome 11. Pei et al. (2012) developed the molecular marker for the gene Rx4 that might be useful in MAS of tomato. Wang et al. (2011) mapped a gene as Rx4 in Hawaii7981 and LA1589, i.e., resistance to race 3. In a later study, this gene was validated being allelic to Rx4 by Zhao et al. (2015). RxopJ4 was identified as resistant locus against bacterial spot in *S. pennellii* accession LA716. Several molecular markers which are tightly associated to RxopJ4 were also reported (Sharlach et al., 2013). Sun et al. (2014) recognized the many QTLs resistance to T_1 to T_4 and *X. gardneri* in PI114490 while Liabeuf (2016) also reported the QTLs resistance to *X. gardneri* in *S. pimpinellifolium* accessions like LA2533 and LA1545 (Table 13.2).

13.15 QTLS IDENTIFICATION FOR BACTERIAL CANKER

Bacterial canker is one of the most destructive bacterial diseases of tomato caused by *Clavibacter michiganensis* (Davis et al., 1984). The resistance source of Bacterial canker was identified in many wild species of tomato that could be integrated into the susceptible tomato cultivar in many countries like the US, Netherlands, and Bulgaria (Sen et al., 2015). MAS facilitated the development of resistance cultivar against many diseases. Yang and Francis (2007) and Sen et al. (2015) identified the molecular markers associated with QTLs conferring the resistance to bacterial canker in *S. peruvianum* accession LA2157 and *S. habrochiates* accession LA407. These markers prepared the platform for transferring the genes resistance to bacterial canker from wild species to susceptible tomato cultivar.

13.16 QTLS IDENTIFICATION FOR BACTERIAL WILT

Bacterial wilt is a vascular disease in tomato caused by *Ralstonia solanacearum* (Jones et al., 2014). It has the highest rank among the serious diseases of tomato having big yield loss in many tropical, sub-tropical, and hot region of the world (Singh et al., 2015). The resistance source against bacterial wilt was reported in *S. pimpinellifolium*, *S. lycopersicum* var. *cerasiforme* and cultivated tomato (Alsam et al., 2017). The highly resistance source was obtained from the three major resistant sources like PI127805A (*S. pimpinellifolium*), CRA66 (*S. lycopersicum* var. *cerasiforme*) and PI129080 (*S. pimpinellifolium*) (Hanson et al., 1998). The resistance is the complex character which are controlled by polygenes. Yang and Francis (2007) identified 7 QTLs on 6 chromosomes (3, 4, 6, 8, 10, and 12) from resistant line Hawaii7996 (obtained from PI127805A) and 3 QTLs on 3 chromosomes (6, 7, and 10) from *S. lycopercisum var cerasiforme* accession L285. The QTL found on chromosome 6 may be used to develop the resistant cultivar against bacterial wilt (Wang et al., 2004). Kim et al. (2018) reported the two QTLs Bwr-6 (Chr-6) and Bwr-12 (Chr-12) tightly linked to SNP markers for developing the resistant cultivars using MAS.

13.17 CONCLUSION

There are many biotic factors such as bacteria, fungi, viruses, nematodes, and insects which cause the significant economic yield in tomato throughout the world. Apart from the chemicals for the management of these stresses, there are many other strategy can be preferred for their management. These strategies include various breeding approaches like MAS, MABB, Gene pyramiding may perform the best role for the development of many tomato varieties resistance to many diseases. The use of resistant cultivars is economically the best strategy, durable, and environmental eco-friendly and showed the great resistance against various biotic stresses. Several QTLs and genes have been recognized in wild species of tomato and could be easily transferred into susceptible tomato cultivars using marker-assisted gene pyramiding. Developed a lot of molecular markers that might be beneficial for the development of the disease-resistant cultivar and save time by reducing the no of progeny by earlier detection in advanced breeding line via the molecular markers.

KEYWORDS

- **bacterial canker**
- **cleaved amplified polymorphic sequence**
- **marker-assisted selection**
- **near-isogenic lines**
- **susceptible cultivar**
- **tomato yellow leaf curl virus**

REFERENCES

Akbar, K., Abbasi, F. M., Sajid, M., Ahmad, M., Khan, Z. U., & Aziz, U. D., (2016). Marker-assisted selection and pyramiding of I1 and Ph3 genes for multiple disease resistance in tomato through PCR analysis. *Int. J. Biosci., 9*(3), 108–113.

Ammiraju, J. S. S., Veremis, J. C., Huang, X., Roberts, P. A., & Kaloshian, I., (2003). The heat-stable, root-knot nematode resistance gene *Mi-9* from *Lycopersicon peruvianum* is localized on the short arm of chromosome 6. *Theor. Appl. Genet., 106*, 478–484.

Ashraf, M. S., & Khan, T. A., (2010). Integrated approach for the management of *Meloidogyne javanica* on eggplant using oil cakes and biocontrol agents. *Arch Phytopathol. Plant Prot., 43*, 609–614.

Ashrafi, H., & Foolad, M., (2015). Characterization of early blight resistance in a recombinant inbred line population of tomato: I. Heritability and trait correlations. *Adv. Stud. Biol., 7*, 131–148.

Aslam, M. N., Mukhtar, T., Hussain, M. A., & Raheel, M., (2017). Assessment of resistance to bacterial wilt incited by *Ralstonia solanacearum* in tomato germplasm. *J. Plant Dis. Prot., 124*, 585–590.

Barone, A., (2003). *Molecular Marker-Assisted Selection for Resistance to Pathogens in Tomato*. Available at: https://pdfs.semanticscholar.org/cb39/fbc9353476b536b24473849070955 1a92de4.pdf (accessed on 1 March 2021).

Brading, P. A., Verstappen, E. C. P., Kema, G. H. J., & Brown, J. K. M., (2002). A gene for gene relationship between wheat and *Mycosphaerella graminicola*, the *Septoria tritici* blotch pathogen. *Phyto. Pathol., 92*, 439–445.

Brouwer, D. J., & St Clair, D. A., (2004). Fine mapping of three quantitative trait loci for late blight resistance in tomato using near-isogenic lines (NILs) and sub NILs. *Theor. Appl. Genet., 108*, 628–638.

Cap, G. B., Roberts, P., & Thomason, I. J., (1993). Inheritance of heat-stable resistance to *Meloidogyne incognita* in *Lycopersicon peruvianum* and its relationship to the *Mi* gene. *Theor. Appl. Genet. 85*, 777–783.

Chaerani, R., & Voorrips, R., (2006). Tomato early blight (*Alternaria solani*): The pathogen, genetics, and breeding for resistance. *J. Gen. Plant Pathol., 72*, 335–347.

Chaerani, R., Smulders, M. J. M., Van, D. L. C. G., Vosman, B., Stam, P., & Voorrips, R. E., (2007). QTL identification for early blight resistance (*Alternaria solani*) in a *Solanum lycopersicum* x *S. arcanum* cross. *Theor. Appl. Genet., 114*, 439–450.

Chunwongse, J., Chunwongse, C., Black, L., & Hanson, P., (2002). Molecular mapping of the Ph-3 gene for late blight resistance in tomato. *J. Hort. Sci. Biotech., 77*, 281–286.

Consuegra, O. G., Gómez, M. P., & Zubiaur, Y. M., (2015). Pyramiding TYLCV and TSWV resistance genes in tomato genotypes. *Rev Protección Veg., 3*(2), 161–164.

Davis, M. J., Gillaspie, A. G., Vidaner, J. A. K., & Harris, R. W., (1984). Clavibacter: A new genus containing some phytopathogenic coryneform bacteria, including *Clavibacter xyli* subsp. xyli sp. nov., subsp. nov. and *Clavibacter xyli* subsp. cynodontis subsp. nov., pathogens that cause ratoon stunting disease of sugarcane and bermudagrass stunting disease. *Int. J. Syst. Bacteriol., 34*, 107–117.

Desaeger, J. A., & Csinos, A. S., (2006). Root-knot nematode management in double-cropped plasticulture vegetables. *J. Nematol., 38*, 59–67.

Devran, Z., & Sogut, M. A., (2010). The occurrence of virulent root-knot nematode populations on tomatoes bearing the Mi gene in protected vegetable growing areas of Turkey. *Phytoparasitica, 38*, 245–251.

El-Sappah, A. H., Islam, M.M., El-Awady H. H., Yan, S., Qi, S., Liu, J., Cheng, G. T., & Liang, Y., (2019). Tomato natural resistance genes in controlling the root-knot nematode. *Genes (Basel), 10*(11), 925.

Foolad, M. R., & Panthee, D. R., (2012). Marker-assisted selection in tomato breeding. *Critical Reviews in Plant Sciences, 31*, 93–123. http://dx.doi.org/10.1080/07352689.2011.616057.

Foolad, M. R., Merk, H. L., & Ashrafi, H., (2008). Genetics genomics and breeding of late blight and early blight resistance in tomato. *Crit. Rev. Plant Sci., 27*, 75–107.

Hanson, P., Lu, S., Wang, J., Chen, W., Kenyon, L., Tan, C., Tee, K. W., et al., (2016). Conventional and molecular marker-assisted selection and pyramiding of genes for multiple disease resistance in tomato. *Scientia Horticulture, 201*, 346–354.

Jablonska, B., Ammiraju, J. S., Bhattarai, K. K., Mantelin, S., Martinez, D. I. O., Roberts, P. A., & Kaloshian, I., (2007). The *Mi-9* gene from *Solanum arcanum* conferring heat-stable resistance to root-knot nematodes is a homolog of *Mi-1. Plant Physiol., 143*, 1044–1054.

Jaiteh, F., Kwoseh, C., & Akromah, R., (2012). Evaluation of tomato genotypes for resistance to root-knot nematodes. *Afr. Crop Sci. J., 20*, 41–49.

Ji, Y., Schuster, D. J., & Scott, J. W., (2007). *Ty-3*, a *Begomovirus* resistance locus near the tomato yellow leaf curl virus resistance locus *Ty-1* on chromosome 6 of tomato. *Mol. Breed., 20*, 271–284.

Ji, Z., Shi, J., & Zeng, Y., (2014). Application of a simplified marker-assisted backcross technique for hybrid breeding in rice. *Biologia (Poland), 69*, 463–468.

Jones, J. B., Lacy, G. H., Bouzar, H., Minsavage, G. V., Stall, R. E., & Schaad, N. W., (2005). Bacterial spot Worldwide distribution, importance and review. *Acta Hortic., 695*, 27–33.

Jones, J. B., Zitter, T. A., Momol, T. M., & Miller, S. A., (2014). *Compendium of Tomato Diseases and Pests* (2nd edn.). APS Press, Minnesota, USA.

Joshi, R. K., & Nayak, S., (2010). Gene pyramiding: A broad-spectrum technique for developing durable stress resistance in crops. *Biotechnol. Genet. Eng. Rev., 5*, 51–60.

Kesavan, V., & Chaudhary, B., (1977). Screening for resistance to Fusarium wilt of tomato. *SABRO J., 9*, 51–65.

Kim, B. Y., Hwang, I. S., Lee, H. J., Lee, J. M., Seo, E. Y., Choi, D., & Oh, C. S., (2018). Identification of a molecular marker tightly linked to wilt resistance in tomato by genome-wide SNP analysis breeding for resistance to tomato bacterial diseases in China. *Theor. Appl. Genet.* doi: 10.1007/s00122-018-3054-1.

Kottapalli, K. R., Narasu, L. M., & Jena, K. K., (2010). Effective strategy for pyramiding three bacterial blight resistance genes into fine-grain rice cultivar, Samba Mahsuri, using sequence-tagged site markers. *Biotechnol. Lett., 32*, 989–996.

Kumar, A., Jindal, S. K., Dhaliwal, M. S., Sharma, A., Kaur, S., & Jain, S., (2019). Gene pyramiding for elite tomato genotypes against ToLCV (*Begomovirus* spp.), late blight (*Phytophthora infestans*) and RKN (*Meloidogyne* spp.) for northern India farmers. *Physiol. Mol. Biol. Plants, 25*(5), 1197–1209.

Kumar, A., Tiwari, K. L., Datta, D., & Singh, M., (2014). Marker-assisted gene pyramiding for enhanced tomato leaf curl virus disease resistance in tomato cultivars. *Biologia Plantarum, 58*(4), 792–797.

Liabeuf, D., Francis, D. M., & Sim, S. C., (2015). Screening cultivated and wild tomato germplasm for resistance to *Xanthomonas gardneri. Acta Hortic., 1069*, 65–70.

Malav, A. K., Kuldeep, I., & Chandrawat, S., (2016). Gene pyramiding: An overview. *Int. J. Curr. Res. Biosci. Plant Biol., 3*, 22–28.

Martin, G. B., Brommonschenkel, S. H., Chunwongse, J., Frary, A., Ganal, M. W., Spivey, R., Wu, T. Y., et al., (1993). Map based cloning of a protein kinase gene conferring disease resistance in tomato. *Science, 262*, 1432–1436.

Miedaner, T., Zhao, Y., Gowda, M., Longin, C. F. H., & Korzun, V., (2013). Genetic architecture of resistance to *Septoria tritici* blotch in European wheat. *BMC Genomics, 14*, 858.

Milligan, S. B., Bodeau, J., Yaghoobi, J., Kaloshian, I., Zabel, P., & Williamson, V. M., (1998). The root-knot nematode resistance gene *Mi* from tomato is a member of the leucine zipper, nucleotide binding, leucine-rich repeat family of plant genes. *Plant Cell, 10*, 1307–1319.

Moreau, P., Thoquet, P., Olivier, J., Laterrot, H., & Grimsley, N., (1988). Genetic mapping of Ph-2, a single locus controlling partial resistance to *Phytophthora infestans* in tomato. *MPMI, 11*, 259–269.

Nowicki, M., Kozik, E. U., & Foolad, M. R., (2013). Late blight of tomato. In: Varshney, R. K., & Tuberosa, R., (eds), *Translational Genomics for Crop Breeding: Biotic Stress* (1st edn., Vol. I). Wiley, Hoboken.

Okorley, B. A., Agyeman, C., Amissah, N., & Nyaku, S. T., (2018). Screening selected solanum plants as potential rootstocks for the management of root-knot nematodes (*Meloidogyne incognita*). *Int. J. Agron., 52*, 455, 456.

Oladokun, J. O., & Mugisa, I., (2019). Exploring MAS: A reliable molecular tool for development of multiple disease resistance in tomato (*Solanum lycopersicum* L.) through gene pyramiding. *International Journal of Environment, Agriculture and Biotechnology, 4*(2), 509–516.

Pedley, K. F., & Martin, G. B., (2003). Molecular basis of Pto-mediated resistance to bacterial speck disease in tomato. *Annu. Rev. Phytopathol., 41*, 215–243.

Pei, C. C., Wang, H., Zhang, J. Y., Wang, Y. Y., Francis, D. M., & Yang, W. C., (2012). Fine mapping and analysis of a candidate gene in tomato accession PI128216 conferring hypersensitive resistance to bacterial spot race T3. *Theor. Appl. Genet., 124*, 533–542.

Pinta, W., Toojinda, T., Thummabenjapone, P., & Sanitchon, J., (2013). Pyramiding of blast and bacterial leaf blight resistance genes into rice cultivar RD6 using marker-assisted selection. *Afr. J. Biotechnol., 12*(28), 4432–4438.

Prasanna, H. C., Sinha, D. P., Rai, G. K., Krishna, R., Sarvesh, P. K., Singh, N. K., Sing, M., & Malathi, V. G., (2014). Pyramiding Ty-2 and Ty-3 genes for resistance to monopartite and bipartite tomato leaf curl viruses of India. *Plant Pathology, 64*(2), 256–264.

Ragimekula, N., Varadarajula, N. N., Mallapuram, S. P., Gangimeni, G., & Reddy, R. K., (2013). Marker-assisted selection in disease resistance breeding. *J. Plant Breed Genet., 1*, 90–109.

Ramyabharathi, S. A., Meena, B., & Raguchander, T., (2012). Induction of chitinase and b-1,3-glucanase PR proteins in tomato through liquid formulated *Bacillus subtilis* EPCO 16 against *Fusarium* wilt. *J. Today's Biol. Sci. Res. Rev. JTBSRR,* 1(1), 50–60.

Rani, C. I., Muthuvel, I., & Veeraragavathatham, D., (2009). Evaluation of 14 tomato genotypes for yield and root-knot nematode resistance parameters. *Pest Technol., 3*, 76–80.

Ranjan, A., Ichihashi, Y., & Sinha, N. R., (2012). The tomato genome: Implications for plant breeding, genomics and evolution. *Genome Biology, 13*, 1.

Reif, J. C., Melchinger, A. E., Xia, X. C., Warburton, M. L., Hoisington, D. A., Vasal, S. K., Beck, D., et al., (2003). Use of SSRs for establishing heterotic groups in subtropical maize. *Theor. Appl. Genet., 107*, 947–957. doi: 10.1007/s00122-003-1333-x.

Ribaut, J. M., & Hoisington, D. A., (1998). Marker assisted selection: New tools and strategies. *Trends Plant Sci., 3*, 236–239.

Robbins, M. D., Masud, M. A. T., Panthee, D. R., Gardner, R. G., Francis, D. M., & Stevens, M. R., (2010). Marker-assisted selection for coupling phase resistance to tomato spotted wilt virus and *Phytophthora infestans* (late blight) in tomato. *Hortscience, 45*(10), 1424–1428.

Rossello, M. A., Descals, E., & Cabrer, B., (1993). Nia epidermoidea, a new marine gasteromycete. *Mycol. Res., 97*(1), 68–70.

Sarfatti, M., Katan, J., Fluhr, R., & Zamir, D., (1989). An RFLP marker in tomato linked to the *Fusarium oxysporum* resistance gene I-2. *Theor. Appl. Genet., 78*, 755–759.

Seah, S., Yaghoobi, J., Rossi, M., Gleason, C. A., & Williamson, V. M., (2004). The nematode-resistance gene, *Mi-1*, is associated with an inverted chromosomal segment in susceptible compared to resistant tomato. *Theor. Appl. Genet., 108*, 1635–1642.

Sen, Y., Derwolf, J. V., Visser, R. G. F., & Heusden, S. V., (2015). Bacterial canker of tomato: Current knowledge of detection, management, resistance, and interactions. *Plant Dis., 99*, 4–13.

Sharlach, M., Dahlbeck, D., Liu, L., Chiu, J., Jimenez-Gomez, J. M., Kimura, S., Koenig, D., et al., (2013). Fine genetic mapping of RXopJ4, a bacterial spot disease resistance locus from *Solanum pennellii* LA716. *Theor. Appl. Genet., 126*, 601–609.

Smith, P. G., (1944). Embryo culture of a tomato species hybrid. *Proc. Am. Soc. Hortic. Sci., 44*, 413–416.

Sorrells, M. E., La-Rota, M., & Bermudez-Kandianis, C. E., (2003). Comparative DNA sequence analysis of wheat and rice genomes. *Genome Res., 13*, 1818–1827.

Sun, H. J., Wei, J. L., Zhang, J. Y., & Yang, W. C., (2014). A comparison of disease severity measurements using image analysis and visual estimates using a category scale for genetic analysis of resistance to bacterial spot in tomato. *Eur. J. Plant Pathol., 139*, 125–136.

Sun, W. Y., Zhao, W. Y., Wang, Y. Y., Pei, C. C., & Yang, W. C., (2011). Natural variation of Pto and Fen genes and marker-assisted selection for resistance to bacterial speck in tomato. *Agric. Sci. China, 10*, 827– 837.

Suresh, S., & Malathi, D., (2013). Gene pyramiding for biotic stress tolerance in crop plants. *Weekly Sci. Res. J., 1*, 2321–7871.

Tanksley, S. D., (1983). Molecular markers in plant breeding. *Plant Molec. Biol. Rep., 1*, 3–8.

Thapa, S. P., Miyao, E. M., Davis, R. M., & Coaker, G., (2015). Identification of QTLs controlling resistance to *Pseudomonas syringae* pv. tomato race 1 strains from the wild tomato, *Solanum habrochaites* LA1777. *Theor. Appl. Genet., 128*, 681–692.

Tuvesson, S., Dayteg, C., Hagberg, P., Manninen, O., Tanhuanpaa, P., Tenhola-Roininen, T., Kiviharju, E., et al., (2007). Molecular markers and doubled haploids in European plant breeding programmes. *Euphytica, 158*, 305–312.

Veremis, J. C., & Roberts, P. A., (1996). Relationship between *Meloidogyne incognita* resistance gene in *Lycopersicon peruvianum* differentiated by heat sensitivity and nematode virulence. *Theor. Appl. Genet., 93*, 950–959.

Vincelli, P., (2016). Genetic engineering and sustainable crop disease management: Opportunities for case-by-case decision making. *Sustain., 8*, 495.

Wang, G. P., Lin, M. B., & Wu, D. H., (2004). Classical and molecular genetics of bacterial wilt resistance in tomato. *Acta Hortic. Sin.*, *31*, 401–407.

Wang, H., Hutton, S. F., Robbins, M. D., Sim, S. C., Scott, J. W., Yang, W. C., Jones, J. B., & Francis, D. M., (2011). Molecular mapping of hypersensitive resistance from tomato cv. Hawaii 7981 to *Xanthomonas perforans* race T3. *Phytopathology, 101*, 1217–1223.

Wang, Y., Yang, W., Zhang, W., Han, Q., Feng, M., & Shen, H., (2013). Mapping of a heat-stable gene for resistance to southern root-knot nematode in *Solanum lycopersicum. Plant Mol. Biol. Rep., 31*, 352–362.

Yaghoobi, J., Yates, J. L., & Williamson, V. M., (2005). Fine mapping of the nematode resistance gene *Mi-3* in *Solanum peruvianum* and construction of an *S. lycopersicum* DNA contig spanning the locus. *Mol. Genet. Genom., 274*, 60–69.

Yang, H. B., Kang, W. H., & Nahm, S. H., (2015). Methods for developing molecular markers. In: *Current Technologies in Plant Molecular Breeding Chapter 2,* (pp. 15–50). Dordrecht: Springer.

Yang, W. C., & Francis, D. M., (2007). Genetics and breeding for resistance to bacterial diseases in tomato: Prospects for marker-assisted selection. In: Razdan, M. K., & Mattoo, A. K., (eds.), *Genetic Improvement of Solanaceous Crops, 1* (pp. 379–419). Tomato Science Publishers, New Hampshire, USA.

Yang, W., & Francis, D. M., (2005). Marker-assisted selection for combining resistance to bacterial spot and bacterial speck in tomato. *J. Amer. Soc. Hort. Sci., 130*(5), 716–721.

Zhao, B. M., Cao, H. P., Duan, J. J., & Yang, W. C., (2015). Allelic tests and sequence analysis of three genes for resistance to *Xanthomonas perforans* race T3 in tomato. *Hortic. Plant J., 1*, 41–47.

Zhou, L., Yuen, G., Wang, Y., Wei, L., & Ji, G., (2016). Evaluation of bacterial biological control agents for control of root-knot nematode disease on tomato. *Crop Prot., 84*, 8–13.

Zhu, M., Sun, Y., Zheng, S., Zhang, X., Wang, T., Ye, Z., & Li, H., (2010). Pyramiding disease resistance genes by molecular marker-assisted selection in tomato. *Acta Horticulturae Sinica, 37*(9), 1416–1422.

CHAPTER 14

MOLECULAR APPROACHES FOR THE POSTHARVEST LOSSES IN TOMATO BY DIFFERENT BIOTIC STRESSES

MD. SHAMIM,[1] MAHESH KUMAR,[1] B. N. SAHA,[2] MD. WASIM SIDDIQUI,[3] ASHUTOSH KUMAR SINGH,[4] DEEPTI SRIVASTAVA,[5] MD. ABU NAYYER,[5] RAJA HUSAIN,[6] S. S. SOLANKEY,[7] and V. B. JHA[8]

[1]Department of Molecular Biology and Genetic Engineering, Dr. Kalam Agricultural College (Bihar Agricultural University, Sabour), Kishanganj, Bihar – 855107, India, E-mail: shamimnduat@gmail.com (Md. Shamim)

[2]Department of Soil Science and Agricultural Chemistry, Dr. Kalam Agricultural College (Bihar Agricultural University, Sabour), Kishanganj, Bihar – 855107, India

[3]Department of Food Science and Post-Harvest Technology, Bihar Agricultural University, Sabour, Bhagalpur – 813210, Bihar, India

[4]Department of Biotechnology and Crop Improvement, College of Horticulture and Forestry, Rani Lakshmi Bai Central Agricultural University, Jhansi, Uttar Pradesh, India

[5]Integral Institute of Agricultural Science and Technology, Integral University, Dasauli, Lucknow – 226021, Uttar Pradesh, India

[6]Department of Agriculture, Himalayan University, Jullang, Itanagar, Arunachal Pradesh – 791111, India

[7]Department of Horticulture (Vegetable and Floriculture), Dr. Kalam Agricultural College (Bihar Agricultural University, Sabour), Kishanganj, Bihar – 855107, India

[8]Department of Plant Breeding and Genetics, Dr. Kalam Agricultural College (Bihar Agricultural University, Sabour), Kishanganj, Bihar – 855107, India

ABSTRACT

Tomato (*Solanum lycopersicum* L.) is one of the most significant horticultural produce and is generally used as fruit as well as vegetable purpose. Loss of postharvest is a significant fact that hampers the production of tomatoes in most growing countries. Such losses bring low returns for farmers, processors, and traders in addition to for the entire country that is suffering in phrases of foreign exchange earnings. Postharvest losses were detected at the time of pre-harvest (on-farm) as well as postharvest (off-farm) condition and create severe problems. On-farm losses are passed on by insufficient harvesting levels, immoderate regulation of heat, incorrect bins for harvesting, poor farm sanitation and irrelevant packaging supplies. Off-farm loss elements include lack of avenue access, insufficient delivery network, lack of producing industrial unit and shortage of correct marketplace information. Using low-price intervention of intermediary creation and molecular approaches will help to reduce postharvest losses in tomato.

14.1 INTRODUCTION

Tomato (*Solanum lycopersicum* L.) is a solitary eaten vegetable and fruit crop within the world. Its recognition curtail from the fact that it could be eaten as a clean or in several of processed forms. Rising plants under greenhouse have several advantages (Jabnoun-Khiareddine et al., 2019). Massive quantities of tomato can be harvested on a small area and it may be continuously harvested due to the fact that the temperature and irrigation is controlled in a greenhouse and it facilitates the growers to supply during off-season markets while clean food expenses are at a best (Yeshiwas and Tolessa, 2018). Storage of tomato is essentially dependent on certain issue that decides its postharvest shelf life (Haile, 2018). These consist of relative humidity of the storage unit, storage temperature, charge of ethylene manufacturing and the rate of respiratory of the fruit. Storage temperature is vital in the renovation of tomatoes postharvest shelf lifestyles. It is immediately connected to the opposite features accountable for tomato fruit weakening. The shelf exists on an average of 7 days at normal temperature and humidity (Znidarcic et al., 2006). Tomatoes saved at the temperature of 10°C maintained good and satisfactory fruit senescence (Ayomide et al., 2019). Practically all post-harvest illnesses of fruit and vegetables are resulting from two important microorganisms, i.e., fungi and bacteria. Postharvest diseases are often categorized in keeping with how contamination is initiated. The dramatic physiological adjustments which arise all through the ripening of tomato are frequently the cause for reactivation of quiescent infections (Quinet et al., 2019). Tomato is known as a climacteric fruit, sense it undertakes a pour in ethylene and respiration manufacturing at the beginning of ripening (Li et al., 2019a). When ripening initiates tomato culmination transit from in part photosynthetic to genuine heterotrophic tissues through the parallel demarcation of chloroplasts into chromoplasts and the dominance of lycopene and carotenoid pigments (Carrari and Fernie, 2006). Presently, the ripening method has advanced to formulate the palatability of tomato fruit to the organisms that devour them and scatter their seeds. When the fruit starts ripening, then different pathways that normally affect the tiers of pigments, acids, sugars, and volatile aroma that are linked with the extra appealing position of tomato. While concurrently tomato tissue goes to soften and degradation allows simpler seed release from the tomato (Matas et al., 2009). The postharvest practices and proper management are the essential paintings in the production of tomato. The postharvest defeat is frequently related to the infection of microorganisms which promotes spoilage of the fruits all through storage, transportation, and advertising. The concerned microbes might also infect fresh tomato fruit at both pre-harvest as well as postharvest stages (Khadka et al., 2017). Irrigation water, fertilizer, compost, soil, and pesticide are the common sources of pre-harvest contamination by microbial populations, whilst postharvest assets encompass soil, cleansing, and water use for treatments, packing shed and transporting equipment (Beuchat, 2002; Izumi, 2010). Dire production and handling practices, i.e., infected water for irrigation purpose and cleansing up and unhygienic managing practices are the general contamination occurred during the different practices. The infected microbes during different harvest practices can produce harmful secondary metabolites and toxins, which might also be the basis of serious fitness problems (Huntanen et al., 1976). Thus the products of microbes are reported as a matter of hazard (FAO, 1998). The microbial stack of the consumable things is the principle measure to decide the postharvest shelf life of the fresh agricultural harvests (Spares et al., 2001). These microbes consist of different types of bacteria (gram-positive and gram negative), fungi (yeasts and molds), viruses, and other parasites on the fruits (Barth et al., 2009). A quantifiable deprivation in the first-rate and amount of tomato from its time of harvest to the end of utilization is named postharvest loss (Ayandiji et al., 2011). The loss in amount entails loss due to the stored amount and that is greater in almost all developing countries (Kitinoja et al., 1999). Different quality namely nutrient contents, caloric composition, acceptability, and edibility are also affected by the microorganism. Tomato losses are encountered in the course of harvesting, managing, garage, processing, packaging, labeling, transportation, and even for the duration of the advertising and marketing of the fruit (Sibomana et al., 2016). Tomato losses that arise because of insufficient storage facilities that leads to quality as well as quantity loss. Losses under primary point of storage can be categorized into microbiological, chemical, biological, physical, mechanical, and physiological losses (Asalfew and Nega, 2020). In the market, a huge amount of harvested tomatoes are vended at intended costs and inappropriate microbial crumble also significantly contributes to the excessive postharvest loss (Arah et al., 2015). Maintenance of excellent quality in the course of promotion is one of the key troubles when harvested fruits are stored at normal temperature. Several fungus and bacteria attacks at the time of tomato storage that result in significant quantity loss (Patel and Patel, 1991). Thus, the above factors extremely vital recourses of postharvest losses of tomato at the time of their delivery to the consumers.

14.2 CAUSES OF POSTHARVEST LOSES IN TOMATO

There are several motives has been recognized for loss after harvest. Most producers have said that insect pests and sicknesses had been the main problems affecting their products, while investors indicated that damage all through delivery became an anxiety. Suitable delicate nature of tomato at the time of harvest requires rigorous practices during a short duration of time to avoid contamination. Fresh tomatoes have a quick put up-harvest life and because of physiological activity, mechanical damage and decay, they are inclined to put up-harvest losses. For lowering the put up-harvest losses, low temperature and refrigerated garage techniques are widely used, which creates delaying in senescence in vegetables in the end preserving their publish-harvest great (Yahya and Mardiyya, 2019). Tohamy et al. (2004) determined that postharvest decay is the principal limiting extension of shelf life in tomato fruits (*Lycopersicon esculentum* Mill.). Srivastava and Tandon (1966) said that a scientific explanation of nine different fungal diseases of tomatoes affected by *Colletotrichum dematium*, *Alternaria tenuis*, *Cladosporium fulvum*, *Malustela aeria*, *Fusarium roseum*, *Rhizopus nigricans* and *Myrothecium roridum* all through storage has been given. All the sicknesses besides Oospora rot were pronounced as new from India and postharvest decay of tomatoes because of *Malustela aeria* become defined as new report. Jones et al. (1993) mentioned that tomato commercialization was limited by way of rotting as a result of *Alternaria alternata* and/or by way of *Botrytis cinerea*. A saprophytic pathogen, namely *Alternaria alternata* causing postharvest losses in tomato at excessive incidence. Oladiran and Iwu (1993) stated that seven fungi connected with fruit rot of tomato namely; *Fusarium equiseti*, *Fusarium chlamydosporum*, *Geotrichum candidum*, *Alternaria solani*, *Acremonium recifei*, *Aspergillus niger*, and *Aspergillus jlavus*. All discussed fungi are pathogenic in nature on the tomato fruits, maximum pathogenic being *Geotrichum candidum* accompanied by using *A. niger*. Sommer et al. (1992) determined that Alterneria spp., Cercospora spp. Colletotrichum spp., *Aspergillus* spp. and *Fusarium* spp. have been accountable for illnesses. It is to some extent similar to postharvest sicknesses signs of tomato fruit. Narain and Rout (1981) said that *C. tenuissimum* changed into related twice with dry rot of tomato fruit. They also first time stated that *Cladosporium* species and *A. alternata* promoted the shriveling of tomato calyces and rachises.

The most commonplace pathogens attacking tomato after harvest are Alternaria species, Alternata, Botrytis cinerea, Colletotrichum spp., Clavibacter, Fusarium spp, Geotrichum candidum, Michiganensis, *Pseudomonas syringae*, *Phoma lycopersici*, *Phytophthora infestans*, *Phytophthora spp.*, *Phomopsis spp.*, *Rhizoctonia. Solani*, *Rhizopus sp.*, *Stemphylium herbarum*, *Sclerotiumr rolfssi*, *Sclerotinia spp* and *Xanthomonas campestris* (Figures 14.1–14.3). To save you towards beating from these types of pathogens and thus hold the fruit quality, inorganic materials, e.g., chlorine is used (100 ppm) during the time of packing for their sanitization. Chlorine eradicates sure gram-effective and gram-bad bacteria, mold, yeasts, viruses, and spores (Gross et al., 2016).

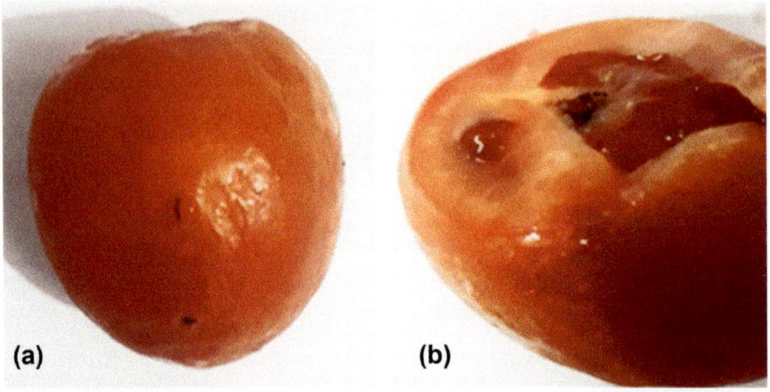

FIGURE 14.1 (a) External sign; and (b) internal black mold rot lesions development in tomato fruit.

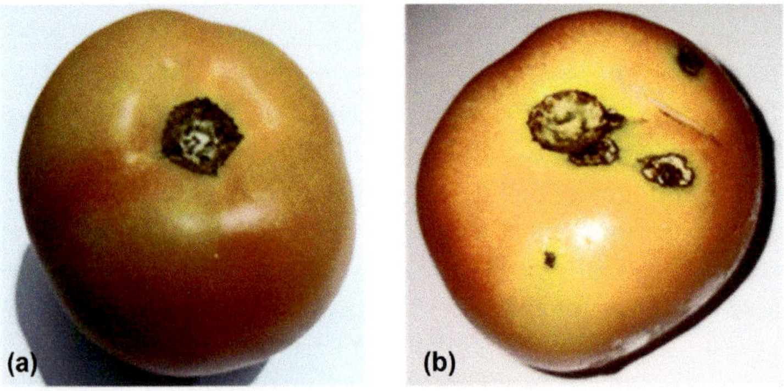

FIGURE 14.2 (a) Fresh tomato without disease; (b) after storage infection of *Colletotrichum lycopersici.*

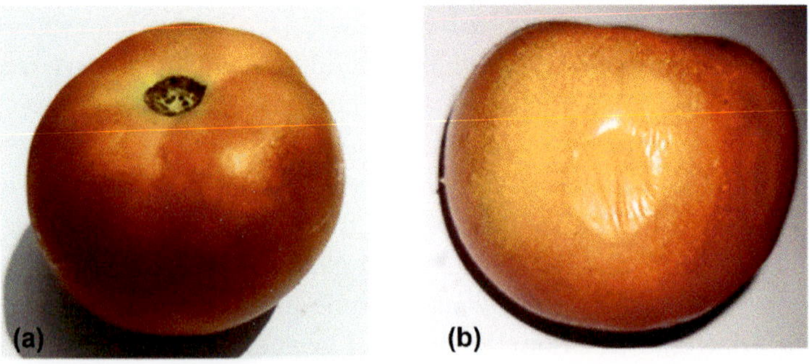

FIGURE 14.3 (a) Fresh tomato without disease; and (b) tomato infected with Rhizopus rot disease.

14.3 BIOLOGICAL LOSSES OF TOMATO FRUITS

Biological losses of tomato production mainly causes by different insects, birds, animals, or rodents, that motive an immediate vanishing of the tomato fruit (Daulty, 1995). In few cases, the tempo of fruit contagion by feathers, feces, and hair of birds and rodents is extraordinarily high. Insects reason defeats in amount by using utilization and quality (Atanda et al., 2011).

14.4 MICROBIOLOGICAL LOSSES OF TOMATO FRUITS

Several microorganisms inclusive of viruses, bacteria, and fungi spoil the stored tomato culmination (Hodges et al., 2011). These microorganisms easily attack sparkling produce and spread rapidly due to the reality that clean produce lack herbal defense mechanism in their tissues and the presence of moisture and vitamins in abundance helps in the growth of microorganism (Tripathi et al., 2013). Hadizadeh et al. (2009) suggested that tomato is susceptible to various put-up harvest diseases brought on by means of numerous pathogenic fungi. A saprophytic pathogen of tomato *Alternaria alternata* is causing (Alternaria rot) postharvest losses at excessive frequency. Pose (2009) suggested that *Alternaria alternata* is a toxigenic fungus, mainly chargeable for black mold of ripe tomato end result, an ailment often causing significant losses of tomatoes, mainly those used for canning purposes. Dal-Bello et al. (2008) pronounced that the fungal pathogen Botrytis cinerea causes intense rots on tomato end result at some point of storage and decreased shelf life. Anthracnose of tomato was ordinarily a disorder of ripe and over-ripe fruit. If left unchecked, the disorder could motive critical losses in yield and marketability (Wani, 2011; Sanoubar and Barbanti, 2017).

A classic pattern of a fungus that has impact the loss of tomato fruit is *Aspergillus niger* and it causes corrosion reputedly inside the fruits at the duration of eight days. The fungus rapidly secrete a protein showing polygalacturonase and cellulase activities (Ajayi et al., 2003). Tomato fruit skin works as a way of safety. However, if they are damaged due to scratch or slash that can provide a door position for microorganisms and thus result in losses (nice and quantity).

The developed green or unripe tomatoes are amazingly impervious to most extreme microorganisms that thought process rot than the ready ones that are more prominent helpless. Natural products with diminished PG were likewise appeared, in lab tests, to have improved protection from *Geotrichum candidum* and *Rhizopus stolonifer*, growths which typically contaminate aging organic product. In this manner, control of PG levels impacts in a few phenotypic modifications which have basic direction on organic product pleasant by means of their results at the pectic substances (Karmakar et al., 1998). Hereditary building has been applied to make plants, final product, and vegetables with new phenotypic qualities for a few goals, as portrayed straightaway. To diminish postharvest organic product misfortunes, tomato perfection with a phenotype of parasites obstruction and natural products with an all-inclusive postharvest time span of usability in examination with regular final product were made (Arah et al., 2016). Aside from the abovementioned, the quantitative and subjective (dietary) misfortunes moreover happen if reasonable distribute collect cures are not applied. Consequently, there might be an earnest need to diminish the critical post-gather loss of harvested tomatoes.

14.5 CAUSES OF POSTHARVEST DISEASE

Right character of the pathogen dispensing postharvest issue is significant to the decision of the perfect infection control procedure. Table 14.1 records some ordinary postharvest diseases and pathogens of tomato. Huge numbers of the parasites which reason postharvest affliction have a place with the phylum Ascomycota and the related Fungi Anamorphici (Fungi Imperfecti). On account of the Ascomycota, the abiogenetic phase of growth (the anamorph) is normally experienced more regularly in postharvest disorders in tomato than the sexual phase of the organism (the teleomorph). These adversaries are broadly apportioned and by and large confined from soil, air, and foodstuff (Thambugala et al., 2017; de Hoog and van der Vegte, 1989). Significant genera of anamorphic postharvest pathogens include Aspergillus, Penicillium, Geotrichum, Botrytis, Fusarium, Alternaria, Colletotrichum, Dothiorella, Lasiodiplodia, and Phomopsis. A portion of those growths likewise structure ascomycete sexual stages. It has been articulated that the absolute best rate reason of the PHLs of tomato organic product are identified with uncommon types of soil-borne phytopathogenic growths (Etebu et al., 2013; Fatima et al., 2009). These species reason sicknesses comprising of early curse (*Alternaria solani*), Sclerotium shrink (Sclerotium rolfsii), Anthracnose (*Colletotrichum* spp.), Damping-off (R. Solani), Tomato shrink (*Fusarium oxysporum*), Phoma decay (*Phoma ruinous*), Fusarium shrivel (*Fusarium oxysporum*), late scourge withering (*Phytophthora capsici*), Rhizopus decay (*Rhizopus stolonifer*) and Septoria leaf spot (*Septoria lycopersici*) (Kleemann et al., 2008; Kumar et al., 2008; Fatima et al., 2009; Ignjatov et al., 2012; Osman, 2015). Genera inside the phylum Basidiomycota are ordinarily not basic causal operators of postharvest affliction, in spite of the fact that parasites comprising of Sclerotium rolfsii and Rhizoctonia solani, which have basidiomycete sexual stages, can cause far-reaching postharvest misfortunes of vegetable vegetation along with tomato (Samuel and Orji, 2015; Barkai-Golan and Paster, 2008). While disorders in view of these pathogens are commonly subject ailments, the improvement of side effects as often as possible quickens after gather. The dominating causal retailers of bacterial delicate decays are different types of Ertuinia, Pseudomonas, Bacillus, Lactobacillus and Xanthomonas. Bacterial delicate decays are basic postharvest sicknesses of numerous vegetables, despite the fact that they're normally of considerably less significance in tomato (Rashid et al., 2016).

TABLE 14.1 Common Pathogens and Diseases of Tomato Causing Postharvest Losses

SL. No	Major Pathogen Causes Postharvest Losses		
	Name of Disease	**Anamorph Stage of Pathogens Causes Postharvest Losses in Tomato**	**Teleomorph Stage of Pathogens Causes Postharvest Losses in Tomato**
1.	Alternaria rot	*Alternaria* spp.	–
2.	Bacterial soft rots	Different species of the following pathogens (*Bacillus* spp., *Erwinia* spp., *Pseudomonas* spp. and *Xanthomonas* spp.)	–
3.	Cladosporium rot	*Cladosporium* spp.	–
4.	Cottony leak	–	*Pythium* spp.
5.	Fusarium rot	*Fursanum* spp.	–
6.	Grey mold	*B. cinerea*	*Botryotinia fuckeliana*
7.	Rhizopus rot	–	*Rhizopus* spp.
8.	Sclerotium rot	*Sclerotium rolfsii* (sclerotial state)	*Athelia rolfsii*
9.	Watery soft rot	–	*Sclerotinia* spp.

14.6 TRADITIONAL STRATEGIES FOR POSTHARVEST DISEASE CONTROL AND PREVENTION

14.6.1 FUNGICIDES

Fungicides are utilized widely for postharvest illness control in leafy foods. Timing of use and kind of fungicide utilized rely ordinarily upon the objective pathogen and when disease happens. The utilization of manufactured fungicides has been an essential strategy for the managing of postharvest waste of tomatoes (Spadaro and Gullino, 2004). For postharvest pathogens which taint produce sooner than collect, region programming of fungicides is regularly fundamental. This may include the rehashed utility of protectant fungicides during the creating season, or potentially vital utility of foundational fungicides. On account of contaminations which emerge all through and after gather, fungicides can be utilized to interfere with pathogen advancement. How hit fungicides are in getting along this relies generally upon the degree to which disease has developed at the time of fungicide programming and how fungicide enters effectively in the host tissue.

In any case, there are expanding worries over fungicide use alongside natural contamination dangers, failure to control parasitic sicknesses because of fungicide opposition, and backbone of fungicide buildups on the tomato (Ippolito and Nigro, 2000). Each one of those requesting circumstances has finished in the search for secure and viable elective procedures for the control of plant pathogens (Liu et al., 2013). Such systems incorporate organic control (counting the microbial enemies) of parasitic pathogens in tomatoes utilizing normally going on microorganisms (Droby et al., 2009). Additionally, this natural control is incredible, solid, and earth benevolent options in contrast to fungicides (Janisiewicz and Korsten, 2002). When all is said in done, fungicides for the administration of wound-attacking pathogens must be completed as fast as conceivable after gather. On the off chance that defilement is very much progressed on the hour of postharvest treatment, control will be difficult to accomplish. The standard strategy with controlling injury pathogens is to hold a definite centralization of the fungicide at the injury site in order to smother (however no longer consistently execute) pathogen improvement until the injury has recuperated. In this sense, the majority of the 'fungicides' which are utilized postharvest are earnestly growths static rather than fungicidal in their movement under standard use. Disinfectants which incorporates sodium hypochlorite might be utilized to slaughter pathogen propagules on the floor of organic product, anyway can't administer pathogens once they have accessed have tissue. Along these lines, bringing down pre-and postharvest utilization of substance fungicides by developing open door the executive's procedures stays a high research need (Droby et al., 2009).

14.6.2 THROUGH MANIPULATION OF THE POSTHARVEST ENVIRONMENT

The capacity to control the postharvest condition manages a very decent chance to delay senescence. Temperature is conceivably the absolute most basic thing impacting jumble improvement after gather. Temperature not best legitimately impacts the expense of pathogen development, yet additionally the charge of tomato aging. The improvement of numerous postharvest sicknesses is firmly identified with natural product readiness, so cures which defer aging have a propensity likewise to delay issue improvement (Asalfew et al., 2020). Low temperature stockpiling of foods grown from the ground is utilized extensively to put off maturing and the improvement of turmoil, despite the fact that the temperatures generally utilized for capacity are not lethal to the pathogen. The general dampness of the capacity condition can impact the advancement of postharvest malady in tomato (Zhou et al., 2014). High mugginess are as often as possible used to restrain water loss of produce. This, however, can expand affliction levels, predominantly if free dampness collects away holders.

14.6.3 PRE-HARVEST FACTORS

A wide assortment of pre gather components impact the advancement of postharvest illness. These incorporate the climate (precipitation, temperature, and so on.), fabricating region, decision of cultivar, social practices (pesticide programming, treatment, water system, planting thickness, pruning, mulching, organic product stowing, and so forth.) and planting material. These variables may have a quick effect on the advancement of ailment by methods for diminishing inoculum assets or by disheartening disease. On the other hand, they will affect the body structure of the produce in a way that effects on sickness advancement after collect. For instance, the utility of specific supplements can likewise improve the 'quality' of the natural product skin so it's miles less in danger of harm after collect and subsequently less powerless to attack with the guide of twisted pathogens in tomato (Arah et al., 2015; Emana et al., 2017).

14.6.4 PREVENTION OF INJURY

Mechanical wounds that emerge eventually of gathering and managing are correct sites for access of pathogens that reason rot in organic products (Jerry et al., 2005). The same number of postharvest pathogens advantage section by means of wounds or contaminate physiologically-harmed tissue, counteraction of injury at all phases at some phase underway, reap, and postharvest adapting to is basic. Wounds can be mechanical (for example, cuts, wounds, and scraped areas), substance (for example, Consumes), organic (for example, bug, fowl, and rat harm) or physiological (for example, chilling injury, warmness injury). Organic products have unnecessary dampness content material which makes them entirely helpless against attacks by method of pathogenic growths that cause decays, making the natural products unfit for human utilization as a result of mycotoxins delivered (Moss, 2002). Wounds might be limited by means of cautious collecting and managing of produce, reasonable bundling of produce, controlling creepy crawly bugs in the field, putting away produce at the upheld temperature and utilizing postharvest cures accurately. Where wounds are available, the procedure of wound recuperating can be expanded in certain occasions through control of the postharvest condition (for example, temperature and moistness) or by means of utility of sure compound cures.

14.6.5 HEAT TREATMENTS

Heat treatment did after gather might be utilized to control certain postharvest ailments (Lurie et al., 1998). Heat works with the guide of both executing the pathogen (and additionally its propagules) or through smothering its pace of advancement following treatment. Be that as it may, wares shift extraordinarily of their physiological resilience of heat treatments. For instance, most mild organic product sorts are really defenseless against heat hurt, uncommonly on the temperatures required to secure infection control. For wares ready to withstand heat cure, warmth might be executed looking like both high-temperature water or warm air. High-temperature water is a more prominent green mechanism for heat move than warm air, but on the other hand, is bound to make injury the ware. The restraint of maturing by the method of heat treatment can be intervened by its effect at the aging hormone, ethylene. Sight-heat treatment of 35–40°C restrains ethylene union inside hours in tomato (Mama et al., 2016; Biggs et al., 1988). High-temperature water is as often as possible utilized in total with fungicides to control postharvest ailments in wares, for example, mango, pawpaw, and rockmelon. Air with the slight high temperatures usually utilized for natural product fly disinfestation in assorted gathered wares additionally can convey some control of postharvest sicknesses in tomato and different organic products (Conway et al., 2004).

14.6.6 *IONIZING RADIATION*

Ionizing radiation is one of the other physical cure that might be utilized after reap to diminish sickness in certain products (Jeonga and Jeong, 2018). Physical cures along with heat or cold have been anticipated as capable cures, there are a couple of restrictions, comprising of their cost, adequacy at non phytotoxic temperatures, absence of preventive action, and low diligence (Lurie and Pedreschi, 2014). The mix of warm water and gamma light synergistically decreased contagious improvement in tomato natural products, bringing about 10.0% and 1.7% contamination rates by method of *Rhizopus stolonifer* and *B. Cinerea*, individually (Barkai-Golan et al., 1993). Vicalvi et al. (2013) utilized ionizing radiation effectively to development the rack presence of tomatoes. The work of ionizing radiation for rot control must be thought about while keeping up at the top of the priority list the solidness among pathogen affectability and host opposition (Liu et al., 2011).

14.7 EMERGING TECHNOLOGIES FOR POSTHARVEST DISEASE CONTROL

Consumer issues is presently increasing over the occurrence of chemical residues in meals have induced the explore the other strategies for non-chemical method of disease controls. Different chemicals used after tomato harvests are of sole subject because they're functional secure to the time of utilization. There are several new techniques to control postharvest losses in tomato are discussed in detail.

14.7.1 BIOLOGICAL CONTROL

The environmental pollution and the non-biodegradable nature creates opportunity for manufacturing/use of other alternative in postharvest losses and other disease control (Migliori et al., 2017). From the available substitutes, biological manipulation by using microorganisms along with their ecological fitness has been identified and presently utilized (Zong et al., 2010; Pal and Gardener, 2006). In current time, significant attention for the use of different microorganisms to manage postharvest losses are invented. There are many microorganisms were isolated and identified from the variety of natural resources, i.e., from fermented meals, surfaces of leaves, fruit, and vegetables. Once the microorganisms isolated, it can be screened (whether they be microorganisms, yeasts or filamentous fungi) with the help of different diverse approaches. The inhibition assay of the identified microorganisms was studied on the selected pathogens causing postharvest losses in tomato. The pathogen activities are affected by the used biocontrol agents with competition for space, mycoparasitism, and nutrients. The microorganisms also emits few agents like antibiotics, antifungal compounds, precarious metabolites, induction of host resistance, participation of the reactive oxygen species (ROS) and the biofilm improvement during the storage time (Dukare et al., 2018; Liu et al., 2013). These biocontrol agents are environment friendly and they also enhance crop production (Khonglah and Kayang, 2018; Shafiq, 2015). Application of such agents involves several practices, i.e., spraying, dipping or drenching, at the time of postharvest period (Di Francesco et al., 2016). Microorganism inhibits the development or boom of pathogens by interfering with the different nutrient uptake or other related cycles is known as antagonism (Rodrigo et al., 2017). There are reports of several of antagonists successfully used to manipulate postharvest losses consist of bacteria, fungi, and yeast (Lledó et al., 2016; Nunes, 2012). There are efficient antagonists that have been accepted as excellent choice to reveal postharvest diseases in citrus fruits against several pathogens, i.e., *Candida guilliermondii, Debaryomyces hanseniis, Trichoderma harzianum, Trichoderma viride, Pythium debaryanum Byssochlamys spectabilis, Gliocladium roseum, Cryptococcus laurentii, Phytophthora cryptogea and Aureobasidium pullulans* (Naglot et al., 2015; Gomathi and Ambikapathy, 2011; De Curtis et al., 2010; Zong et al., 2010; Agrios, 2004).

14.7.2 CONSTITUTIVE AND INDUCED HOST RESISTANCE

Plants acquire several biochemical and structural defense mechanisms which shield against the different pathogen. Among the defense response by hosts, there are two types of defense mechanisms are in the region before onset of the pathogen (i.e., Constitutive resistance), while others are effectively activated in response to infection (i.e., induced resistance). However, in tomato fruits there is little reports against the postharvest losses has been reported in comparison to the related fruit plants (Morrissey and Osbourn, 1999). Phytoalexins are the simplest antimicrobial produced in response to pathogen invasion by the microorganism, however, in some cases, they may be elicited by positive chemical (Arruda et al., 2016). For example, non-ionizing ultraviolet-C radiation is thought to result in the manufacturing of phytoalexins in numerous crops. Carrot slices treated with ultraviolet rays induce the production of 6-methoxymellen which is inhibitory to *Sclerotinia sclerotiorum* and *Botrytis cinerea*. UV treatments and their effects on citrus have also been accounted. Chitosan, a natural compound present within the cell wall of several fungi. Several other elicitors of host defense responses present in the host are also investigated. Chitosan can motivate some important defense signaling including manufacturing of chitinase, accumulation of phytoalexins and increase in lignification in tomato (Lafontaine and Benhamou, 1996).

A large variety of different defense-related chemical compounds (e.g., salicylic acid (SA), phosphonates, and methyl jasmonate) and virtual biochemical action (e.g., heat) can stimulated the host defenses in harvested losses (Díaz et al., 2002; Guo et al., 2015; Tsai et al., 2019).

14.7.3 THROUGH ENHANCEMENT OF BIOCHEMICAL FACTORS

There are several report of the improvement of anthocyanin content, a natural pigment compound can substantially make better shelf existence in tomato (Zhang et al., 2013). The anthocyanin are located on the tomato chromosomes and the expression of corresponding genes encoding transcription factors, Delila (Del) and Roseal (Ros1), shows multiplied expression of all the related genes dedicated to the biosynthesis of anthocyanin to develop red color tomato fruit (Butelli et al., 2008). However, the purple color tomato had less shelf life in comparison to red fruit color tomato.

Accordingly, each fruit softening timing and duration of ripening as well as pathogen contamination affect the shelf life of tomato fruits. Purple tomato fruit from Del/Ros1 tomato lines have regular shape, size, and sort of seeds show less shelf life. However, red fruit shows delayed ripening time to purple fruit. This is obvious from the advent of the crimson fruit both on the vine and all through submit harvest garage and from a reduced level of fungal contamination underneath either condition. Anthocyanin rich red tomato showed 2-fold more shelf life, increased resistance to pathogens, and slower ripening at late stages. These trends are directly related to the anthocyanin's developments in tomato fruits. There are research work has been initiated for the breeding of anthocyanin contents in tomato fruits by traditional plant breeding approaches (Bassolino et al., 2013; Povero, 2011). Several new tomato varieties were containing better quantities of anthocyanins in fruits are produced by utilizing the wild tomato species (Gonzali et al., 2009). Especially, the dominant gene Aft (Anthocyanin fruit), transferred from *Solanum chilense* (Jones et al., 2003; Georgiev, 1972), which grants anthocyanin amassing in the fruit, and the recessive gene atv (atroviolacea) derived from *Solanum cheesmaniae* (L. Riley) Fosberg (Rick et al., 1968), had been introgressed in *S. Lycopersicum*. The anthocyanin biosynthetic pathway activator R2R3 MYB coded by Aft gene probable has been particularly identified (Boches and Myers, 2007; Schreiber et al., 2011). Recently a gene namely atv also documented which encodes a R3 MYB repressor of the pathway of pigment production in fruits (Colanero et al., 2018; Cao et al., 2017). Additionally, cross between the Aft and atv tomato lines were carried out and extremely pigmented fruit (Aft/Aft atv/atv genotype) has been observed (Gonzali et al., 2009;

Povero et al., 2011; Scott and Myers, 2012; Borghesi et al., 2016). In high anthocyanin containing tomato varieties accumulates more anthocyanin in fruit peel (Povero et al., 2011). Tomatoes rich in anthocyanin pigment have a longer shelf life as compared to wild-type as well as multiple resistance to necrotrophic pathogens and slower ripening after harvest (Bassolino et al., 2013; Zhang et al., 2013).

14.7.4 THROUGH INDUCED MUTATIONS AND MUTATION BREEDING

In tomato, a single gene creates different characteristics of ripening, and mutations can be created to acquire the information of ripening-related characteristics. The single gene that affects more than one character (Pleiotropic) in tomatoes consists of ripening-inhibitor (rin), colorless non-ripening (Cnr), green-ripe (Gr), never-ripe (Nr), and excessive-pigment (hp-1 and hp-2). Characterization of diverse ripening-related mutations in tomato has been possible for the use of molecular techniques together with positional cloning, mapping of mutated loci and candidate genes. There are two mated lines namely Cnr and Rin (recessive and dominant respectively) are identified, which efficiently obstruct the ripening progression in tomato. The discussed process are responsible for the breakdown to manufacture high ethylene or reply to exogenous ethylene (Manning et al., 2006; Vrebalov et al., 2002) which is responsible for the ripening during the ripening stage. The ethylene receptor gene (Nr and Gr mutation) has been observed to encode a novel element of ethylene signaling molecules. GR gene was identified by positional cloning approach from a dominant ripening mutation lines for the establishment of self-life enhancement in tomato fruits (Barry and Giovannoni, 2006). Kopeliovitch et al. (1979) worked on numerous ripening gene mutants, inclusive of non-ripening (nor), alcobaca (alc), in no way ripe, and ripening inhibitor (rin) to increase traces and cultivars with delayed ripening thru interruption of the ethylene signaling pathway.

14.7.5 MARKER ASSISTED SELECTION (MAS)

Genetic improvement of fruit quality by exploration of the available genetic diversity within the accessible germplasm is a reasonable and environmentally safe choice. Integrating molecular methods with conventional breeding to beautify fruit best could notably enhance the postharvest shelf life of tomatoes. The ranges of genetic variation inside the cultivated tomato detected by most molecular markers are very small, that's a burden for marker-assisted selection (MAS) with admire to a wide range of vital fruit satisfactory tendencies.

Several tomato hybrids with improved shelf life were advanced by using ripening mutants' lines and agronomically advanced varieties with the help of molecular markers. The developed varieties has some important QTL, and these QTLs has been transferred for the enhancement of shelf life, yield, and different other fruit qualities (Table 14.2). Yogendra and Gowda (2013) reported simple sequence repeat (SSR) markers-based screening of F_2 population from the best executing F_1 hybrid (alc × 'Vaibhav') for the mapping of important quantitative trait loci (QTLs) responsible for better fruit firmness and shelf life. Among different crosses, nine crosses showed three fruit ripening gene mutants of *S. lycopersicum* (non-ripening, alcobaca (alc), and ripening inhibitor) and three Indian agronomically superior cultivars (Sankranti, Pusaruby, and Vaibhav) with poor shelf life. The hybrid offspring developed from alc × Vaibhav had the highest prolonged shelf existence (up to 40 days) as compared with that of different hybrids and varieties. Fruit quality traits and yield parameters in F_2 progenies of alc × 'Vaibhav' have been assessed. A broad variety of genetic diversity becomes determined for fruit firmness (0.55–10.65 lbs/cm^2) and shelf life (5–106 days). Molecular markers identified the QTLs related to tomato fruit firmness in several segregating populations (Fulton et al., 2000; Walley and Seymour, 2006) an on chromosomes 2, 3, 4, and 8 the QTLs for fruit firmness has been identified by RFLP (restriction fragment length polymorphism) marker (Tanksley et al., 1996). Similarly, Fulton et al. (2000) also identified fruit firmness QTLs on chromosomes 1, 3, 4, 6, 9, and 11 in segregating tomato population.

14.7.6 IMPORTANT GENE AND QTLS FRO SHELF LIFE IN TOMATO

Important QTLs from wild tomatoes *Solanum pimpinellifolium, Solanum peruvianum, Solanum esculentum* and their crosses with cultivated tomato

TABLE 14.2 Some Important QTLs for the Postharvest Losses in Tomato

Important Traits Related to Postharvest	Wild Species Used	Mapping Population Used	Genes/QTLs	Chromosome Location	References
Ripening	*L. esculentum*	F_2	2 QTLs	5, 12	Doganlar et al. (2000)
	L. pennellii	F_2	Many loci	All chromosome	Kinzer et al. (1990); Slater et al. (1985)
	L. pimpinellifolium	Backcross (BC_1)	3QTLs	2, 8, 9	Grandillo and Tanksley (1996b)
	L. pimpinellifolium	Backcross BC_2 and BC_3)	4 QTLs	2, 4, 7, 8	Tanksley et al. (1996)
	L. peruvianum	Backcross (BC_3, BC_4	4 QTLs	2, 3, 8, 9	Fulton et al. (1997)
Fruit ripening (*Colorless non-rip.*)	*L. cheesmanii*	F_2	*Cnr*	2	Tor et al. (2002)
Fruit ripening (*never-ripe*	*L. cheesmanii*	F_2	*Nr*	9	Yen et al. (1995)
Fruit ripening (*non-ripening*)	*L. pennellii, L. cheesmanii*	F2	Nor, Rin	5, 10	Moore et al. (2002); Giovannoni et al. (1999); Vrebalov et al. (2002)
Fruit ripening (polygalacturonase)	*L. pimpinellifolium*	Backcross (BC_1)	TOM6	10	Kinzer et al. (1990)
Fruit ripening (*uniform rip.*)	*L. pimpinellifolium*	Backcross (BC_1)	*u*	10	Kinzer et al. (1990)

Source: Adapted and modified from: Foolad (2007).

for enhanced self-life (Table 14.2) and firmness has been identified by using BC_2F_1 and BC_3 mapping populations (Tanksley et al., 1996; Fulton et al., 1997). There are important QTLs for tomato firmness have been also recognized from *Solanum parviflorum* (LA2133), *Solanum hirsutum* (LA1777), and *Solanum pennellii* (LA1657) by using backcross QTL strategy (Frary et al., 2004; Bernacchi et al., 1998). Seventy-five QTLs also reported by using 152 lines of *S. hirsutum* BC_2F_2 individuals (Okmen et al. (2011) for tomato phenotypes and firmness. Chapman et al. (2012) also reported five QTLs that were responsible for tomato firmness, and identified QTLs are located within 8.6 Mb regions and linked with the TG453 and TG567 markers. QTLs for fruit firmness have been diagnosed via hand squeezing method by Okmen et al. (2011); Frary et al. (2004); Bernacchi et al. (1998); Fulton et al. (1997); and Tanksley et al. (1996) for the improvement of tomato fruit firmness. The QTLs related with other postharvest factor, i.e., pericarp puncturing also identified for the improvement of different lines of tomato (Liu et al., 2015; Chapman et al., 2012).

14.8 BIOTECHNOLOGICAL APPROACHES

The physiological, biochemical, and molecular developments of the tomato as well as other vegetables are directly related to the postharvest garage of greens to maintain the fine and shelf existence of postharvest products. These trends are genetically decided and may be manipulated the usage of genetic breeding and biotechnology. It has been discovered in published research effects that the manipulated genes have potential and may be used to develop excellent postharvest crop plant life. Biotechnological programs, which include genetic engineering, have a first-rate capacity to engineer the fruits which not on time ripening characters. In 1994 Calgene, United States of America had got first popularity of commercial sale of a genetically engineered transgenic tomato product 'Flavr-savr' tomato which had behind schedule ripening trait; culmination with stepped forward shelf lifestyles which were suitable for food processing can be developed (the use of down-regulation/change of ethylene metabolism and manipulating cell wall metabolism).

14.8.1 EMERGING TRANSGENIC TECHNOLOGIES

Alterations of fruit softening and of the general firmness were attained mainly by the modification of different enzymes which may associate with the cell wall. These enzymes include polygalacturonase (Langley et al., 1994), ß-galactosidase (Smith et al., 2002), pectin methylesterase (Tieman and Handa, 1994) and expansin (Brummell et al., 1999) and related genes actively involved in controlling softening and fruit firmness in tomato. A transgenic line of tomato with the gene fw2.2, a poor regulator of cellular division has been achieved via genetic transformation (Frary et al., 2000). The transgenic tomato lines produced normal tomato without affecting any cellular length in pericarp, placenta tissues with normal weight (Liu et al., 2003).

Two examples of a hit metabolic engineering modifying fruit flavor in tomato relayed on heterologous single-gene expression to introduce untypical traits in tomato. In the primary example, a biologically lively thaumatin, a sweet recognizing, flavor-enhancing protein from the African plant *Thaumatococcus daniellii* Benth became articulated in transgenic tomatoes that produced sweeter fruits with a selected after-taste (Bartoszewski et al., 2003). In the second example, the lemon basil geraniol synthase (GES) gene changed into overexpressed under the manipulation of strong fruit ripening definite tomato polygalacturonase promoter (PG). GES encodes the enzyme liable for the production of geraniol from GDP and its expression precipitated the plastidial terpenoid biosynthetic flux to divert, leading to a reduced lycopene accumulation and to dramatic changes in the aroma and normal flavor of the transgenic culmination (Davidovich-Rikanati et al., 2007). Expressing yeast spermidine synthase (ySpdSyn) gene, which bureaucracy the polyamine spermidine, constitutively or beneath the manage of the promoter of the fruit specific gene E8 have led to accelerated shelf-life of tomato and is because of the buildup of lycopene, an antioxidant, combined with the anti-senescence outcomes of polyamine (Nambeesan et al., 2010). Liu et al. (2016) determined that genetic manipulation of SlMSI1 and RIN successfully extended the fruit shelf existence by using *regulating* the fruit-ripening genes in tomato.

14.8.2 RNAI

RNAi technique, conferring delayed ripening in tomato has been pronounced the use of ACO gene. With the help of RNAi technology, transgenic cultivars of tomato have also been produced by changing ethylene biosynthesis with an anti-ACO gene (Nath et al., 2006). Researchers additionally said that transgenic tomato culmination had a prolonged shelf existence of as a minimum a hundred and twenty days (Xiong et al., 2005). Najat et al. (2018) stated that lowland transgenic RNAiACO1-21 capable of put off the softening technique and in addition storage at low temperature extended the postharvest lifestyles and maintained the satisfactory of the tomato fruit. Similarly, GRAS gene known as SlFSR (fruit shelf-lifestyles regulator) detected and this gene increased duration of fruit ripening. The mutant of S1FSR gene appreciably decreased inside the tomato mutant rin (ripening inhibitor) and enhanced the self-line of tomato (Zhang et al., 2018)). Suppression of SlFSR with the help of RNAi resulted in abridged expression of numerous cell wall modification-related genes. These genes finally decreased the activities of polygalacturonase), tomato β-galactosidase, cellulase, and β-D-xylosidase and appreciably extended fruit shelf-life. In addition, over expression of SlFSR in mutant rin tomato provided upward thrust to up-regulated expression of multiple mobile wall modification-associated genes, including TBG4, PG, XYL1, CEL2, PL, MAN1, PE, XTH5, and EXP1 drastically reduced the fruit shelf-life. These results disclose a numeral of the genetic systems and their correlation in fruit cell wall metabolism and endorse that the SlFSR gene is another ability biotechnological goal for the manipulate of tomato fruit shelf-life.

14.8.3 MI RNA

At present, new, and critical additives of complicated gene-regulatory networks which controlling plant improvement are referred to as microRNAs (miRNAs) (Jones-Rhoades et al., 2006). However, miRNAs, and their objects have been diagnosed inside the wide variety of plants. There are several work has been executed in terms of their involvement in fruit improvement and ripening. In recent times, scientists discriminated a preset of miRNAs and their dreams from tomatoes that were conserved with the stage alternate from vegetative to reproductive growth (Yin et al., 2008; Zhang et al., 2008). In addition, a recent era of excessive throughput pyrosequencing has obtrusive miRNAs s targeting genes which can be engaged in vegetable fruit ripening (Moxon et al., 2008; Shukla et al., 2008).

14.8.4 CRISPR APPROACH

A superior technique to generate genetic unevenness in a particular and focused approach, the genome-enhancing method CRISPR/Cas (clustered frequently interspaced quick palindromic repeats (CRISPR) related proteins), has strained the eye of plant breeders. A guiding RNA-based genome editing CRISPR/Cas device is used that control a nuclease to produce a double-strand break (DSB) in the region of target DNA. Further, this enzyme activating the mobile renovation of the systems and finally lead to the deletions or insertions of nucleotides in the concerned region (Koltun et al., 2018). In tomato, the CRISPR-Cas9 device become successfully applied first time in 2014 by using argonaute seven approaches. The results altered into knocked wiry phenotypes; in which first leaves of mutants' tomato had leaflets without petioles and in the end fashioned leaves lacking laminae (Brooks et al., 2014). *Botrytis cinerea* is an airborne plant pathogen that reasons gray mold disease, resulting in serious economic losses in each pre- and postharvest stages. For *B. cinerea* losses, CRISPR-Cas9 technology with Mitogen-activated protein kinase 3 (MAPK3) used to confer resistance (Zhang et al., 2018). Regulation of ripening has also been examined by site-directed mutagenesis, wherein a sequence of mutations of a RIN gene in tomato has investigated (Ito et al., 2015). Further lncRNA1459 gene mutants with deficient-ripening fruit production has also been investigated in tomato (Li et al., 2018). Genome alteration by site-directed mutagenesis has also been applied in tomato lines for the development of new characters. Quality attribute in tomato also modified with the help of CRISPR/Cas9 in ALC gene. The nucleotide substitute of the tomato sequence of the homologous restore template (replacement of thymine by using adenine in position 317 of the coding sequence) has been utilized (Table 14.3). In the T_1 generation of tomato, it turned into viable to produce recessive homozygous alc mutants freed from the CRISPR/Cas9 additives which presented outstanding garage performance (Yu et al., 2017). Nonaka et al. (2017) investigated GABA (gamma-aminobutyric acid) accumulation in tomato fruit, and found increased SlGAD3 and SlGAD2 genes, through the CRISPR/Cas9 system. The intake of GABA-enriched meals tomato in daily existence can convey anti-hypotensive effects and could be an interesting path to prevent high blood pressure in humans as well as postharvest losses reduction.

TABLE 14.3 Genome Editing Technologies to Study Tomato Fruit Ripening, and CRISPR/Cas9

SL. No.	Genome Editing Tools	Gene-Edited	Character Coded	References
1.	CRISPR/Cas9	Ripening-inhibitor (RIN)	Fruit ripening	Ito et al. (2015, 2017)
2.	CRISPR/Cas9	Pectate lyase (PL)	Fruit firmness	Uluisik et al. (2016)
3.	CRISPR/Cas9	Long-non coding RNA (lncRNA1459)	Fruit ripening	Li et al. (2018)

Source: Adapted and modified from: Martin-Pizarro and Pose (2019).

14.9 CONCLUSIONS

A wide type of biotic stress (virus, bacteria, and fungus, etc.), causes postharvest losses in tomato and their related products. The pathogens invade the tomato harvest throughout the year via the wounding and other opening in the fruits. Improvement approaches for postharvest losses is basic to make a stride returned and recall the creation and postharvest managing structures. Numerous pre-harvest factors also easily affect losses during harvesting periods. Thus advancement of postharvest issue is urgently needed to minimize the losses in tomato for the better yield and quality purpose. Chemical fungicides are commonly utilized for the postharvest loss's purpose in tomato. Thus there is an urgent need to develop new tools with the help of biotechnology to minimize the losses of tomato after harvest. This will stand to exciting challenge to meet the new techniques for the 21st century.

KEYWORDS

- **colorless non-ripening**
- **gamma aminobutyric acid**
- **gene silencing**
- **marker-assisted selection**
- **microRNAs**
- **postharvest losses**
- **ripening gene**

REFERENCES

Agrios, G. N., (2004). *Plant Pathology* (5th edn.). Molecular plant pathology.

Ajayi, A. A., Olutiola, P. O., & Fakunle, J. B., (2003). Studies on polygalacturonase associated with the deterioration of tomato fruits (*Lycopersicon esculentum* Mill.) infected by *Botryodiplodia theobromae* PAT. *Sci. Focus, 5,* 68–77.

Arah, I. K., Ahorbo, G. K., Anku, E. K., Kumah, E. K., & Amaglo, H., (2016). Postharvest handling practices and treatment methods for tomato handlers in developing countries: A mini-review. *Adv. Agric., 8.*

Arah, I. K., Ernest, K. K., Etornam, K. A., & Harrison, A., (2015). An overview of postharvest losses in tomato production in Africa: Causes and possible prevention strategies. *J. Biol. Agric. Healthcare*, *5,* 78–88.

Arruda, R. L., Paz, A. T. S., Bara, M. T. F., Côrtes, M. V. C. B., De Filippi, M. C. C., & Da Conceição E. C., (2016). An approach on phytoalexins: Function, characterization and biosynthesis in plants of the family *Poaceae. Cienc. Rural., 46,* 1206–1216.

Asalfew, G. K., & Nega, M., (2020). Postharvest loss of tomato (*Solanum Lycopersicum* L) in Ethiopia: A review. *Acta Sci. Agricul., 4*, 01–06.

Atanda, S. A., Pessu, P. O., Agoda, S., Isong, I. U., & Ikotun, I., (2011). The concepts and problems of postharvest food losses in perishable crops. *Afr. J. Food Sci., 5,* 603–613.

Ayandiji, A., Adeniyo, O. R., & Omidiji, D., (2011). Determinant postharvest losses among tomato farmers in Imeko-Afon local government area of Ogun State, Nigeria. *Global J. Sci. Front. Res., 11*, 5975–5896.

Ayomide, O. B., Ajayi, O. O., & Ajayi, A. A., (2019). Advances in the development of a tomato postharvest storage system: Towards eradicating postharvest losses. *J. Physics: Conference Series 1378*, 022064.

Barkai-Golan, R., & Paster, N., (2008). *Mycotoxins in Fruits and Vegetables*. Academic Press.

Barkai-Golan, R., Padova, R., Ross, I., Lapidot, M., Davidson, H., & Copel, A., (1993). Combined hot water and radiation treatments to control decay of tomato fruits. *Sci. Horticul., 56*, 101–105.

Barry, C. S., & Giovannoni, J. J., (2006). Ripening in the tomato green-ripe mutant is inhibited by ectopic expression of a protein that disrupts ethylene signaling. *Proc. Natl. Acad. Sci., 103*(20), 7923–7928.

Barth, M., Hankinson, T. R., Zhuang, H., & Breidt, H., (2009). Microbiological spoilage of fruits and vegetables. In: Sperber, WHaMD, (ed.), *Compendium of the Microbiological Spoilage of Foods and Beverages* (pp. 135–183). Springer Science+Business Media, Berlin, Germany.

Bartoszewski, G., Niedziela, A., Szwacka, M., & Niemirowicz-Szczytt, K., (2003). Modification of tomato taste in transgenic plants carrying a thaumatin gene from *Thaumatococcus daniellii* benth. *Plant Breed., 122,* 347–351.

Bassolino, L., Zhang, Y., Schoonbeek, H. J., Kiferle, C., Perata, P., & Martin, C., (2013). Accumulation of anthocyanins in tomato skin extends shelf life. *New Phytol., 200,* 650–655.

Bernacchi, D., Beck-Bunn, T., Eshed, Y., Lopez, J., Petiard, V., Uhlig, J., Zamir, D., & Tanksley, S., (1998). Advanced backcross QTL analysis in tomato. I. Identification of QTLs for traits of agronomic importance from *Lycopersicon hirsutum. Theor. Appl. Genet., 97*, 381—397.

Beuchat, L. R., (2002). Ecological factors influencing survival and growth of human pathogens on raw fruits and vegetables. *Microb. Infection., 4*, 413–423.

Biggs, M. S., Woodson, W. R., & Handa, A. K., (1988). Biochemical basis of high temperature inhibition of ethylene biosynthesis in ripening tomato fruits. *Physiol. Plant., 72*, 572–578.

Boches, P., & Myers, J., (2007). The anthocyanin fruit tomato gene (Aft) is associated with a DNA polymorphism in a MYB transcription factor. *Hortsci., 42*, 856.

Borghesi, E., Ferrante, A., Gordillo, B., RodrõÂguez-Pulido, F., Cocetta, G., Trivellini, A., Mensuali-Sodi, A., et al., (2016). Comparative physiology during ripening in tomato rich-anthocyanins fruits. *Plant Growth Regul., 80*, 207–214.

Brooks, C., Nekrasov, V., Lippman, Z. B., & Van, E. J., (2014). Efficient gene editing in tomato in the first generation using the clustered regularly interspaced short palindromic repeats/CRISPR-associated9 system. *Plant Physiol., 166*, 1292–1297.

Brummell, D. A., Harpster, M. H., Civello, P. M., Palys, J. M., Bennett, A. B., & Dunsmuir, P., (1999). Modification of expansin protein abundance in tomato fruit alters softening and cell wall polymer metabolism during ripening. *The Plant Cell, 11*, 2203–2216.

Butelli, E., Titta, L., Giorgio, M., Mock, H. P., Matros, A., Peterek, S., Schijlen, E. G., et al., (2008). Enrichment of tomato fruit with health-promoting anthocyanins by expression of select transcription factors. *Nat. Biotechnol., 26*, 1301–1308.

Cao, X., Qiu, Z., Wang, X., Van, G. T., Liu, X., Wang, J., et al., (2017). A putative R3 MYB repressor is the candidate gene underlying atroviolacium, a locus for anthocyanin pigmentation in tomato fruit. *J. Exp. Bot., 68*, 5745–5758.

Carrari, F., & Fernie, A. R., (2006). Metabolic regulation underlying tomato fruit development. *J. Exp. Bot., 57*, 1883–1897.

Chapman, N. H., Bonnet, J., Grivet, L., Lynn, J., Graham, N., Smith, R., Sun, G., et al., (2012). High-resolution mapping of a fruit firmness-related quantitative trait locus in tomato reveals epistatic interactions associated with a complex combinatorial locus. *Plant Physiol., 159*, 1644–1657.

Colanero, S., Perata, P., & Gonzali, S., (2018). The *atroviolacea* Gene encodes an R3-MYB protein repressing anthocyanin synthesis in tomato plants. *Front Plant Sci., 9*, 830.
Dal Bello, G., Mónaco, C., Rollan, M. C., Lampugnani, G., Arteta, N., Abramoff, C., Ronco, L., & Stocco, M. (2008). Biocontrol of postharvest grey mould on tomato by yeasts. Journal of Phytopathology, 156, 257–263.
Dauthy, M. E., (1995). *Fruit and Vegetable Processing* (pp. 1–6). FAO.
Davidovich-Rikanati, R., Sitrit, Y., Tadmor, Y., Iijima, Y., Bilenko, N., Bar, E., Carmona, B., et al., (2007). Enrichment of tomato flavor by diversion of the early plastidial terpenoid pathway. *Nat. Biotech., 25*, 899–901.
De Curtis, F., Lima, G., Vitullo, D., et al., (2010). Biocontrol of *Rhizoctonia solani* and *Sclerotium rolfsii* on tomato by delivering antagonistic bacteria through a drip irrigation system. *Crop Protection, 29*, 663–670.
De Hoog, G. S., & Van, D. V. W. H. B., (1989). Retroconis, a new genus of ascomycetous hyphomycetes. *Studies in Mycology, 31*, 99–105.
Di Francesco, A., Martini, C., & Mari, M., (2016). Biological control of postharvest diseases by microbial antagonists: How many mechanisms of action? *Eur. J. Plant Pathol., 145*, 711–717.
Díaz, J., Ten, H. A., & Van, K. J. A., (2002). The role of ethylene and wound signaling in resistance of tomato to Botrytis cinerea. *Plant Physiol., 129,* 1341–1351.
Doganlar, S., Tanksley, S. D., & Mutschler, M. A., (2000). Identification and molecular mapping of loci controlling fruit ripening time in tomato. *Theor. Appl. Genet., 100,* 249–255.
Dukare, A. S., Paul, S., Nambi, V. E., et al., (2018). Exploitation of microbial antagonists for the control of postharvest diseases of fruits: A review. *Crit. Rev. Food Sci. Nut.*, 1–16.
Emana, B., Afari-Sefa, V., Nenguwo, N., Ayana, A., Kebede, D., & Mohammed, H., (2017). Characterization of pre- and postharvest losses of tomato supply chain in Ethiopia. *Agric. Food Secure., 6*, 3.
FAO, Food Quality and Safety Systems, (1998). *A Training Manual on Food Hygiene and the Hazard Analysis and Critical Control Point (HACCP) System*. Publishing Management Group, FAO Information Division, Rome, Italy.
Fatima, N., Batool, H., Sultana, V., et al., (2009). Prevalence of postharvest rot of vegetables and fruits in Karachi, Pakistan. *Pak. J. Botany, 41*, 3185–3190.
Foolad, M. R., (2007). Molecular mapping, marker-assisted selection and map-based cloning in tomato. In: Varshney, R. K., & Tuberosa, R., (eds.), *Genomics Assisted Crop Improvement: Genomics Applications in Crops* (Vol. 2, pp. 307–356).
Frary, A., Fulton, T. M., Zamir, D., & Tanksley, S. D., (2004). Advanced backcross QTL analysis of a *Lycopersicon esculentum L.* pennellii cross and identification of possible orthologs in the Solanaceae. *Theor. Appl. Genet., 108*, 485–496.
Frary, A., Nesbitt, T. C., Grandillo, S., Van, D. K. E., Cong, B., Liu, J., Meller, J., et al., (2000). fw2.2: A quantitative trait locus key to the evolution of tomato fruit size. *Science, 289*, 85–88.
Fulton, T. M., Beck-Bunn, T., Emmatty, D., Eshed, Y., Lopez, J., Petiard, V., Uhlig, J., Zamir, D., & Tanksley, S. D., (1997). QTL analysis of an advanced backcross of *Lycopersicon peruvianum* to the cultivated tomato and comparison with QTLs found in other wild species. *Theor. Appl. Genet., 95*, 881–894.

Fulton, T. M., Beck-Bunn, T., Emmatty, D., Eshed, Y., Lopez, J., Petiard, V., Uhlig, J., Zamir, D., & Tanksley, S. D., (1997). QTL analysis of an advanced backcross of *Lycopersicon peruvianum* to the cultivated tomato and comparisons with QTLs found in other wild species. *Theor. Appl. Genet., 95*, 881–894.

Fulton, T. M., Grandillo, S., Beck-Bunn, T., Fridman, E., Frampton, A., Lopez, J., Petiard, V., et al., (2000). Advanced backcross QTL analysis of a *Lycopersicon esculentum* x *Lycopersicon parviflorum* cross. *Theor. Appl. Genet., 100*, 1025–1042.

Georgiev, C., (1972). Anthocyanin fruit (Af). *Rep. Tomato Genet. Coop., 22*, 10.

Giovannoni, J. J., Yen, H., Shelton, B., Miller, S., Vrebalov, J., Kannan, P., Tieman, D., et al., (1999). Genetic mapping of ripening and ethylene-related loci in tomato. *Theor. Appl. Genet., 98*, 1005–1013.

Gomathi, S., & Ambikapathy, V., (2011). Antagonistic activity of fungi against *Pythium debaryanum* (Hesse) isolated from chili field soil. *Ad. Applied Sci. Res., 2*, 291–297.

Gonzali, S., Mazzucato, A., & Perata, P., (2009). Purple as a tomato: Towards high anthocyanin tomatoes. *Trends Plant Sci., 14*, 237–241.

Grandillo, S., & Tanksley, S. D., (1996b). QTL analysis of horticultural traits differentiating the cultivated tomato from the closely related species *Lycopersicon pimpinellifolium. Theor. Appl. Genet., 92*, 935–951.

Gross, K. C., Wang, C. Y., & Salveit, M., (2016). *The Commercial Storage of Fruits, Vegetables, and Florist and Nursery Stocks*. Beltsville Maryland, united states department of agriculture.

Guo, M., Feng, J., Zhang, P., Jia, L., & Chen, K., (2015). Postharvest treatment with trans-2-hexenal induced resistance against *Botrytis cinerea* in tomato fruit. *Aust. Plant Pathol., 44*, 121–128.

Hadizadeh, I., Pivastegan, B., & Hamzehzarghani, H., (2009). Antifungal activity of essential oils from some medicinal plants of Iran against *Alternaria alternata. Am. J. Appl. Sci., 6*, 857–861.

Haile, A., (2018). Shelf life and quality of tomato (*Lycopersicon esculentum* mill.) fruits as affected by different packaging materials. *Afr. J. Food Sci., 12*, 21–27.

Hodges, R. J., Buzby, J. C., & Bennett, B., (2011). Postharvest losses and waste in developed and less developed countries: Opportunities to improve resource use. *The J. Agricul. Sci., 149*(S1), 37–45.

Huntanen, C. N., Naghski, J., Custer, C. S., & Russel, R. W., (1976). Growth and toxin production by *Clostridium botulinum* in moldy tomato juice. *Appl. Environ. Microb., 32*, 711–715,

Ignjatov, M., Milosevic, D., Nikolic, Z., Gvozdanović-Varga, J., Jovičić, D., & Zdjelar, G., (2012). *Fusarium oxysporum* as causal agent of tomato wilt and fruit rot. *Pestic. Phytomed. (Belgrade), 27*, 25–31.

Ito, Y., Yokoi, N., Endo, M., Mikami, M., & Toki, S., (2015). CRISPR/Cas9-mediated mutagenesis of the RIN locus that regulates tomato fruit ripening. *Biochem. Biophys. Res. Comm., 6*, 76–82.

Izumi, H., (2010). Development of technologies for safe fresh and fresh-cut produce in Japan, *Acta Horticulturae, 875*, 229–236.

Jabnoun-Khiareddine, H., Abdallah, R. A. B., Daami-Remadi, M., Nefzi, A., & Ayed, F., (2019). Grafting tomato cultivars for soil borne disease suppression and plant growth and yield improvement. *J. Plant Pathol. Microbiol., 10*, 1.

Jeong, M. A., & Jeong, R. D., (2018). Applications of ionizing radiation for the control of postharvest diseases in fresh produce: Recent advances. *Plant Pathol., 67*, 18–29.

Jerry, A. B., Steven, A. S., & Michael, M., (2005). *Guide to Identifying and Controlling of Post-Harvest Tomato Diseases in Florida* (p. 131). Department of Horticultural Sciences, University of Florida, Gainesville.

Jones, C. M., Mes, P., & Myers, J. R., (2003). Characterization and inheritance of the anthocyanin fruit (Aft) tomato. *J. Hered., 94*, 449–456.

Jones, J. B., Jones, J. P., Stall, R. E., & Zitter, T. A., (1993). *Compendium of Tomato Diseases* (p. 73). The Amer. Phytopathol. Society. USA.

Jones-Rhoades, W., Matthew, D., Bartel, P., & Bartel, B., (2006). MicroRNAs and their regulatory roles in plants. *Annu. Rev. Plant Biol., 57*, 19–53.

Khadka, R. B., Marasini, M., Rawal, R., Gautam, D. M., & Acedo, A. L. Jr., (2017). Effects of variety and postharvest handling practices on microbial population at different stages of the value chain of fresh tomato *(Solanum lycopersicum)* in western Terai of Nepal. *Bio-Med Res. Int.*, 7148076. https://doi.org/10.1155/2017/7148076.

Khonglah, D., & Kayang, H., (2018). Antagonism of indigenous fungal isolates against *Botrytis cineria* the causal of gray mold disease of tomato (*Solanum lycopersicum* L.). *Int. J. Curr. Res. Life Sci., 7*, 806–812.

Kinzer, S. M., Schwager, S. J., & Mutschler, M. A., (1990). Mapping of ripening-related or specific cDNA clones of tomato (*Lycopersicon esculentum*). *Theor. Appl. Genet., 79*, 489–496.

Kitinoja, L., & Gorny, J. R., (1999). *Postharvest Technology for Small-Scale Produce Marketers: Economic Opportunities, Quality and Food Safety.* University of California, Davies, Postharvest Hort. Ser., 21.

Kleemann, R., Zadelaar, S., & Kooistra, T., (2008). Cytokines and atherosclerosis: A comprehensive review of studies in mice. *Card. Res., 79*, 360–376.

Koltun, A., Corte, L. E. D., Mertz-Henning, L. M., & Gonçalves, L. S. A., (2018). Genetic improvement of horticultural crops mediated by CRISPR/Cas: A new horizon of possibilities. *Horti. Bras., 36*, 290–298.

Kopeliovitch, E., Mizrahi, Y., Rabinowitch, H. D., & Kedar, N., (1979). Effect of the fruit-ripening mutant genes rin and nor on the flavor of tomato fruit. *J. Amer. Soc. Hort. Sci., 107*, 361–364.

Kramer, M., Sanders, R., Bolkan, H., Waters, C., Sheeny, R. E., & Hiatt, W. R., (1992). Postharvest evaluation of transgenic tomatoes with reduced levels of polygalacturonase: Processing, firmness and disease resistance. *Postharvest Biol. Tech., 1*, 241–255.

Kumar, V., Haldar, S., Pandey, K. K., Singh, R. P., Singh, A. K., & Singh, P. C., (2008). Cultural, morphological, pathogenic and molecular variability amongst tomato isolates of *Alternaria solani* in India. *World J. Microb. Biotech., 24*, 1003–1009.

Lafontaine, J. P., & Benhamou, N., (1996). Chitosan treatment: An emerging strategy for enhancing resistance of greenhouse tomato plants to infection by *Fusarium oxysporum* f. sp. *radicis-lycopersici*. *Biocontrol Sci. Technol., 6*, 111–124.

Langley, K. R., Martin, A., Stenning, R., Murray, A. J., Hobson, G. E., Schuch, W. W., & Bird, C. R., (1994). Mechanical and optical assessment of the ripening of tomato fruit with reduced polygalacturonase activity. *J. Sci. Food Agric., 66*, 547–554.

Li, J., Tao, X., Li, L., Mao, L., Luo, Z., Khan, Z. U., & Ying, T., (2016a). Comprehensive RNA-Seq analysis on the regulation of tomato ripening by exogenous auxin. *PloS One, 11*, e0156453.

Li, R., Fu, D., Zhu, B., Luo, Y., & Zhu, H., (2018). CRISPR/Cas9-mediated mutagenesis of lncRNA1459 alters tomato fruit ripening. *The Plant J., 94*, 513–524.

Liu, C., Cai, L., Han, X., & Ying, T., (2011). Temporary effect of postharvest UV-C irradiation on gene expression profile in tomato fruit. *Gene., 15*, 56–84.

Liu, D. D., Zhou, L. J., Fang, M. J., Dong, Q. L., An, X. H., You, C. X., & Hao, Y. J., (2016). Polycomb-group protein SlMSI1 represses the expression of fruit ripening genes to prolong shelf life in tomato. *Sci. Rep., 6*, 31806.

Liu, J., Cong, B., & Tanksley, S. D., (2003). Generation and analysis of an artificial gene dosage series in tomato to study the mechanisms by which the cloned quantitative trait locus fw2.2 controls fruit size. *Plant Physiol., 132*, 292–299.

Liu, J., Sui, Y., Wisniewski, M., et al., (2013). Review: Utilization of antagonistic yeasts to manage postharvest fungal diseases of fruit. *Int. J. Food Microb., 167*, 153–160.

Liu, L., Song, Y., Zheng, Z., & Li, J. M., (2015). QTL mapping of fruit firmness with an Introgression Line population derived from the wild tomato species *Solanum pennellii* LA0716. *J. Plant Gent. Res., 16*, 323–329 (in Chinese).

Lledó, S., Rodrigo, S., Poblaciones, M. J., & Santamaría, O., (2016). Biomass yield, nutritive value and accumulation of minerals in *Trifolium subterraneum* L. as affected by fungal endophytes. *Plant and Soil, 405*, 197–210.

Lurie, S., & Pedreschi, R., (2014). Fundamental aspects of postharvest heat treatments. *Horticulture Res., 1*, 14030.

Lurie, S., (1998). Postharvest heat treatments. *Post. Biol. Tech., 14*, 257–269.

Mama, S., Yemer, J., & Woelore, W., (2016). Effect of hot water treatments on shelf life of tomato (*Lycopersicon esculentum* Mill). *J. Nat. Sci. Res., 6*, 69–77.

Manning, K., Tör, M., Poole, M., Hong, Y., Thompson, A. J., King, G. J., Giovannoni, J. J., & Seymour, G. B. A., (2006). Naturally occurring epigenetic mutation in a gene encoding an SBP-box transcription factor inhibits tomato fruit ripening. *Nat. Genet., 38*, 948–952.

Matas, A. J., Gapper, N. E., Chung, M. Y., Giovannoni, J. J., & Rose, J. K., (2009). Biology and genetic engineering of fruit maturation for enhanced quality and shelf-life. *Curr. Opin. Biotechnol., 20*, 197–203.

Migliori, C. A., Salvati, L., Di Cesare, L. F., Lo Scalzo, R., & Parisi, M., (2017). Effects of preharvest applications of natural antimicrobial products on tomato fruit decay and quality during long-term storage. *Sci. Hortic., 222*, 193–202.

Moore, S., Vrebalov, J., Payton, P., & Giovannoni, J., (2002). Use of genomics tools to isolate key ripening genes and analyze fruit maturation in tomato. *J. Exp. Bot., 53*, 2023–2030.

Morrissey, J. P., & Osbourn, A. E., (1999). Fungal resistance to plant antibiotics as a mechanism of pathogenesis. *Microbiol. Mol. Biol. Rev., 63*, 708–724.

Moss, M. O., (2002). Mycotoxin review journal on *Aspergillus penicillium. Mycologist, 16*, 116–119.

Moxon, S., Jing, R., Szittya, G., Schwach, F., Pilcher, R. L. R., Moulton, V., & Dalmay, T., (2008). Deep sequencing of tomato short RNAs identifies microRNAs targeting genes involved in fruit ripening. *Genome Res., 18*, 1602–1609.

Naglot, A., Goswami, S., Rahman, I., et al., (2015). Antagonistic potential of native *Trichoderma viride* strain against potent tea fungal pathogens in north east India. *Plant Path. J., 31*, 278–289.

Najat, M. E., Maizom, H., Zamri, Z., & Alhdad, G. M., (2018). Comparative study of quality changes in lowland transgenic RNAiACO1(T2) tomato fruit during storage at ambient and low temperature. *Int. J. Chem. Tech. Res., 10*, 75–83.

Nambeesan, S., Datsenka, T., Ferruzzi, M. G., Malladi, A., Mattoo, A. K., & Handa, A. K., (2010). Overexpression of yeast spermidine synthase impacts ripening, senescence and decay symptoms in tomato. *Plant J., 63*, 836–847.

Narain, A., & Rout, G. B. A., (1981). tomato rot caused by *Cladosporium tenuissimum. Ind. Phytopathol., 34*, 237–238.

Nath, P., Trivedi, P. K., Sane, V. A., & Sane, A. P., (2006). Role of ethylene in fruit ripening. In: Nafees, A., & Khan, (eds.), *Ethylene Action in Plants* (pp. 151–184). Springer Berlin: Heidelberg.

Nonaka, S., Arai, C., Takayama, M., Matsukura, C., & Ezura, H., (2017). Efficient increase of γ-aminobutyric acid (GABA) content in tomato fruits by targeted mutagenesis. *Scientific Reports, 7*, 7057.

Nunes, C. A., (2012). Biological control of postharvest diseases of fruit. *European Journal of Plant Pathol., 133*, 181–196.

Okmen, B., Sigva, H. O., Gurbuz, N., Ulger, M., Frary, A., & Doganlar, S., (2011). Quantitative trait loci (QTL) analysis for antioxidant and agronomically important traits in tomato (*Lycopersicon esculentum*). *Turk. J. Agric. For., 35*, 501–514.

Oladiran, A. O., & Iwu, L. N., (1993). Studies on the fungi associated with tomato fruit rots and effects of environment on storage. *Mycopathologia, 121*, 157–161.

Osman, A. O. A., (2015). *Antifungal Evaluation of Some Plants Extracts and Fungicide Against (Fusarium oxysporum f.sp. lycopersici) Causal Agent Wilt of Tomato* (pp. 1–73). Sudan University of Science and Technology College of Graduate Studies.

Pal, K. K., & Gardener, B. M., (2006). Biological control of plant pathogens. *The Plant Health Instructor*, 1–25. https://doi.org/10.1094/PHI-A-2006-1117-02. Biological.

Patel, R. B., & Patel, G. S., (1991). Postharvest diseases of the tomato (*Lycopersicon esculentum*) fruits and their control. *Ind. J Agric. Res., 25*, 173–176.

Pose, G., Patriarca, A., Kyanko, V., Pardo, A., & Fernández, P. V., (2009). Effect of water activity and temperature on growth of *Alternaria alternata* on a synthetic tomato medium. *Int. J. Food Microbiol., 135*, 60-63.

Povero, G., (2011). *Physiological and Genetic Control of Anthocyanin Pigmentation in Different Species*. PhD Thesis, Vrije Universiteit of Amsterdam. Available from: https://research.vu.nl/ws/portalfiles/portal/42207803 (accessed on 1 March 2021).

Povero, G., Gonzali, S., Bassolino, L., Mazzucato, A., & Perata, P., (2011). Transcriptional analysis in high-anthocyanin tomatoes reveals synergistic effect of Aft and Atv genes. *J. Plant Physiol., 168*, 270–279.

Quinet, M., Angosto, T., Yuste-Lisbona, F. J., Blanchard-Gros, R., Bigot, S., Martinez, J. P., & Lutts, S., (2019). Tomato fruit development and metabolism. *Front. Plant Sci., 10*, 1554.

Rashid, T. S., Sijam, K., Awla, H. K., Saud, H. M., & Kadir, J., (2016). Pathogenicity assay and molecular identification of fungi and bacteria associated with diseases of tomato in Malaysia. *Am. J. Plant Sci., 7*, 949–957.

Rick, C., Reeves, A., & Zobel, R., (1968). Inheritance and linkage relations of four new mutants. *Rep. Tomato Genet. Coop, 18*, 34–35.

Rodrigo, S., Santamaria, O., Halecker, S., Llied, S., & Stadler, M., (2017). Antagonism between *Byssochlamys spectabilis* (anamorph *Paecilomyces variotii*) and plant pathogens: Involvement of the bioactive compounds produced by the endophyte. *Ann. Appl. Biol., 171*, 1–13.

Samuel, O., & Orji, M. U., (2015). Fungi associated with the spoilage of postharvest tomato fruits sold in major markets in Awka, Nigeria. *Uni. J. Microb. Res., 3*, 11–16.

Sanoubar, R., & Barbanti, L., (2017). Fungal diseases on tomato plant under greenhouse condition. *Eur. J. Biol. Res*., *7*, 299–308.

Sapers, G. M., Miller, R. L., Pilizota, V., & Mattrazzo, A. M., (2001). Antimicrobial treatments for minimally processed cantaloupe melon. *J. Food Sci., 66*, 345–349.

Schreiber, G., Reuveni, M., Evenor, D., Oren-Shamir, M., Ovadia, R., Sapir-Mir, R., Bootbool-Man, A., et al., (2011). Anthocyanin from *Solanum chilense* is more efficient in accumulating anthocyanin metabolites than its *Solanum lycopersicum* counterpart in association with the anthocyanin fruit phenotype of tomato. *Theor. Appl. Genet., 124*, 295–308.

Scott, J., & Myers, J., (2012). *Purple Tomato Debuts as 'Indigo Rose'*. Available from: http://extension.oregonstate.edu/gardening/purple-tomato-debuts-indigo-rose (accessed on 1 March 2021).

Shafiq, S. A., (2015). Antagonistic activity of probiotic and seaweed extract against vegetative growth for some fungi and zearalenone production. *World J. Pharm. Res., 4*, 1577–1585.

Shukla, L. I., Chinnusamy, V., & Sunkar, R., (2008). The role of microRNAs and other endogenous small RNAs in plant stress responses. *Biochimica et Biophysica Acta (BBA)-Gene Regulatory Mechanisms, 1779*, 743–748.

Sibomana, M. S., Workneh, T. S., & Audain, K., (2016). A review of postharvest handling and losses in the fresh tomato supply chain: A focus on Sub-Saharan Africa. *Food Security, 8*, 389–404.

Sinha, S. R., Singha, A., Faruquee, M., Jinu, M. A. S., Rahaman, M. A., Alam, M. A., & Kader, M. A., (2019). Postharvest assessment of fruit quality and shelf life of two elite tomato varieties cultivated in Bangladesh. *Bull. Natl. Res. Cent., 43,* 185.

Slater, A., Maunders, M. J., Edwards, K., Schuch, W., & Grierson, D., (1985). Isolation and characterization of cDNA clones for tomato polygalacturonase and other ripening-related proteins. *Plant Mol. Biol., 5*, 137–147.

Smith, D. L., Abbott, J. A., & Gross, K. C., (2002). Down-regulation of tomato beta-galactosidase 4 results in decreased fruit softening. *Plant Physiol., 129*, 1755–1762.

Sommer, N. F., Fortlage, R. J., & Edwards, D. C., (1992). Postharvest diseases of selected commodities. In: Kader, A. A., (eds), *Postharvest Technology of Horticultural Crops* (p. 136). University of California, Division of agriculture and natural resources, USA.

Srivastava, M. P., & Tandon, R. N., (1966). Postharvest diseases of tomato in India. *Mycopathologia et Mycologia Applicata., 29*, 254–264.

Tanksley, S. D., Grandillo, S., Fulton, T. M., Zamir, D., Eshed, Y., Petiard, V., Lopez, J., & Beck-Bunn, T., (1996). Advanced backcross QTL analysis in a cross between an elite processing line of tomato and its wild relative *L. pimpinellifolium*. *Theor. Appl. Genet., 92*, 213–224.

Thambugala, K. M., Wanasinghe, D. N., Phillips, A. J. L., Camporesi, E., Bulgakov, T. S., Phukhamsakda, C., Ariyawansa, H. A., et al., (2017). Mycosphere notes 1–50, grass (Poaceae) inhabiting dothideomycetes. *Mycosphere, 8*, 697–796.

Tieman, D. M., & Handa, A. K., (1994). Reduction in pectin methyltransferase activity modifies tissue integrity and cation levels in ripening tomato (*Lycopersicon esculentum* Mill.) fruits. *Plant Physiol., 106*, 429–436.

Tohamy, M. R. A., Helal, G. A., Ibrahim, K. I., & El-Aziz, S. A. A., (2004). Control of postharvest tomato fruit rot 11 using heat treatments. *Egyp. J. Phytopath., 32*, 129–138.

Tor, M., Manning, K., King, G. J., Thompson, A. J., Jones, G. H., Seymour, G. B., & Armstrong, S. J., (2002). Genetic analysis and FISH mapping of the colorless non-ripening locus of tomato. *Theor. Appl. Genet., 104*, 165–170.

Tripathi, A., Sharma, N., Sharma, V., & Alam, A., (2013). A review on conventional and non-conventional methods to manage postharvest diseases of perishables. *Researcher, 5*, 6–19.

Tsai, W. A., Weng, S. H., Chen, M. C., Lin, J. S., & Tsai, W. S., (2019). Priming of plant resistance to heat stress and tomato yellow leaf curl Thailand virus with plant-derived materials. *Front. Plant Sci., 10*, 906.

Vrebalov, J., Ruezinsky, D. M., Padmanabhan, V., White, R., Medrano, D., Drake, R., Schuch, W., & Giovannoni, J., (2002). A MADS-box gene necessary for fruit ripening at the tomato ripening-inhibitor (*rin*) locus. *Science, 296*, 343.

Vrebalov, J., Ruezinsky, D., Padmanabhan, V., White, R., Medrano, D., Drake, R., Schuch, W., & Giovannoni, J., (2002). A MADS-box gene necessary for fruit ripening at the tomato ripening-inhibitor (rin) locus. *Science, 296*, 343–346.

Walley, C., & Seymour, G., (2006). Investigating the polygenetic nature of texture traits in tomato fruit. *VI International Solanaceae Conference*. University of Wisconsin, Madison.

Wani, A. H., (2011). An overview of the fungal rot of tomato. *Mycopathology, 9*, 33–38.

Xiong, A. S., Yao, Q. H., Peng, R. H., Li, X., Han, P. L., & Fan, H. Q., (2005). Different effects on ACC oxidase gene silencing triggered by RNA interference in transgenic tomato. *Plant Cell Rep., 23*(9), 639–646.

Yen, H. C., Lee, S., Tanksley, S. D., Lanahan, M. B., Klee, H. J., & Giovannoni, J. J., (1995). The tomato *Never-ripe* locus regulates ethylene-inducible gene expression and is linked to a homologue of the *Arabidopsis ETR1* gene. *Plant Physiol., 107*, 1343–1353.

Yeshiwas, Y., & Tolessa, K., (2018). Postharvest quality of tomato (*Solanum lycopersicum*) varieties grown under greenhouse and open field conditions. *Int. J. Biotech. Mol. Biol. Res., 9*, 1–6.

Yin, J. Q., Zhao, R. C., & Morris, K. V., (2008). Profiling microRNA expression with microarrays. *Trends Biotechnol., 26*, 70–76.

Yogendra, K. N., & Gowda, P. H. R., (2013). Phenotypic and molecular characterization of a tomato (*Solanum lycopersicum* L.) F-2 population segregation for improving shelf life. *Genet. Mol. Res., 12*, 506–518.

Zhang, J., Zeng, R., Chen, J., Liu, X., & Liao, Q., (2008). Identification of conserved microRNAs and their targets from *Solanum lycopersicum* mill. *Gene., 423*, 1–7.

Zhang, L., Zhu, M., Ren, L., Li, A., Chen, G., & Hu, Z., (2018). The *SlFSR* gene controls fruit shelf-life in tomato. *J. Exp. Bot., 69*, 2897–2909.

Zhang, Y., Butelli, E., De Stefano, R., Schoonbeek, H. J., Magusin, A., Pagliarani, C., Wellner, N., et al., (2013). Anthocyanins double the shelf life of tomatoes by delaying overripening and reducing susceptibility to gray mold. *Curr. Biol., 23*, 1094–1100.

Zhou, B., Luo, Y., Nou, X., Yang, Y., Wu, Y., & Wang, Q., (2014). Effects of postharvest handling conditions on internalization and growth of *Salmonella enterica* in tomatoes. *J. Food Prot., 77*, 365–370.

Znidarcic, D., & Pozrl, T., (2006). Comparative study of quality changes in tomato cv. 'Malike' (*Lycopersicon esculentum* Mill.) whilst stored at different temperatures. *Acta Agricul. Sloven., 87*, 235–243.

Zong, Y., Liu, J., Li, B., Qin, G., & Tian, S., (2010). Effects of yeast antagonists in combination with hot water treatment on postharvest diseases of tomato fruit. *Biological Control, 54*, 316–321.

INDEX

C

Q

R

U

V